Stratospheric Ozone Depletion and Climate Change

Foreword

Few fields in atmospheric sciences have experienced such a stormy develop-ment as atmospheric chemistry. When I first entered the field some 40 years ago, it was still thought that stratospheric ozone was regulated by four reac-tions, after Chapman in 1930, involving only the oxygen allotropes O, O_2 and O_3. We now know that a complete stratospheric reaction scheme con-tains more than 100 reactions in the gas phase, the liquid phase and on the surface of particulate matter. These reactions involve chemical compounds, which have their origins in human activities, such as anthropogenic nitrogen and halogen oxides (NO_x, ClO_x, BrO_x). Together with dynamics and trans-port, they create a mix of processes that determine the observed distribution of stratospheric ozone. Despite the complexity of such processes, numerical models are becoming increasingly successful in describing the state and the evolution of the stratospheric ozone layer. However, reliable predictions of the future of the ozone layer are complicated by the fact that the atmosphere is changing due to climate change, which impacts the future evolution of the ozone layer. *Vice versa*, the depletion of stratospheric ozone has caused important changes in tropospheric climate and will continue to do so for decades to come. This ambitious book successfully brings together all aspects necessary for an in-depth understanding of the interaction of stratospheric ozone depletion and climate change.

Starting with a description of the past and future development of the source gases that drive stratospheric change, followed by a discussion of the relevant gas-phase and heterogeneous chemistry, a comprehensive presentation is given of the current state and past development of the polar and mid-latitude stra-tospheric ozone layer. Furthermore, three chapters deal with the interaction between climate change and stratospheric ozone depletion as well as model predictions about the future development of the stratospheric ozone layer. The book concludes with a consideration of the possible effects on the ozone layer of the now widely discussed concept of "geoengineering", (i.e. a deliberate,

Stratospheric Ozone Depletion and Climate Change
Edited by Rolf Müller
© Royal Society of Chemistry 2012
Published by the Royal Society of Chemistry, www.rsc.org

man-made, enhancement of the stratospheric sulphur layer). The chapters in this book were contributed by top researchers in the chemistry and dynamics of the middle atmosphere so that the book is highly recommended both for young academics specializing in the field and for advanced researchers.

Paul J. Crutzen

Preface

ROLF MÜLLER

Institute for Energy and Climate Research (IEK-7), Forschungszentrum
Jülich, Jülich, Germany

*"The Anthropocene could be said to have started in the late eighteenth century,
when analyses of air trapped in polar ice showed the beginning of growing
global concentrations of carbon dioxide and methane."* (Paul J. Crutzen, 2002)

Today, the activities of mankind have a greater effect on the composition of the
atmosphere for a number of important atmospheric trace species than emissions from natural sources; examples include sulphur dioxide and nitric oxides.
Further, growing fossil-fuel consumption and agricultural activities have
caused substantial increases in greenhouse gases; in particular carbon dioxide,
methane and nitrous oxide, leading to climate change on the timescale of
decades to millennia. Moreover, atmospherically long-lived chemical compounds that have no natural analogues have been emitted into the atmosphere
as a result of anthropogenic activities, the most notable ones being the chlorofluorocarbons, which are the cause of stratospheric ozone depletion in general
and the Antarctic ozone hole in particular. For several years, even the possible
necessity of an active manipulation of the Earth's system to counteract climate
change ("geo-engineering") has been widely discussed. For these reasons, Paul
J. Crutzen suggested that the term "Anthropocene" should be applied to the
present, in many ways human-dominated, geological epoch.

   Two of the most striking examples of the impact of mankind on the atmosphere and the whole Earth-system are climate change, driven by anthropogenic emissions of greenhouse gases (most notably carbon dioxide) and the
appearance of the "ozone hole", driven by anthropogenic emissions of

Stratospheric Ozone Depletion and Climate Change
Edited by Rolf Müller
© Royal Society of Chemistry 2012
Published by the Royal Society of Chemistry, www.rsc.org

chlorofluorocarbons and related chemicals. In recent decades it has become increasingly clear that the two problems are intimately related. First, climate change is affecting stratospheric temperatures and circulation patterns that impact the physical and chemical processes which determine the state of the stratospheric ozone layer. Second, the depletion of the stratospheric ozone layer and most notably the Antarctic ozone hole has a significant effect on tropospheric climate. And the effect of changes in tropospheric winds caused by the ozone hole have likely influenced carbon dioxide uptake in the southern ocean. Third, the chlorofluorocarbons are also potent greenhouse gases, so that the controls on chlorofluorocarbons in the Montreal Protocol for the protection of the ozone layer and its amendments and adjustments have also helped to protect climate by avoiding an enhanced warming that would have occurred otherwise.

The aim of this book is to present the scientific basis for understanding stratospheric ozone depletion and its interplay with climate change. The book is aimed at filling the gap between advanced textbooks and scientific assessments, which focus on communicating policy-relevant, recent advances in scientific understanding of these issues.

I would like to thank all of the contributors to the chapters in this book for supporting this project and for their great effort in putting together chapters at the edge of the most recent research. I would also like to thank the staff at the Royal Society of Chemistry, especially Gwen Jones, for their important contribution in bringing this book to completion. It is hoped that the book will be helpful to both established researchers in the field and young scientists embarking on a career in this fascinating field.

Contents

Chapter 1 Introduction   **1**
Rolf Müller

  1.1 The Stratospheric Ozone Layer   1
      1.1.1 Early Observations of Stratospheric Ozone   1
      1.1.2 The Chemistry of Stratospheric Ozone   3
      1.1.3 The Distribution of Ozone in the Stratosphere   7
  1.2 Anthropogenic Influence on the Stratospheric
      Ozone Layer   9
      1.2.1 Increase in Halogen Source Gases in
           the Atmosphere   10
      1.2.2 Upper Stratospheric Ozone Depletion   13
  1.3 Polar Stratospheric Ozone Depletion   15
      1.3.1 The Antarctic Ozone Hole   15
      1.3.2 Arctic Ozone Depletion   16
      1.3.3 Chemical Mechanisms of Polar Ozone
           Depletion   18
  1.4 The Future of the Stratospheric Ozone Layer   21
      1.4.1 Projections of Future Stratospheric Ozone
           Recovery   21
      1.4.2 The World Avoided by the Montreal Protocol   24
  Acknowledgements   26
  References   26

Chapter 2 Source Gases that Affect Stratospheric Ozone   **33**
Stephen A. Montzka

  2.1 Introduction   33
  2.2 Longer-lived Halogenated Source Gases   38

Stratospheric Ozone Depletion and Climate Change
Edited by Rolf Müller
© Royal Society of Chemistry 2012
Published by the Royal Society of Chemistry, www.rsc.org

2.2.1   The Timescales and Processes that Remove
        Ozone-depleting Substances from the
        Atmosphere   38
2.2.2   The Relative Contribution of Human *vs.*
        Natural Sources to Ozone-depleting
        Halogen   40
2.2.3   Measuring and Interpreting Modern-day
        Changes in the Atmospheric Abundance
        of Ozone-depleting Substances   46
2.2.4   Long-lived Halogen-containing Gases
        Emitted from Natural Processes   50
2.2.5   Systematic Changes in Total Tropospheric
        Chlorine and Bromine from Long-lived
        ODSs   52
2.3   Very Short-lived Substances: Accounting for
      All of the Chlorine and Bromine in the
      Stratosphere   54
2.4   Past and Future Changes in Total Atmospheric
      Halogen Loading   59
2.5   Non-halogenated Gases that Affect Stratospheric
      Ozone Chemistry   62
      2.5.1   Methane (CH$_4$)   63
      2.5.2   Nitrous Oxide (N$_2$O)   65
      2.5.3   Sulfur Compounds   66
2.6   The Contributions of Ozone-depleting Gases
      to Changes in Climate   67
Acknowledgements   69
References   69

Chapter 3   **Stratospheric Halogen Chemistry**   **78**
            Marc von Hobe and Fred Stroh

3.1   Introduction   78
3.2   A Brief History of Halogen Chemistry   79
3.3   Overview of Stratospheric Halogen Chemistry,
      Abundances and Partitioning   83
      3.3.1   Abundances   84
      3.3.2   Partitioning   85
      3.3.3   Chlorine *versus* Bromine Chemistry   87
3.4   Halogen Catalyzed Ozone Loss Cycles in the
      Stratosphere   89
      3.4.1   The ClO Dimer Cycle   92
      3.4.2   The ClO/BrO Cycle   93
3.5   ClOOCl Photolysis   93

3.6 Halogen Chemistry in the Upper Troposphere
and Lowermost Stratosphere (UTLS) 95
    3.6.1 The Inorganic Bromine Budget 96
    3.6.2 Chlorine Chemistry in the Tropopause
        Region 98
Acknowledgements 99
References 99

**Chapter 4 Polar Stratospheric Clouds and Sulfate Aerosol Particles:
Microphysics, Denitrification and Heterogeneous
Chemistry** **108**
Thomas Peter and Jens-Uwe Grooß

4.1 Historical Overview 108
4.2 Distribution and Composition of PSCs 112
    4.2.1 LiDAR-based PSC Classification 113
    4.2.2 Antarctic PSC Diversity 115
    4.2.3 Arctic Measurements of STS and NAT 117
4.3 Nucleation Mechanisms of NAT-PSCs 119
    4.3.1 Ice-assisted NAT Nucleation 121
    4.3.2 NAT Nucleation Without Ice 124
    4.3.3 Summary on NAT Nucleation 125
4.4 Simulation of NAT-Rock Formation
    and Denitrification 128
    4.4.1 Simple Fixed Grid Simulations 129
    4.4.2 Lagrangian Modeling with Constant Volume
        or Ice-induced NAT Nucleation 130
4.5 Heterogeneous Reaction Rates on PSCs and Cold
    Sulfate Aerosols 131
4.6 Which Type of Particles—Solid NAT *vs.* Liquid
    STS—Control Chlorine Activation? 135
4.7 Outlook 138
Acknowledgements 139
References 139

Chapter 5 Ozone Loss in the Polar Stratosphere **145**
Neil R. P. Harris and Markus Rex

5.1 Introduction 145
5.2 Antarctic Ozone Loss 149
    5.2.1 Main Features of the Antarctic Ozone Hole 149
    5.2.2 Chemical Ozone Loss in the Antarctic Vortex 150
5.3 Arctic Ozone Loss 154
    5.3.1 Natural Variability in Stratospheric
        Ozone over the Arctic 154

5.3.2   Chemical Ozone Loss in the Arctic Vortex   155
5.3.3   Interannual Variability in Ozone Loss in
         the Arctic Vortex   160
5.4   Summary   164
Acknowledgements   165
References   165

Chapter 6   Mid-latitude Ozone Depletion   169
M. P. Chipperfield

6.1   Introduction   169
6.2   Observations of Past Mid-latitude Changes   172
      6.2.1   Column Ozone   172
      6.2.2   Ozone Profile   174
6.3   Understanding of Mid-latitude Ozone Depletion   174
      6.3.1   Chemical Processes   174
      6.3.2   Dynamical Contributions   180
      6.3.3   Other Factors Affecting Mid-latitude Ozone 183
      6.3.4   Assessment Model Calculations   183
6.4   Summary   186
Acknowledgements   186
References   187

Chapter 7   Impact of Polar Ozone Loss on the Troposphere   190
N. P. Gillett and S.-W. Son

7.1   Introduction   190
7.2   Antarctic   192
      7.2.1   Ozone Depletion Effect on the Stratosphere 192
      7.2.2   Effects of Stratospheric Ozone Depletion on
               the Troposphere and Ocean   192
               7.2.2.1   Tropospheric Circulation   192
               7.2.2.2   Surface Climate   197
               7.2.2.3   Ocean   199
      7.2.3   Mechanisms of Tropospheric Response to
               Stratospheric Ozone Depletion   201
               7.2.3.1   Radiative Effects   201
               7.2.3.2   Zonal-mean Dynamics   202
               7.2.3.3   Zonal Asymmetry Effects   205
      7.2.4   Future Changes   206
7.3   Arctic   207
7.4   Summary   208
Acknowledgements   209
References   209

**Chapter 8   Impact of Climate Change on the Stratospheric
Ozone Layer   214**
Martin Dameris and Mark P. Baldwin

8.1   Introduction   214
8.2   Impact of Enhanced Greenhouse Gas Concentrations
      on Radiation and Chemistry   216
      8.2.1   Past Temperature Changes   217
      8.2.2   Expected Future Temperature Changes   220
      8.2.3   Temperature Ozone Feedback   221
8.3   Impact of Enhanced Greenhouse Gas Concentrations
      on Stratospheric Dynamics   222
      8.3.1   Importance of Atmospheric Waves   223
      8.3.2   The Brewer-Dobson Circulation and Mean
              Age of Air   225
      8.3.3   The Role of Sea Surface Temperatures   231
8.4   Coupling of the Stratosphere and the Troposphere in a
      Changing Climate   233
      8.4.1   Stratosphere-troposphere Coupling   234
      8.4.2   The Tropical and the Extra-tropical
              Tropopause Layer   238
      8.4.3   Expected Future Changes   240
8.5   Concluding Remarks   241
Acknowledgements   243
References   243

Chapter 9   Stratospheric Ozone in the 21st Century   253
D. W. Waugh, V. Eyring and D. E. Kinnison

9.1   Introduction   253
9.2   Models and Simulations   254
      9.2.1   Chemistry-climate Models   254
      9.2.2   Simulations   256
      9.2.3   Evaluation   259
9.3   Changes in Major Factors Affecting Stratospheric
      Ozone   260
      9.3.1   Stratospheric Halogens   260
      9.3.2   Temperature   261
      9.3.3   Transport   262
      9.3.4   Other Factors   264
9.4   Projections of the Behavior of Ozone   265
      9.4.1   Tropical Ozone   265
      9.4.2   Mid-latitude Ozone   267
      9.4.3   Springtime Polar Ozone   269

9.5   Summary and Concluding Remarks   272
Acknowledgements   273
References   273

**Chapter 10   Impact of Geo-engineering on Stratospheric Ozone
and Climate   279**
Simone Tilmes and Rolando R. Garcia

10.1   Motivation for Proposed Geo-engineering
Approaches   279
10.2   Impact of Major Volcanic Eruptions on Climate
and Ozone   281
10.2.1   Impact on Climate and Stratospheric
Dynamics   282
10.2.2   Impact on Stratospheric Ozone   282
10.3   Impact of Geo-engineering on Climate and Ozone   284
10.3.1   Impact of Geo-engineering on Surface
Temperature and Precipitation   284
10.3.2   Impact on Atmospheric Temperatures and
Dynamics   286
10.3.3   Impact of Geo-engineering on Stratospheric
Ozone   287
10.3.4   Impact of Geo-engineering on Polar Ozone   290
Acknowledgements   295
References   295

Subject Index   299

CHAPTER 1
Introduction

ROLF MÜLLER

Institute for Energy and Climate Research (IEK-7), Forschungszentrum Jülich, 52425 Jülich, Germany

1.1 The Stratospheric Ozone Layer

1.1.1 Early Observations of Stratospheric Ozone

Regular measurements of stratospheric ozone started in 1924, when G. M. B. Dobson[1] designed a spectrograph to measure the total ozone column that was suitable for routine outdoor use. Dobson's new instrument allowed regular measurements to be made over extended time periods. The success of the measurement program initiated by Dobson is obvious by the fact that today the unit for the total atmospheric column of ozone is called the Dobson unit (DU). Dobson's first observations at Oxford in 1924–1925 showed a marked annual variation of ozone and a strong day-to-day variability that was closely connected to meteorological conditions.[1,2] The Dobson instrument network was extended to worldwide measurements until end of 1927.[1] In the USSR, the first measurements of total ozone commenced in 1933; later, in 1959, the M-83 ozonometer was developed, which became the basis of the USSR ozone station network.[3] Up to today, ground-based total ozone measurements are essential for long-term monitoring of the ozone content of the atmosphere.[4]

The early total ozone measurements, however, did not allow estimates of the altitude profile of the stratospheric ozone concentration to be made, so that in the late 1920s, the ozone layer was still assumed to be located in the upper stratosphere.[5,6] However, in 1933 Götz *et al.*,[7] based on so-called "Umkehr" measurements with a Dobson spectrograph, realized what is well known today,

Stratospheric Ozone Depletion and Climate Change
Edited by Rolf Müller
© Royal Society of Chemistry 2012
Published by the Royal Society of Chemistry, www.rsc.org

BOX 1.1: Units of Measurement

In atmospheric chemistry, the most unambiguous way of expressing the abundance of a species in the gas phase is as the ratio of the number of moles of the species to the number of moles of air (a mole fraction). Throughout this book, the abbreviations ppm (parts per million), ppb (parts per billion), ppt (parts per trillion) and the term mixing ratio are used to denote a mole fraction in dry air. Atmospheric measurement data are also sometimes expressed as parts per million by volume or ppmv (and, similarly, as ppbv and pptv), for example, as an approximation of a mole fraction. Because mole fractions and pressures being considered in atmospheric chemistry are quite low, inter-molecular interactions are small and gases frequently behave in ways well approximated by the ideal gas law. As a result, a mole fraction of 300 ppb, for example, is nearly equivalent to 300 ppbv. For an ideal gas, the molar mixing ratio is *exactly* equivalent to the volume mixing ratio, so that to denote mole fractions also the abbreviations ppmv, ppbv, and pptv are used. The rationale behind adding the "v" is frequently to avoid confusion between molar and mass mixing ratios. (When mass mixing ratios are used in the literature, the abbreviations ppmm, ppbm, and pptm are commonly employed). However, for non-ideal gases at sufficiently high mixing ratios, *e.g.*, CO_2 in modern ambient air, the volume correction factor for real gas behaviour is about 0.993–0.995.[10] In this case, the molar mixing ratio is not equal to the volume mixing ratio and the difference in the order of one ppm can be scientifically significant.

The term "concentration" is also used throughout this document to express a concept similar to mole fraction and molar mixing ratio. Concentrations, however, are an expression of the number of particles or molecules per unit area, so are dependent on the total pressure of an air parcel. To convert a molar mixing ratio μ_X of a species X to a concentration [X] in units of particles per cubic centimetre, the mixing ratio is multiplied by the density of air M in units of molecules per cubic centimetre, which can be expressed as:

$$M = \frac{n}{V} N_A \tag{1.1}$$

where n is number of moles, V is volume, and $N_A = 6.0221 \times 10^{23}$ molecules mol^{-1} is the Avogadro constant. When pressure p and temperature T are expressed in units of hPa and K, respectively, (as is common in the atmospheric sciences) a useful relation results for M in units of molecules per cubic centimetre is

$$M = 7.243 \times 10^{18} \frac{p}{T} \tag{1.2}$$

The total thickness of the ozone layer in a column of air is measured in Dobson units (DU), one DU is defined as 2.687×10^{16} ozone molecules per square centimetre. The DU describes the thickness of a layer of pure ozone if the total amount of ozone in a column of the atmosphere were brought to standard conditions (1013.25 hPa, 0 °C). For example, an ozone column of 300 DU brought down to the surface of the Earth would occupy a 3 mm thick layer of pure ozone. It is sometimes useful to express the concentration of ozone in DU per km rather than in molecules per cubic centimetre (see *e.g.*, Figure 1.3). From the definition of the Dobson unit one obtains 1 DU/km $\approx 3.717 \times 10^{12}$ molecules per cubic centimetre.

namely that "the average height [of the ozone in the atmosphere] at Arosa now appears to be about 20 km". Soon thereafter, the first balloon-borne measurements of the solar UV spectrum in the stratosphere[8] and the ozone profiles deduced from these measurements independently confirmed the Umkehr observations. The first measurements above balloon altitudes were made by a UV spectrograph mounted on a rocket in 1946.[6,9]

1.1.2 The Chemistry of Stratospheric Ozone

When Dobson and Harrison[2] published their first report on their column ozone measurements in 1926, the formation mechanism of stratospheric ozone was still unclear. Today it is well established that stratospheric ozone is produced by the photolysis of molecular oxygen (O_2) at ultraviolet wavelengths below 242 nm,

$$R1: \quad O_2 + h\nu \rightarrow 2O;$$

where $h\nu$ denotes an ultraviolet photon. The atomic oxygen (O) produced in reaction R1 reacts rapidly with molecular oxygen to form ozone (O_3)

$$R2: \quad O + O_2 + M \rightarrow O_3 + M;$$

where M denotes a collision partner (N_2 or O_2) that is not affected by the reaction. Ozone is photolyzed rapidly

$$R3: \quad O_3 + h\nu \rightarrow O + O_2.$$

The dissociative absorption of short-wave solar radiation by ozone is not only relevant for photochemistry but also constitutes the dominant source of heating in the stratosphere.

Through reactions R2 and R3, ozone and O establish a rapid photochemical equilibrium. Therefore, instead of considering ozone and atomic oxygen as separate species, the sum of ozone and O is often considered and referred to as "odd oxygen", or, alternatively, as the "odd oxygen family". It is denoted by the symbol O_x. This concept is useful because the sum of the family members is produced and destroyed much more slowly than the individual members. Throughout the stratosphere (up to about 50 km altitude) ozone constitutes the vast majority of odd oxygen. Through the reaction

$$R4: \quad O + O_3 \rightarrow 2O_2$$

both an O atom and an ozone molecule are lost. Because O and ozone are in rapid photochemical equilibrium, the loss of one oxygen atom effectively implies the loss of an ozone molecule, i.e., R4 destroys two molecules of odd oxygen. Reactions R1-R4 were proposed by Chapman[11] in 1930 as the first photochemical theory for the formation of ozone and are therefore referred to as the "Chapman reactions".

However, destruction of ozone by reaction R4 alone cannot explain the observed ozone abundances in the stratosphere. Today it is established that in the mid-latitudes and in the tropics the stratospheric ozone production through reaction R1 is largely balanced by the destruction in catalytic cycles of the form

$$R5: \quad XO + O \quad \rightarrow \quad X + O_2$$

$$R6: \quad X + O_3 \quad \rightarrow \quad XO + O_2$$

$$C1: \text{Net: } O + O_3 \quad \rightarrow \quad 2O_2$$

where the net reaction is identical to reaction R4. It is important to note that the catalyst X is not used up in the reaction cycle. The most important cycles of this type in the stratosphere involve reactive nitrogen ($X = NO$), originally proposed by Crutzen,[12] and hydrogen ($X = H$, OH) radicals, originally proposed by Bates and Nicolet[13] with a mesospheric focus and by Hampson[14] with a stratospheric focus. Stolarski and Cicerone[15] introduced the possibility of chlorine-catalyzed ozone loss (*via* cycle C1 with $X = Cl$).

Because of their great reactivity, the species X and XO are referred to as "active" and are commonly considered together as a so-called "chemical family", in analogy to the odd oxygen family O_x. Thus, $NO + NO_2$ is referred to as active nitrogen (NO_x), $H + OH + HO_2$ as active hydrogen (HO_x), and $Cl + ClO$ as active chlorine (ClO_x). Under polar winter conditions, the dimer of ClO, Cl_2O_2 is also part of the active chlorine family so that $ClO_x = Cl + ClO + 2 Cl_2O_2$. However, most of the atmospheric nitrogen and chlorine not tied up in very long-lived gases (like N_2O and chlorofluorocarbons) is not prevailing in active form as NO_x or ClO_x. Rather, most nitrogen and chlorine is bound in so-called "reservoir species"; HNO_3 and N_2O_5 in the case of nitrogen and HCl, $ClONO_2$, and HOCl in the case of chlorine. The sum of $NO_x + HNO_3 + 2 N_2O_5$ is referred to as NO_y and the sum of $ClO_x + HCl + ClONO_2 + HOCl$ as Cl_y or "total inorganic chlorine".

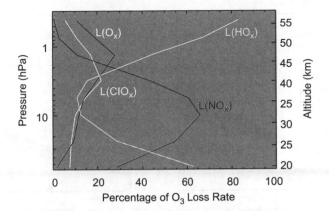

Figure 1.1 The vertical distribution of the relative importance of the individual contributions to ozone loss in the gas-phase by the HO_x, ClO_x, and NO_x cycles as well as the Chapman loss cycle (R4), denoted by the symbol O_x. The calculations are based on HALOE satellite measurements and are for overhead sun (23°S, January) and for total inorganic chlorine (Cl_y) in the stratosphere corresponding to 1994 conditions. (Figure courtesy of Jens-Uwe Grooß, adapted from IPCC/TEAP 2005).[16]

Because of the strong increase of O with altitude, the rates of the catalytic ozone loss cycles increase strongly between 25 and 40 km; the same is true for the rate of photolytic ozone production through reaction R1.[17] The relative importance of the cycles for ozone loss varies considerably with altitude. Between 25 and 40 km the NO_x cycle is the dominant ozone loss process, whereas above 45 km HOx-catalysed ozone loss dominates (Figure 1.1). Gas-phase reactions causing ozone loss through the ClOx cycle (which also depends on the stratospheric chlorine loading) peak at 40 km. The HOx catalyzed ozone loss dominates below about 25 km (Figure 1.1) where the concentration of O deceases strongly because a HO_x-catalyzed cycle exists (C2) which only involves ozone and does not require O to be present.

$$R7: \quad OH + O_3 \quad \rightarrow \quad HO_2 + O_2$$

$$\underline{R8: \quad HO_2 + O_3 \quad \rightarrow \quad OH + 2O_2}$$

$$C2: Net: O_3 + O_3 \quad \rightarrow \quad 3O_2$$

Two different chemical regimes exist for stratospheric ozone: the upper stratosphere and the lower stratosphere. In the upper stratosphere, the ozone distribution is largely determined by the balance between production from the photolysis of molecular oxygen (R1) and destruction *via* the catalytic cycles involving hydrogen, nitrogen and halogen radical species discussed above (see Section 1.1.2). In the upper stratosphere, a reduction in temperature slows the

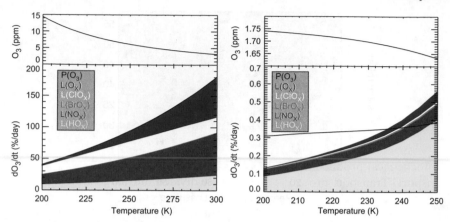

Figure 1.2   The contribution of the various ozone loss cycles in the gas-phase to the ozone loss rate, together with the rate of ozone production *via* reaction R1 $P(O_3)$, at 45°N for equinox conditions (end of March) as a function of temperature (bottom panels). The estimates were obtained from short (20 day) runs of a chemical box model, starting from climatological values of ozone; the ozone mixing ratios at the end of the run are shown in the top panels. Left-hand panels show conditions for 40 km altitude (2.5 hPa), where the climatological temperature is about 250 K; right-hand panels show conditions for 20 km (55 hPa), where the climatological temperature is about 215 K. Reaction rate constants were taken from Sander *et al.*[18] The production and loss rates are for the simulated ozone value. Note that at 40 km ozone is in steady state (*i.e.*, production equals the sum of all chemical loss terms) and thus ozone is under photochemical control. At 20 km ozone production does not equal chemical loss, implying that transport also has a strong influence on ozone concentrations. At 20 km the simulated ozone value after 20 days remains close to its initial climatological value of 1.70 ppm. (Figure courtesy of Jens-Uwe Grooß, adapted from IPCC/TEAP 2005).[16]

destruction rate of ozone (Figure 1.2). The rate of both the ozone destruction cycles and of ozone production via reaction R1 is substantially faster in the upper stratosphere than in the lower stratosphere; therefore, chemical equilibrium is reached rapidly in the upper stratosphere, which is not the case in the lower stratosphere (Figure 1.2).

In the lower stratosphere, in addition to gas-phase chemistry, reactions on aerosol and cloud particles (*i.e.*, heterogeneous reactions) become important. Throughout the lower stratosphere, a layer of aerosol particles exists which consist of sulphuric acid and water, the so-called Junge layer.[19] In the polar stratosphere, in winter, polar stratospheric clouds (PSCs) form.[20,21] (Chapter 4). The distribution of the radicals (and the partitioning of the nitrogen, hydrogen and halogen species between radicals and the reservoir species which do not destroy ozone) are affected by heterogeneous chemistry. In the mid-latitudes, reactions on aerosol surfaces convert active nitrogen to the HNO_3 reservoir, making mid-latitude ozone less vulnerable to active nitrogen

(X = NO in cycle 1), but increase the efficiency of chlorine-catalysed (X = Cl in cycle 1) ozone loss.[22,23] Heterogeneous reaction at low temperatures are of particular importance in the chemical mechanisms causing polar ozone loss[24,21] (see Section 1.3.3 and Chapter 4).

1.1.3   The Distribution of Ozone in the Stratosphere

The distribution of ozone in the stratosphere is governed by three processes: photochemical production, photochemical destruction by catalytic cycles, and transport. Transport processes are typically divided into large-scale advection and mixing processes on smaller scales. The large-scale circulation of the stratosphere, with rising motion at low latitudes followed by poleward motion and descent at high latitudes, systematically transports ozone poleward and downward (Figure 1.3). This circulation in the stratosphere is referred to as the "Brewer–Dobson circulation" because such a circulation was originally suggested by Brewer[26] based on water vapour measurements in the stratosphere and by Dobson *et al.*[27] based on column ozone measurements. Because of the short photochemical lifetime of ozone in the upper stratosphere (see Section 1.1.2 above), the Brewer–Dobson circulation has little effect on the ozone distribution there (Figure 1.3). However, in the lower stratosphere, the

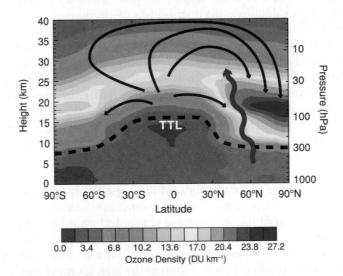

Figure 1.3   The meridional cross section of ozone concentration[25] (in DU per km) during northern hemisphere winter (January to March). The dashed line denotes the approximate location of the tropopause, and TTL stands for tropical tropopause layer. The black arrows indicate the Brewer–Dobson circulation in northern hemisphere winter, and the wiggly red arrow represents planetary waves that propagate from the troposphere into the stratosphere. (Figure reproduced from Box 1.2 in IPCC/TEAP 2005).[16]

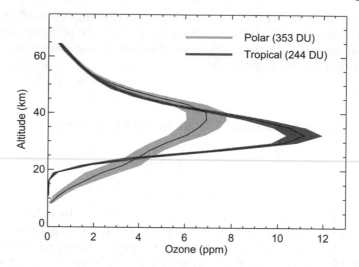

Figure 1.4 The vertical profile of ozone mixing ratio against altitude for polar
(equivalent latitude 72.5°N) and tropical (equivalent latitude 2.5°N)
conditions in March. The ozone data are from a climatology deduced
from satellite measurements.[29] (Figure adapted from Müller, 2010).[30]

photochemical lifetime of ozone is long (several months or longer)[28] so that
transport processes dominate the distribution of ozone.

Maximum ozone mixing ratios occur in the tropics between 30 km
(Figure 1.4). In this altitude region, ozone is very short-lived and is essentially
in photochemical equilibrium; that is, rapid photochemical loss is balanced by
rapid production[17] (see Figure 1.2). Under these conditions, transport timescales are slow compared to chemical timescales. Therefore, the altitude region
30–40 km in the deep tropics cannot serve as a source region for the extra-
tropical stratosphere. The region where the ozone mixing ratios that are
exported to the extra-tropics are determined, is the transition region between
the area of chemical control and the area where ozone is controlled by
transport.[31]

Transport of ozone from the high latitudes poleward is important in the
extra-tropical lower stratosphere, where ozone can accumulate on the time
scale of a season. Variations in the ozone concentration in this region, between
the tropopause and 20–25 km altitude, control changes in total column ozone
abundance (Figure 1.3). Below about 25 km, ozone concentrations and mixing
ratios in the extra-tropics are greater than in the high latitudes and (Figures 1.3
and 1.4). Thus, the ozone column at high and mid-latitudes are greater than in
the tropics (Figure 1.5).

The Brewer–Dobson circulation is driven by planetary wave activity which is
strongest in winter.[34] Further, because of the asymmetric distribution of the
topography and land-sea thermal contrasts that force planetary waves, planetary wave activity is stronger in the northern hemisphere than in the southern

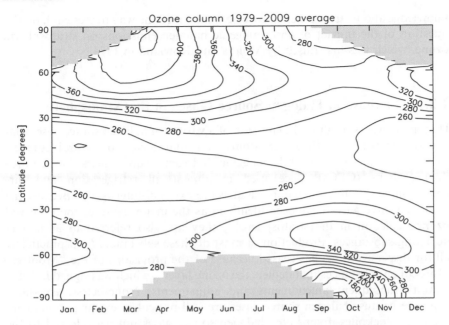

Figure 1.5 A climatology of total column ozone as a function of latitude and month. Data were taken from version 2.7 of the Bodeker Scientific combined ozone database[32,33] (courtesy of Greg Bodeker) which provides daily total column ozone fields from 1979 to present day. (Plot courtesy of Jens-Uwe Grooß).

hemisphere. The stronger planetary wave activity in the northern hemisphere causes the Brewer–Dobson circulation to be stronger during the northern hemisphere winter than during the southern hemisphere winter. Therefore, ozone builds up in the extra-tropical lower stratosphere during winter and spring, with a greater build-up occurring in the northern hemisphere.[35] The ozone then decays photochemically during the summer when transport is weaker and strong NO_x-driven ozone loss occurs at the poles.[36,37] The development of total ozone over the year as a function of latitude (Figure 1.5) reflects the seasonality of the Brewer–Dobson circulation.

1.2 Anthropogenic Influence on the Stratospheric Ozone Layer

The first concern about the impact of anthropogenic activities on the ozone layer was formulated in the late 1950s: the possible impact of nuclear weapons tests on the ozone layer.[6] Later, in the early 1970s, attention was focused on the effect a planned fleet of hundreds of supersonic aircraft might have on the stratospheric ozone layer.[38,39] Research programs directed at assessing the impact of supersonic transport on stratospheric ozone greatly improved knowledge about stratospheric processes and paved the way for research on the question of the impact of anthropogenic halogen emissions on stratospheric ozone, which was first raised by Molina and Rowland in 1974.[40]

1.2.1 Increase in Halogen Source Gases in the Atmosphere

Human activities result in the emission of a variety of halogen source gases that contain chlorine and bromine atoms and that have no natural sources. Important examples of anthropogenic halogen source gases are chloro-fluorocarbons (CFCs), once used in almost all refrigeration and air-conditioning systems, and halons, used as fire-extinguishing agents. Because the halogen source gases have been identified as the major cause of the observed ozone depletion in the stratosphere[4,21] they are also referred to as ozone-depleting substances. Production of most of these substances (and practically all for dispersive uses) has ceased because of the provisions of the "Montreal Protocol on substances that deplete the ozone layer", which was signed in 1987, and its subsequent amendments and adjustments. Nonetheless, emissions continue because halogen source gases are still present in existing equipment, chemical stockpiles, foams *etc.*; halogen source gases not yet released to the atmosphere are referred to as "banks".

Without the Montreal Protocol, production, consumption, and thus emission, of ozone-depleting substances would have continued with an annual growth rate of about 3%. As a result, the stratospheric halogen loading would have increased by 2030 by about a factor of ten compared to only natural sources (Figure 1.6). Model studies[41,42] suggest that, had this happened, stratospheric ozone would have been strongly depleted globally with erythemal UV radiation more than doubling in the mid-latitudes in Northern hemisphere summer by the middle of this century (Section 1.4.2). However, both the Montreal Protocol itself and the later London amendments in 1990 would only have slowed the growth of the stratospheric halogen burden. Not until the amendments and adjustments signed in Copenhagen in 1992 do the projections of the future halogen loading indicate the decrease (Figure 1.6) required for a true recovery of the stratospheric ozone layer.

The most abundant naturally emitted chlorine source gas is methyl chloride. Methyl chloride is present in the troposphere in globally averaged concentrations of about 550 ppt and accounts for about 16% of the chlorine loading of the stratosphere today[4] (Figure 1.7). At the end of the 21st century, when the abundance of anthropogenic chlorine source gases (e.g. CFCs) will have been greatly reduced as a consequence of the Montreal Protocol, methyl chloride is expected to account for a large fraction of the remaining stratospheric chlorine.

The emission of halogen source gases to the atmosphere ultimately leads to stratospheric ozone depletion. The first step is the photochemical breakdown of the source gases. Because of the great chemical stability of most source gases, this breakdown only occurs at appreciable rates in the upper stratosphere, where there is high-energy solar radiation. The chlorine atoms released in this

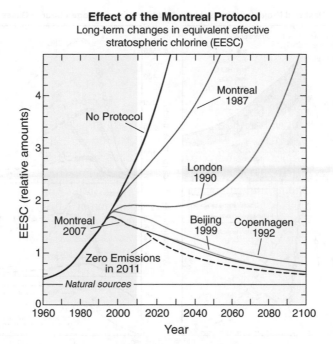

Effect of the Montreal Protocol
Long-term changes in equivalent effective
stratospheric chlorine (EESC)

Figure 1.6 Projections of the future stratospheric halogen loading expressed as equivalent effective stratospheric chlorine (EESC) values (see Chapter 2) are shown for the mid-latitude stratosphere for (1) no Protocol provisions, (2) the provisions of the original 1987 Montreal Protocol and some of its subsequent amendments and adjustments, and (3) zero emissions of ozone depleting substances starting in 2011. The city names and years indicate where and when changes to the original 1987 Protocol provisions were agreed upon.[4] (Figure taken from WMO 2011).[4]

way from the source gases are mostly converted to reservoir species, the most important being HCl and $ClONO_2$. Because the reservoir species themselves do not cause ozone depletion, for ozone depletion to occur chlorine must be liberated from the reservoirs and converted into an active form. This occurs through gas-phase processes in the upper stratosphere and through heterogeneous chemistry in the polar lower stratosphere in winter.[21]

An important measure of the potential for ozone depletion in the stratosphere due to the presence of halogen-containing (ozone-depleting) source gases in the stratosphere is the so-called "equivalent effective stratospheric chlorine".[4,45] Equivalent effective stratospheric chlorine values are calculated by summing over adjusted amounts of all chlorine and bromine source gases. The adjustments are designed to account for the different rates of decomposition of the source gases and the greater per-atom effectiveness of bromine in depleting ozone compared to chlorine (Chapter 2).

In the latter half of the 20th century up until the 1990s, equivalent effective stratospheric chlorine values increased steadily and rapidly (Figure 1.7). As a

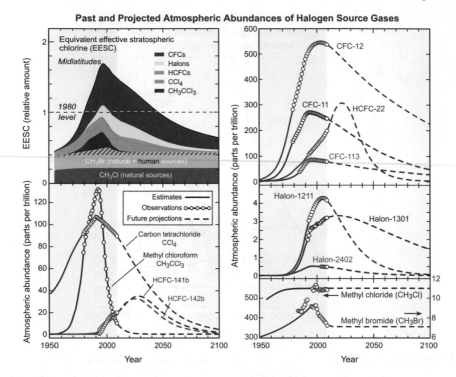

Past and Projected Atmospheric Abundances of Halogen Source Gases

Figure 1.7 The development of the stratospheric halogen loading measured as the equivalent effective stratospheric chlorine (Chapter 2). The rise of equivalent effective stratospheric chlorine in the 20th century has slowed and reversed in the past decade (top left panel). Values are derived from individual halogen source gas abundances obtained from measurements, historical estimates of abundances, and projections of future abundances. Equivalent effective stratospheric chlorine levels as shown here for mid-latitudes will return to 1980 values around 2050. The return to 1980 values will occur around 2065 in polar regions.[43] The year 1980 is used here as a reference for a relatively unperturbed ozone layer as usually done in ozone assessments.[44] A decrease in equivalent effective stratospheric chlorine abundance results from decreases in the abundance of individual halogen source gases which follows from reductions in their emissions. Total summed emissions and atmospheric concentrations have decreased and will continue to decrease given international compliance with the provisions of the Montreal Protocol. The changes in the atmospheric abundance of individual gases at the Earth's surface shown in the panels were obtained using a combination of direct atmospheric measurements, estimates of historical abundance, and future projections of abundance. The abundances of most CFCs in the troposphere, along with those of carbon tetrachloride and methyl chloroform, have decreased in the past decade. The concentrations of HCFCs, which are used as CFC substitutes, will continue to increase in the coming decades. The abundances of some halon compounds will also continue to grow in the future while halons still present in existing equipment (so called "banks") are released to the atmosphere. Smaller relative decreases are expected for methyl bromide in response to production and use restrictions because it has substantial natural sources. Methyl chloride has large natural sources and is not regulated under the Montreal Protocol. (Figure taken from WMO 2011).[4]

result of the regulations of the Montreal Protocol, the long-term increase in equivalent effective stratospheric chlorine slowed, reached a peak in the mid-1990s and began to decrease thereafter. With a delay of a few years, the stratospheric abundance of halogen source gases follows the changes observed in the troposphere. The concentrations of all major chlorine-containing source gases decrease now in the troposphere, the longest-lived source gas, CFC-12, reached its peak approximately in 2000.[4] Although measurements in the stratosphere are much sparser, declining growth rates of CFC-12 in the stratosphere have been reported.[46] The start of a reduction in equivalent effective stratospheric chlorine values means that, as a result of the Montreal Protocol, the total stratospheric concentration of ozone-depleting halogen and thus the potential for stratospheric ozone depletion has begun to decrease.

1.2.2 Upper Stratospheric Ozone Depletion

The possibility of chlorine-catalyzed ozone loss was first related to the accumulation of anthropogenic ozone-depleting substances (the CFCs) in the atmosphere by Molina and Rowland in 1974.[40] In an early model study Crutzen[47] predicted that enhanced levels of chlorine in the stratosphere would lead to a depletion of upper stratospheric ozone via cycle C1 with $X = Cl$ (Figure 1.8).

$$R5: \quad ClO + O \quad \rightarrow \quad Cl + O_2$$

$$\underline{R6: \quad Cl + O_3 \quad \rightarrow \quad ClO + O_2}$$

$$C1: Net : O + O_3 \quad \rightarrow \quad 2O_2$$

This prediction was subsequently confirmed by a variety of model studies.[48] Based on the scientific evidence provided by those studies, the United States, Canada, Sweden, and Norway moved to ban the use of CFCs in aerosol spray cans during the late 1970s. However, it took until 1987 for the Montreal Protocol to be signed, the first international agreement on the protection of the ozone layer with legally binding controls on halogen source gases. Following country ratification, the Montreal Protocol entered into force in 1989 and is now ratified by all 196 United Nations members.

The predictions of the early studies were remarkably far-sighted. Those early studies already stated that CFCs accumulate in the lower atmosphere, with photolysis in the middle and upper stratosphere being the major sink for CFCs. And it was already understood in 1974 that the accumulation of CFCs in the atmosphere constitutes a long-term problem, on a timescale of many decades. Most importantly, it was predicted that enhanced levels of stratospheric chlorine would lead to a decline of ozone in the upper stratosphere by the catalytic ozone loss cycle C1 (with $X = Cl$); correctly, as we know today.

Today, more than 35 years after the first scientific studies were published linking the accumulation of anthropogenic CFCs in the atmosphere with the danger of future stratospheric ozone loss,[40,47] ozone decline in the altitude region between 30 and 50 km is observed by a variety of ground-based and space-borne instruments since many years (Chapter 6). The observed altitude variation of ozone loss in the upper stratosphere—peak percentage losses at about 40 km—was correctly predicted by the first model studies,[47] (Figure 1.8, panel a). Overall, the understanding of upper stratospheric chemistry, as expressed in current models, is consistent with observations.[4,17,44]

Further, as early as in the late 1970s, it was predicted that an important factor influencing upper stratospheric ozone is temperature so that increasing CO_2 in the stratosphere should have an impact on ozone.[51,52] Because the rates of the gas-phase ozone destruction cycles decrease with lower temperatures (Figure 1.2), a cooling of the stratosphere leads to an increase of ozone concentrations above about 25 km.[16] Increases in CO_2 in the atmosphere are expected to cool the stratosphere and a cooling of the stratosphere, at a rate of 0.5–1.5 K/decade from 1979–2005, is indeed observed[53] (Chapter 8).

The relation of the observed decline of upper stratospheric ozone and the increase of the stratospheric chlorine loading is shown schematically in Figure 1.8. A significant reduction in the concentration of ozone was observed during the 1980s and 1990s, with the largest losses at 40 km altitude (Figure 1.8, panel b, see also Chapter 6). A similar pattern of the decline in upper

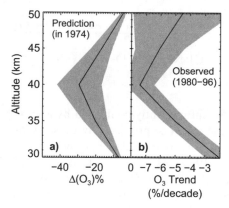

Figure 1.8   Upper stratospheric ozone trend and ozone loss rates caused by chlorine catalysed reaction cycles for mid-latitudes of the Northern Hemisphere. Panel a) shows the percentage reduction in ozone concentrations predicted to occur by Crutzen[47] due to the build-up of CFCs. The calculation assumed a growth of the stratospheric chlorine loading to a level of 5.3 ppb in the uppermost stratosphere, which was never reached in the atmosphere (see Figure 1.7). Panel b) shows the observed reduction of upper stratospheric ozone for the latitude range 30° to 50°N, between 1980 and 1996, derived from SAGE, SBUV and Umkehr measurements.[49,50] The shaded areas indicate the range of uncertainty. (Figure adapted from WMO 1999).[50]

stratospheric ozone due to the build-up of anthropogenic chlorine was predicted by the first modelling studies (Figure 1.8, panel a). These calculations were based on a projected increase of Cl_y to a level of 5.3 ppb, a value never reached in the contemporary stratosphere (Figure 1.7). The decline of upper stratospheric ozone, observed for 1980–1996, has slowed substantially over the last decade and there is evidence that this decline can be attributed to the decline of halogen source gases in the atmosphere.[44,54,55]

1.3 Polar Stratospheric Ozone Depletion

1.3.1 The Antarctic Ozone Hole

In 1985, Farman, Gardiner, and Shanklin reported that in Antarctic spring strongly reduced total ozone values occurred at the British Antarctic Survey station at Halley.[56,57] (The data from the original publication updated to present time are shown in Chapter 5, Figure 5.3) This phenomenon, soon referred to as the Antarctic "ozone hole",[6,58] is one of the most striking examples of the direct impact of human activities on the atmosphere. The discovery by Farman *et al.*, which was based on a time series of measurements by classical Dobson instruments started by Dobson himself in 1956,[1] was soon confirmed by satellite measurements which showed that the ozone depletion extended over roughly the entire Antarctic continent.[58] A satellite measurement of total ozone on 5 October 2006, one of the largest and deepest ozone holes ever observed is shown in Figure 1.9. The term "ozone hole" for the phenomenon of extremely low total ozone values in Antarctic spring was first used by Stolarski *et al.*:[58] "The deep minimum, or hole". But of course, the Antarctic ozone hole is not a true hole. Some column ozone always remains; values of a about 100 DU are found even in the core of the ozone hole (Figure 1.9).

When ozone measurements in mid-winter and spring are compared (Figure 5.5 in Chapter 5), the signature of the ozone hole becomes obvious through substantially lower ozone partial pressures in October in the altitude range (14–20 km) where the ozone maximum occurs during mid-winter. More recently, satellite measurements[59,60] have provided additional, detailed information on the vertical ozone distribution and on chemical ozone loss in the ozone hole. Nonetheless, ozonesonde observations remain important as they provide unique information. Measurements of extremely low ozone mixing ratios (below 0.1 ppm) are not possible with satellite instruments but such observations could potentially provide insights into future ozone layer recovery.[61,62]

Although chemical destruction of ozone is established as the fundamental reason for the appearance of the ozone hole, dynamical processes play an important role in its formation as well. In fall, a large vortex forms over Antarctica as the polar stratosphere radiatively cools creating a thermal contrast between polar and mid-latitude air. As the cooler polar air descends, mid-latitude air moves poleward and a strong wind jet forms, when the Coriolis

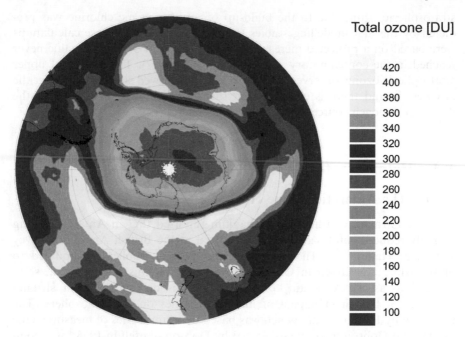

Total ozone [DU]

420
400
380
360
340
320
300
280
260
240
220
200
180
160
140
120
100

Figure 1.9   Total ozone over the southern hemisphere on 5 October 2006. A value of 220 DU is commonly used as the definition for the edge of the ozone hole. (Data from the Ozone Monitoring Instrument [OMI], plot courtesy of Markus Rex).

force deflects the air eastward. Because of this wind jet, the Antarctic polar vortex is strongly isolated from the mid-latitude stratosphere.[63] If the vortex were not isolated, mixing with mid-latitude air would make it difficult to maintain both low enough temperatures and the perturbed chlorine chemistry required for the chemical destruction of ozone.

In recent years, it has emerged that the occurrence of the Antarctic ozone hole has also important consequences for tropospheric climate. Both model simulations and observations show that the Antarctic ozone hole is the cause of much of the poleward shift of the southern hemisphere middle latitude jet in the troposphere in summer observed since about 1980. And this poleward shift of the tropospheric jet has been linked with a range of observed changes in the southern hemisphere troposphere, including decreasing CO_2 uptake over the southern ocean (Chapter 7).[4]

1.3.2   Arctic Ozone Depletion

In many respects, the Arctic wintertime stratosphere resembles its Antarctic counterpart; it exhibits a cold polar vortex separating the air enclosed in it from mid-latitude air. Strong diabatic descent throughout the winter transports air

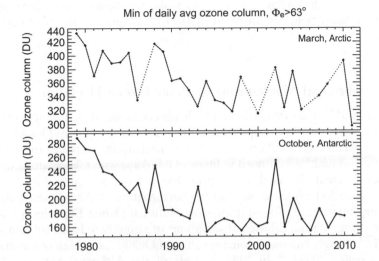

Figure 1.10 Time series of minimum of daily average column ozone poleward of 63° equivalent latitude for March in the Arctic (top panel) and October in the Antarctic (bottom panel). Winters in which the vortex broke up before March (1987, 1999, 2001, 2006, and 2009) are not shown for the Arctic time series (dashed lines in top panel). Data were taken from version 2.7 of the Bodeker Scientific combined ozone database (courtesy of Greg Bodeker). (Figure adapted and updated from Müller *et al.*;[33] figure courtesy of Jens-Uwe Grooß).

from the upper stratosphere and partly from the mesosphere to the lower stratosphere.[64] However, the Arctic polar vortex is warmer, smaller and more variable than the Antarctic vortex. The reason for this difference is a stronger planetary wave activity in the northern hemisphere that is caused by the different (less zonally symmetric) distribution of land masses in the northern hemisphere (see also Section 1.1.3).[34]

The stronger dynamical variability in the Arctic compared to the Antarctic leads to a stronger interannual variability in both chemical loss of ozone and in dynamical supply of ozone-rich air to high latitudes (Chapter 5). Therefore, compared to the Antarctic, Arctic ozone abundances in winter and spring are more variable (Figure 1.10). Nonetheless, in particularly cold winters and in winters with an enhanced burden of volcanic aerosol substantial chemical loss of ozone has been observed in the Arctic[65–68] (Chapter 5) and has led to Arctic column ozone losses of up to 30% and local ozone losses in the lower stratosphere exceeding 50%. In dynamically active and therefore warm Arctic winters, however, the estimated chemical ozone loss has been very small. The Arctic winter 2010/2011 was unusually cold and, most importantly for chemical ozone destruction, a cold and stable polar vortex persisted into spring, until April 2011. The interplay between continuing heterogeneous processing and increasing abundance of sunlight in spring led to an extremely strong chemical ozone loss in the Arctic vortex, probably the largest loss since regular satellite measurements commenced in 1979 (see Figure 1.10, cover page, and Chapter 5).

1.3.3   Chemical Mechanisms of Polar Ozone Depletion

Farman et al.[56] had already linked their observations of Antarctic ozone loss with the increase of anthropogenic CFCs in the atmosphere. Quickly, alternative explanations based on very different mechanisms were put forward,[69–71] but were eventually abandoned in favor of a linkage with CFCs, albeit through a rather different chemical mechanism than the one Farman et al. had proposed.[21] The first evidence for a strong perturbation of the Antarctic chlorine chemistry was already found during the National Ozone Expedition (NOZE) in 1986 through ground-based observations of strongly reduced column values of HCl,[72] strongly enhanced column values of OClO,[73] and high concentrations of ClO below 20 km.[74] In 1987, as part of the Airborne Antarctic Ozone Experiment (AAOE),[75] aircraft measurements in the stratosphere provided even clearer evidence that the cause of the Antarctic ozone hole is indeed a strongly perturbed chlorine chemistry. Likewise, a perturbed chlorine chemistry was found in measurements in the Arctic stratosphere.[76] Since that time, in both polar regions in winter, strongly enhanced ClO mixing ratios enhanced OClO, and strongly depleted HCl and $ClONO_2$, are regularly detected from balloon, aircraft, and remote sensing experiments.[4,21]

One year after the discovery of the ozone hole, Solomon et al.[24] proposed that the reaction of HCl and $ClONO_2$ on the surfaces of polar stratospheric clouds (PSCs) constituted the key initiation step leading to the perturbed polar chlorine chemistry and the subsequent greatly accelerated ozone loss. Later in the same year, Toon et al.[77] and Crutzen and Arnold[78] suggested that PSCs were crystalline particles consisting of nitric acid trihydrate (NAT) rather than ice as was previously thought (Chapter 4). NAT particles may exist in the polar stratosphere at significantly higher temperatures (≈ 195 K) than ice particles.[79] Toon et al.[77] further proposed (correctly as we know today) that sedimentation of PSC particles may remove active nitrogen species from the polar stratosphere and that this process contributes to bringing about a perturbed chlorine chemistry in the polar stratosphere.

The suggestion that NAT particles prevailed in the winter polar stratosphere was shown to be correct much later through balloon-borne mass spectrometer measurements in the Arctic.[80] The characteristic radii of NAT PSCs were originally assumed to be 0.5–3.0 μm.[81,82] However, from January to March 2000, large NAT particles with radii of 20–40 μm (often referred to as "NAT rocks") were detected by in situ measurements on high-flying aircraft.[83] This observation demonstrated that NAT particles can reach sedimentation velocities large enough to allow HNO_3 to be removed from the polar stratosphere thereby causing a "denitrification" of the polar stratosphere. There are also liquid PSC particles, super-cooled ternary solutions (STS) consisting of liquid

$H_2O/HNO_3/H_2SO_4$ aerosol particles and, when temperatures drop several degrees Kelvin below the frost point, ice crystals may form (Chapter 4).[20,21]

Both in the Arctic and Antarctic, temperatures reach minimum values in the lower stratosphere in winter. However, average minimum temperatures in the Antarctic are much lower (by about 10 K) than in the Arctic.[4] Therefore, the PSC period is much longer (five to six months) in the Antarctic than in the Arctic, where in warm winters practically no PSC formation occurs.[84] Average minimum values over Antarctica are as low as 185 K in July and August, so that ice PSCs frequently form. In the Arctic, in contrast, ice PSC formation (and PSC formation in general) is often only possible when large amplitude temperature excursions occur caused by mountain waves.[85,86] However, the 2009–2010 Arctic winter was unusually cold from mid-December until the end of January, and was one of only a few winters from the past 52 years with synoptic-scale regions of temperatures below the ice frost point. Consequently, an unusually large number of ice PSCs was detected by the CALIPSO satellite instrument.[87]

Today, a variety of heterogeneous reactions of importance to stratospheric chemistry are known (Chapter 4) the most important ones for polar heterogeneous chlorine activation being:

$$R9: \quad HCl + ClONO_2 \rightarrow Cl_2 + HNO_3$$

$$R10: \quad HCl + HOCl \rightarrow Cl_2 + H_2O$$

In the first years after the discovery of the ozone hole, it was thought that stratospheric heterogeneous reactions in the polar regions occur only on solid surfaces.[20,21] Heterogeneous reactions on the ubiquitous stratospheric sulphate aerosol, which is non-crystalline, were known to be important for mid-latitude chemistry,[88–90] but were not thought to be of great relevance for polar chlorine activation.

From laboratory measurements of heterogeneous reaction rates of chlorine compounds on stratospheric sulphate aerosol[91–93] and theoretical studies[94–96] it emerged that reactions on stratospheric sulphate aerosol and on liquid PSC particles are very effective for heterogeneous chlorine activation.[20,21] The effect of chlorine activation on the stratospheric sulphate aerosol is particularly pronounced after strong volcanic eruptions that lead to a substantial increase in the stratospheric sulphate loading. Consequently, enhancements of polar ozone loss is found after strong volcanic eruption (Chapter 10).[67,68,97,98] Today, detailed information from laboratory studies on the reaction probabilities of stratospheric species is available for liquid aerosol and PSC particles.[93,96,99]

These reaction probabilities are frequently strongly temperature dependent (being relevant only at low temperatures), with the important exception of the reaction of N_2O_5 with H_2O (Chapter 4), which is important at all temperatures occurring in the stratosphere. The strong increase of the reaction probabilities on liquid sulphate aerosol particles with temperature means that the onset of chlorine activation, under most conditions, is controlled by reactions on liquid sulphate aerosol particles before a significant depletion of gas-phase HNO_3 occurs, *i.e.*, before NAT or STS particles form (Chapter 4).[97,100]

The Cl_2 formed in the heterogeneous reaction R9 and R10 photolyses rapidly in sunlit air and forms ClO. Further, reaction R9 and related heterogeneous reactions (Chapter 4) suppress the concentration of NO_2 by forming HNO_3. If NO_2 concentrations were not suppressed, the released ClO would readily reform the $ClONO_2$ reservoir.[101,102] Thus, rapid chlorine-catalysed ozone loss requires first the heterogeneous release of chlorine from the HCl and $ClONO_2$ reservoirs ("activation") while the suppression of NO_2 in the gas phase slows down "deactivation", so that chlorine can remain in an active form for a longer time. The production of Cl_2 in heterogeneous chlorine activation implies that sunlight is required to release Cl and start the catalytic cycles.

In the polar regions, in winter and spring, the conventional ozone loss cycles (C1) are not effective because of the lack of atomic oxygen (O). Under these conditions, two different catalytic cycles dominate the catalytic chemical ozone loss; the efficiency of both cycles depends on the concentration of ClO (Chapter 3). The most important cycle was proposed by Molina and Molina:[103]

$$R11: \quad ClO + ClO + M \;\rightarrow\; Cl_2O_2 + M$$

$$R12: \quad Cl_2O_2 + h\upsilon \;\rightarrow\; Cl + ClOO \;(\lambda < 400\,nm)$$

$$R13: \quad ClOO + M \rightarrow Cl + O_2 + M$$

$$\underline{R14: \quad 2(Cl + O_3 \;\rightarrow\; ClO + O_2)}$$

$$C3: \quad Net: \quad 2O_3 \;\rightarrow\; 3O_2$$

where hυ denotes a photon and M a collision partner (N_2 or O_2). The second cycle also depends on BrO concentrations and thus on the stratospheric bromine loading:[69,104]

$$R15: \quad ClO + BrO \rightarrow Cl + Br + O_2$$

$$R14: \quad Cl + O_3 \rightarrow ClO + O_2$$

$$\underline{R16: \quad Br + O_3 \rightarrow BrO + O_2}$$

$$C4: \quad Net: \quad 2O_3 \rightarrow 3O_2$$

The net result of both cycle C3 and cycle C4 is the destruction of two ozone molecules. Like cycle C1, cycles C3 and C4 are catalytic: chlorine (Cl) and bromine (Br) are not lost in the reaction cycle. In the stratosphere, chlorine is much more abundant than bromine (160 times). Nonetheless, cycle C4 is important as bromine atoms are about 60 times more efficient than chlorine atoms in chemically destroying ozone.[4,105]

Sunlight is necessary to maintain a large ClO abundance through the photolysis of Cl_2O_2. However, in contrast to the rather shortwave radiation required to produce atomic oxygen, the species that is essential for cycle C1, the photolysis of Cl_2O_2 proceeds at rather long wavelengths and thus under the conditions of low sun found at the poles in spring (Chapter 3). Cycles C3 and C4 account for the majority of the ozone loss observed in late winter/early spring in the polar stratosphere. Under cold polar vortex condition, with high ClO abundances, the rate of ozone destruction can reach substantial values, up to 2–3% per day.[106–108] Outside the polar regions both cycles are of reduced importance; cycle C3 is negligible because it is only effective at the low polar temperatures in winter and spring, and cycle C4 is of minor significance because of the much lower ClO concentrations outside the polar regions (Chapter 6). In summary, large ozone loss rates can only occur when air is both sufficiently cold as well as sunlit, *i.e.* conditions that largely prevail in spring in the polar regions when ozone depletion is indeed observed.

1.4 The Future of the Stratospheric Ozone Layer

1.4.1 Projections of Future Stratospheric Ozone Recovery

A recovery of the ozone layer from the effect of anthropogenic halogen source gas emissions is expected in the later decades of the 21st century (Chapter 9). Recovery will occur as halogen source gases that cause ozone depletion (*i.e.*, ozone-depleting substances) decrease in the future because of the success of the Montreal Protocol and its adjustments and amendments in reducing global production and consumption of these substances (Figures 1.7 and 1.6). However, the atmosphere will not return to pre-1960 or to pre-1980 conditions; the influence of climate change could either accelerate or delay ozone recovery (Chapter 8). As the amount of ozone-depleting substances in the atmosphere declines towards the end of this century (Chapter 2), the impact of climate change on future ozone is increasing. Climate change is expected to have an influence on stratospheric ozone through the projected changes in the stratospheric circulation and stratospheric temperatures (Chapter 8). Moreover, the chemical composition of stratospheric air will change throughout the 21st century, for example, the expected future increases in N_2O are likely the dominant emission into the atmosphere throughout the 21st century of a compound which causes ozone depletion.[109]

Chemistry-climate models are used both to assess past changes in the global ozone distribution and to project how ozone is expected to respond to the decline in ozone depleting substances and to climate change in different geographical regions in future decades (Chapter 9). Figure 1.11 shows projections of total ozone until 2100 from a group of chemistry-climate models[4] that take into account the influences of changes in ozone-depleting substances in the

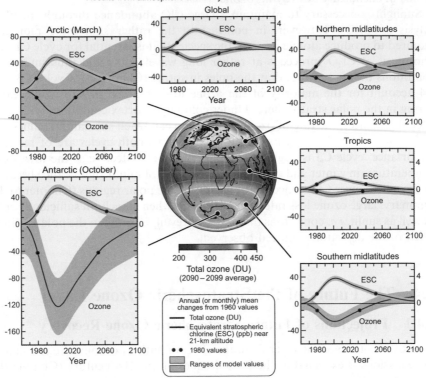

Total Ozone and Equivalent Stratospheric Chlorine
Results from atmospheric chemistry-climate models for 1960 to 2100

Figure 1.11 Long-term changes in ozone and the stratospheric halogen loading
measured by the equivalent stratospheric chlorine (ESC). Chemistry-
climate models are used to make projections of total ozone amounts that
account for the effects of changes in the stratospheric halogen loading
and climate change. Regional and global projections are shown for total
ozone and ESC for the period 1960–2100, referenced to 1960 values. The
globe in the centre shows average total ozone projections for the last
decade of the 21st century. Total ozone depletion increased after 1960 as
ESC values steadily increased throughout the stratosphere. ESC values
have peaked and are now in a slow decline (Chapter 2). All the projections show maximum total ozone depletion around 2000, coincident with
the highest abundances of ESC. Thereafter, as ESC slowly declines. Total
ozone increases, except in the tropics. In all the projections except the
Antarctic and the tropics, total ozone returns to 1960 values by the
middle of the century, which is earlier than expected from the decrease in
ESC alone. The earlier returns are attributable to climate change, which
influences total ozone through changes in stratospheric transport and
temperatures (Chapter 8). In the tropics, by contrast, climate change
causes total ozone to remain below 1960 values throughout the century.
In the Antarctic, the effect of climate change is smaller than in other
regions. As a result, Antarctic total ozone in springtime mirrors the
changes in ESC, with both closely approaching 1960 values at end of the
century. The dots on each curve mark the occurrences of 1980 values of
total ozone and ESC. Note that the equal vertical scales in each panel
allow direct comparisons of ozone and ESC changes between regions.
(Figure taken from WMO 2011).[4]

atmosphere (that are changes in the stratospheric halogen loading) and climate on ozone for a variety of geographical regions. The stratospheric halogen loading, or the amount of ozone-depleting substances in the stratosphere is measured in Figure 1.11 by the equivalent stratospheric chlorine (ESC), which is deduced from the results of chemistry-climate models (Chapter 2).

Global total ozone is projected to return to 1980 levels in the coming decades (2020–2050) and to 1960 levels around the middle of this century (2040–2080). However, stratospheric halogen levels (ESC) return to 1960 values only near the end of the century (Figure 1.11). The results from chemistry-climate models suggest that the early return of total ozone to 1960 values compared to the return of ESC is primarily a result of upper stratosphere cooling and the strengthened Brewer–Dobson circulation in the stratosphere (Chapter 8).

Total ozone changes are greatest in the Antarctic in October, in austral spring (Section 1.3.1 and Chapter 5). Results from chemistry-climate models indicate that changes in stratospheric halogen levels have the largest impact on Antarctic ozone depletion in the past and in the coming decades. Changes in climate parameters are of lesser importance. Therefore, total ozone changes follow changes in stratospheric halogen levels (ESC): as ESC increases, ozone decreases; as ESC decreases, ozone increases (Figure 1.11). Antarctic total ozone is projected to return to 1980 levels after about 2050 and thus later than in any other region, and yet slightly earlier than when the stratospheric halogen loading is projected to return to 1980 levels. The slightly earlier return of ozone to 1980 levels results primarily from a cooling of the upper stratosphere and the consequent increases in upper stratospheric ozone.[4]

Total ozone depletion in the Arctic in boreal spring (March) is considerably smaller than in the Antarctic. In contrast to the Antarctic, Arctic ozone changes do not closely mirror changes in stratospheric halogen levels (ESC). After the middle of the century, Arctic total ozone is projected to increase to values above those expected from ESC reductions alone because of the strengthening of the Brewer–Dobson circulation in the stratosphere and the enhanced stratospheric cooling associated with increases in CO_2 (Chapters 8 and 9). By 2100 Arctic total ozone values are projected to lie well above both 1960 and 1980 values (Figure 1.11). The large range in projections compared to extrapolar regions is caused by the strong year-to-year variability of meteorological conditions in Arctic winter and spring (Chapter 5). The strong natural year-to-year variability in the Arctic also makes it difficult to obtain accurate model simulations for this region, in particular for polar temperatures and the transport barrier at the polar vortex edge. Thus, most current models underestimate present-day Arctic ozone depletion.[4,28,110,111]

In the northern and southern mid-latitudes, the annual averages of total ozone changes are much smaller than the springtime losses in polar regions. Both mid-latitude regions show similar return dates of ESC to 1960 and 1980 conditions. Like in the Arctic, total ozone is projected to return to 1960 and 1980 conditions much sooner than ESC does (Figure 1.11). In the northern mid-latitudes, a return of total ozone values to 1980 values is projected to occur between 2015 and 2030, whereas ESC only returns to close to 1980 values by

2050.[4] Total ozone in southern mid-latitudes is projected to return to 1980 values between 2030 and 2040 while ESC is projected to develop similarly as in the northern hemisphere.[4] Further, in southern mid-latitudes, the 1960 return date for total ozone is somewhat later (2055) and the maximum ozone depletion observed near 2000 is greater than in northern mid-latitudes because of the influence of the Antarctic ozone hole, where stratospheric air, strongly depleted in ozone, is transported to southern mid-latitudes in spring when the polar vortex dissipates. The more rapid return of total ozone in northern and southern mid-latitudes compared with ESC is caused again by the influence of climate change induced trends in transport and upper stratospheric temperatures in the model projections.[4]

In the tropics, total ozone changes are smaller than in the mid-latitudes and in the polar regions (Figure 1.11), because ozone is less sensitive to stratospheric halogen levels in the tropical stratosphere. In contrast to the extra-tropics, chemistry-climate models project total ozone to remain below 1960 values throughout the 21st century. The evolution of total ozone in the tropics is determined by the balance between ozone increases in the upper stratosphere and decreases in the lower stratosphere. Increase of ozone in the upper stratosphere is caused by the declining stratospheric halogen burden (Chapter 2) and a slowing of ozone destruction due to decreases in temperatures (Chapter 6) caused by increasing greenhouse gases (Chapters 8 and 9). Decrease of ozone in the lower stratosphere is caused by the strengthening of the stratospheric Brewer-Dobson circulation (Chapters 8 and 9), which reduces the time for ozone production in the upwelling air in the tropical stratosphere.

1.4.2 The World Avoided by the Montreal Protocol

Recently, in model studies, the question was addressed, what would have been the future of the stratospheric ozone layer, had no legally binding controls on halogen source gas emissions been put into effect?[41,42] If this had happened, the production of ozone-depleting substances would have grown at an annual rate of about 3% (Figure 1.6). Newman et al.[42] used a fully coupled radiation-chemical-dynamical model to simulate a hypothetical future world that would have developed under these circumstances. In their "world avoided" simulation, 17% of the globally averaged column ozone is destroyed by 2020, and 67% is destroyed by 2065 in comparison to 1980.[42] Similarly, Morgenstern et al.[41] find for a non-regulated growth of halogen source gases to a level of 9 ppb Cl_y until 2030 a much larger ozone depletion than has occurred hitherto. Further, Newman et al.[42] find that large ozone depletions in the polar region are to extend into a year-round phenomenon in contrast to the seasonal ozone loss that is observed in the Antarctic ozone hole from the early 1980s to the present day (see Section 1.3.1 and Chapter 5). Very large temperature reductions are found in the simulation[42] in response to circulation changes and decreased shortwave radiation absorption by ozone. Moreover, simulated

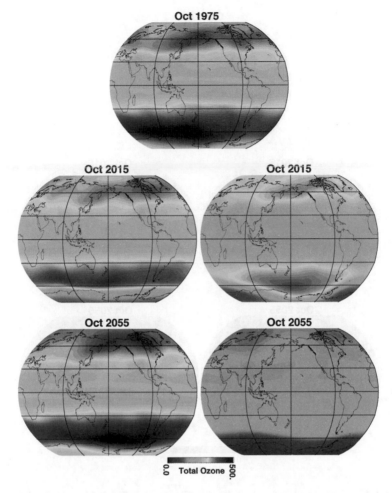

Figure 1.12   Simulated total ozone column (in DU) for the years 1975, 2015, and 2055 from the model study by Newman *et al.* 2009.[42] (Figure courtesy of Paul Newman).

ozone levels in the tropical lower stratosphere remain constant until about 2053 and then collapse to near zero by 2058 (Figure 1.12). This is because the same heterogeneous chemical processes as currently observed in the Antarctic ozone hole (Chapter 4) are triggered in this simulation[42] by 2058, when tropical temperatures drop to low enough values. The strong tropical cooling is caused by an increase of the tropical upwelling in the model.[42] An increase in tropical upwelling in a hypothetical world without a Montreal Protocol has also been reported by Morgenstern *et al.*[41] In response to simulated ozone reductions, ultraviolet radiation increases,[42] more than doubling the erythermal radiation in the northern summer mid-latitudes by 2060 (Figure 1.13).

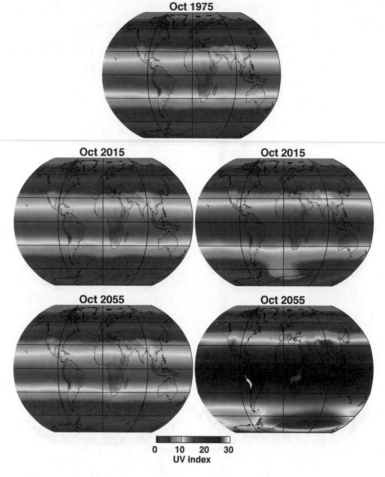

Figure 1.13   The simulated UV index for the years 1975, 2015, and 2055 from the
model study by Newman *et al.* 2009.[42] (Figure courtesy of Paul
Newman).

Acknowledgements

I thank Steve Montzka, Paul Newman and Fred Stroh for helpful comments on
this chapter. I also thank Jens-Uwe Grooß, Paul Newman, and Markus Rex for
providing figures. Furthermore, I am grateful to J. Carter-Sigglow for revising
the grammar and style of the manuscript.

References

1. G. M. B. Dobson, *Appl. Opt.*, 1968, **7**, 387–405.
2. G. M. B. Dobson and D. N. Harrison, *Proc. R. Soc.* London A, 1926,
   110, 660–693.

3. G. P. Gushchin, *Izv. A. N. Fiz. Atmos. Ok.*, 1995, **31**, 6–9.
4. WMO *Scientific assessment of ozone depletion: 2010, Global Ozone Research and Monitoring Project-Report No. 52*, Geneva, Switzerland, 2011, p. 516.
5. G. P. Brasseur, in *Climate Variability and Extremes During the Past 100 Years,* ed. S. Brönnimann *et al.*, Springer, 2008, vol. 33, pp. 303–316.
6. R. Müller, *Meteorol. Z.*, 2009, **18**, 3–24, DOI: 10.1127/0941-2948/2009/353.
7. F. W. P. Götz, G. M. B. Dobson and A. R. Meetham, *Nature*, 1933, **132**, 281.
8. E. Regener and V. H. Regener, *Phys. Z.*, 1934, **35**, 788–793.
9. F. S. Johnson, J. D. Purcell and R. Tousey, *J. Geophys. Res.*, 1951, **56**, 583–59.
10. R. F. Weiss, *Marine Chem.*, 1974, **2**, 203–215.
11. S. Chapman, *Mem. Roy. Soc.*, 1930, **3**, 103–109.
12. P. J. Crutzen, *Q. J. R. Meteorol. Soc.*, 1970, **96**, 320–325.
13. D. R. Bates and M. Nicolet, *J. Geophys. Res.*, 1950, **55**, 301–327.
14. J. Hampson, *Photochemical behaviour of the ozone layer*, Canadian armament research and development establishment technical report, Nr. 1627, 1964, p. 280.
15. R. S. Stolarski and R. J. Cicerone, *Canad. J. Chem.*, 1974, **52**, 1610–1615.
16. IPCC/TEAP, *Special Report on Safeguarding the Ozone Layer and the Global Climate System: Issues Related to Hydrofluorocarbons and Perfluorocarbons*, Cambridge University Press, Cambridge, United Kingdom, and New York, USA, 2005, p. 478.
17. P. J. Crutzen, J.-U. Grooß, C. Brühl, R. Müller and J. M. Russell III, *Science*, 1995, **268**, 705–708.
18. S. P. Sander, R. R. Friedl, D. M. Golden, M. J. Kurylo, R. E. Huie, V. L. Orkin, G. K. Moortgat, A. R. Ravishankara, C. E. Kolb, M. J. Molina and B. J. Finlayson-Pitts, *Evaluation number 14, NASA Panel for Data Evaluation, JPL Publication 02-25, Jet Propulsion Laboratory*, California Insitute of Technology, Pasadena, California, 2003.
19. C. E. Junge, J. E. Manson and C. W. Chagnon, *Science*, 1961, **133**, 1478–1479, DOI: 10.1126/science.133.3463.1478-a.
20. T. Peter, *Ann. Rev. Phys. Chem.*, 1997, **48**, 785–822.
21. S. Solomon, *Rev. Geophys.*, 1999, **37**, 275–316, DOI: 10.1029/1999RG900008.
22. J. M. Rodriguez, M. K. W. Ko and S. N. D, *Nature*, 1991, **352**, 134–137, DOI: 10.1038/352134a0.
23. D. W. Fahey, S. R. Kawa, E. L. Woodbridge, P. Tin, J. C. Wilson, H. H. Jonsson, J. E. Dye, D. Baumgardner, S. Borrmann and D. W. Toohey, *Nature*, 1993, **363**, 509–514.
24. S. Solomon, R. R. Garcia, F. S. Rowland and D. J. Wuebbles, *Nature*, 1986, **321**, 755–758.
25. J. P. F. Fortuin and H. Kelder, *J. Geophys. Res.*, 1998, **103**, 31709–31734.
26. A. W. Brewer, *Q. J. R. Meteorol. Soc.*, 1949, **75**, 351–363.
27. G. M. B. Dobson, A. W. Brewer and B. M. Cwilong, *Proc. R. Soc. London A*, 1946, **185**, 144–175.

28. D. Sankey and T. G. Shepherd, *J. Geophys. Res.*, 2003, **108**, DOI: 10.1029/2002JD002799.
29. J.-U. Grooß and J. M. Russell, *Atmos. Chem. Phys.*, 2005, **5**, 2797–2807.
30. R. Müller, Tracer-tracer relations as tool for research on polar ozone loss, *Forschungszentrum Jülich*, 2010, vol. 58, p. 116.
31. R. R. Garcia and S. Solomon, *J. Geophys. Res.*, 1985, **90**, 3850–3868.
32. G. E. Bodeker, H. Shiona and H. Eskes, *Atmos. Chem. Phys.*, 2005, **5**, 2603–2615.
33. R. Müller, J.-U. Grooß, C. Lemmen, D. Heinze, M. Dameris and G. Bodeker, *Atmos. Chem. Phys.*, 2008, **8**, 251–264.
34. T. G. Shepherd, *Chem. Rev.*, 2003, **103**, 4509–4532, DOI: 10.1021/cr020511z.
35. A. F. Tuck, T. Davies, S. J. Hovde, M. Noguer-Alba, D. W. Fahey, S. R. Kawa, K. K. Kelly, D. M. Murphy, M. H. Proffitt, J. J. Margitan, M. Loewenstein, J. R. Podolske, S. E. Strahan and K. R. Chan, *J. Geophys. Res.*, 1992, **97**, 7883–7904.
36. J. C. Farman, R. J. Murgatroyd, A. M. Silnickas and B. A. Thrush, *Q. J. R. Meteorol. Soc.*, 1985, **111**, 1013–1025.
37. D. W. Fahey and A. R. Ravishankara, *Science*, 1999, **285**, 208–210.
38. H. Johnston, *Science*, 1971, **173**, 517–522.
39. P. J. Crutzen, *J. Geophys. Res.*, 1971, **76**, 7311–7327.
40. M. J. Molina and F. S. Rowland, *Nature*, 1974, **249**, 810–812.
41. O. Morgenstern, P. Braesicke, M. M. Hurwitz, F. M. O'Connor, A. C. Bushell, C. E. Johnson and J. A. Pyle, *Geophys. Res. Lett.*, 2008, **35**, DOI: 10.1029/2008GL034590.
42. P. A. Newman, L. D. Oman, A. R. Douglass, E. L. Fleming, S. M. Frith, M. M. Hurwitz, S. R. Kawa, C. H. Jackman, N. A. Krotkov, E. R. Nash, J. E. Nielsen, S. Pawson, R. S. Stolarski and G. J. M. Velders, *Atmos. Chem. Phys.*, 2009, **9**, 2113–2128, DOI: 10.5194/acp-9-2113-2009.
43. P. A. Newman, J. S. Daniel, D. W. Waugh and E. R. Nash, *Atmos. Chem. Phys.*, 2007, **7**, 4537–4552.
44. WMO, *Scientific assessment of ozone depletion: 2006, Global Ozone Research and Monitoring Project-Report No. 50*, Geneva, Switzerland, 2007, p. 572.
45. J. S. Daniel, S. Solomon and D. L. Albritton, *J. Geophys. Res.*, 1996, **100**, 1271–1285.
46. A. Engel, U. Schmidt and D. S. McKenna, *Geophys. Res. Lett.*, 1998, **25**, 3319–3322, DOI: 10.1029/98GL02520.
47. P. J. Crutzen, *Geophys. Res. Lett.*, 1974, **1**, 205–208.
48. WMO, *Scientific assessment of ozone depletion: 1985, Report No. 16*, Geneva, Switzerland, 1986, p. 1095.
49. N. Harris, R. Hudson and C. Phillips (ed.), *Assessment of Trends in the Vertical Distribution of Ozone*, WMO Ozone Research and Monitoring Project Report No. 43, 1998, p. 289.
50. WMO, *Scientific assessment of ozone depletion: 1998, Global Ozone Research and Monitoring Project-Report No. 44*, Geneva, Switzerland, 1999.

51. K. S. Groves, S. R. Mattingly and A. F. Tuck, *Nature*, 1978, **273**, 711–715, DOI: 10.1038/273711a0.

52. J. D. Haigh and J. A. Pyle, *Nature*, 1979, **279**, 222–224, DOI: 10.1038/279222a0.

53. W. J. Randel, K. P. Shine, J. Austin, J. Barnett, C. Claud, N. P. Gillett, P. Keckhut, U. Langematz, R. Lin, C. Long, C. Mears, A. Miller, J. Nash, D. J. Seidel, D. W. J. Thompson, F. Wu and S. Yoden, *J. Geophys. Res.*, 2009, **114**, DOI: 10.1029/2008JD010421.

54. M. J. Newchurch, E. S. Yang, D. M. Cunnold, G. C. Reinsel, J. M. Zawodny and J. M. Russell, *J. Geophys. Res.*, 2003, **108**, DOI: 10.1029/2003JD003471.

55. W. Steinbrecht, H. Claude, F. Schönenborn, I. S. McDermid, T. Leblanc, S. Godin, T. Song, D. P. J. Swart, Y. J. Meijer, G. E. Bodeker, B. J. Connor, N. Kämpfer, K. Hocke, Y. Calisesi, N. Schneider, J. de la Nöe, A. D. Parrish, I. S. Boyd, C. Brühl, B. Steil, M. A. Giorgetta, E. Manzini, L. W. Thomason, J. M. Zawodny, M. McCormick, J. M. Russell, P. K. Bhartia, R. S. Stolarski and S. M. Hollandsworth-Frith, *J. Geophys. Res.*, 2006, **111**, DOI: 10.1029/2005JD006454.

56. J. C. Farman, B. G. Gardiner and J. D. Shanklin, *Nature*, 1985, **315**, 207–210.

57. A. E. Jones and J. D. Shanklin, *Nature*, 1995, **376**, 409–411.

58. R. S. Stolarski, A. J. Krueger, M. R. Schoeberl, R. D. McPeters, P. A. Newman and J. C. Alpert, *Nature*, 1986, **322**, 808–811.

59. K. Hoppel, R. Bevilacqua, D. Allen, G. Neduluha and C. Randall, *Geophys. Res. Lett.*, 2003, **30**, DOI: 10.1029/2002GL016899.

60. S. Tilmes, R. Müller, J.-U. Grooß, R. Spang, T. Sugita, H. Nakajima and Y. Sasano, *J. Geophys. Res.*, 2006, **111**, DOI: 10.1029/2005JD06260.

61. S. Solomon, R. W. Portmann, T. Sasaki, D. J. Hofmann and D. W. J. Thompson, *J. Geophys. Res.*, 2005, **110**, DOI: 10.1029/2005JD005917.

62. S. Solomon, R. W. Portmann and D. W. J. Thompson, *Proc. Natl. Acad. Sci.*, 2007, **104**, 445–449.

63. M. R. Schoeberl and D. L. Hartmann, *Science*, 1991, **251**, 46–52.

64. A. F. Tuck, *J. Geophys. Res.*, 1989, **94**, 11687–11737.

65. S. Tilmes, R. Müller, J.-U. Grooß and J. M. Russell, *Atmos. Chem. Phys.*, 2004, **4**, 2181–2213.

66. F. Goutail, J.-P. Pommereau, F. Lefèvre, M. V. Roozendael, S. B. Andersen, B.-A. Kåstad-Høiskar, V. Dorokhov, E. Kyrö, M. P. Chipperfield and W. Feng, *Atmos. Chem. Phys.*, 2005, **5**, 665–677.

67. M. Rex, R. J. Salawitch, H. Deckelmann, P. von der Gathen, N. R. P. Harris, M. P. Chipperfield, B. Naujokat, E. Reimer, M. Allaart, S. B. Andersen, R. Bevilacqua, G. O. Braathen, H. Claude, J. Davies, H. De Backer, H. Dier, V. Dorokov, H. Fast, M. Gerding, S. Godin-Beekmann, K. Hoppel, B. Johnson, E. Kyrö, Z. Litynska, D. Moore, H. Nakane, M. C. Parrondo, A. D. Risley Jr., P. Skrivankova, R. Stübi, P. Viatte, V. Yushkov and C. Zerefos, *Geophys. Res. Lett.*, 2006, **33**, DOI: 10.1029/2006GL026731.

68. S. Tilmes, R. Müller, R. J. Salawitch, U. Schmidt, C. R. Webster, H. Oelhaf, J. M. Russell III and C. C. Camy-Peyret, *Atmos. Chem. Phys.*, 2008, **8**, 1897–1910.

69. K. K. Tung, M. K. W. Ko, J. Rodriguez and N. D. Sze, *Nature*, 1986, **333**, 811–814.

70. J. D. Mahlman, H. Levy II and W. J. Moxim, *J. Geophys. Res.*, 1986, **91**, 2687–2707.

71. L. Callis and M. Natarajan, *J. Geophys. Res.*, 1986, **91**, 10771–10780.

72. C. B. Farmer, G. C. Toon, P. W. Schaper, J.-F. Blavier and L. L. Lowes, *Nature*, 1987, **329**, 126–130.

73. S. Solomon, G. H. Mount, R. W. Sanders and A. L. Schmeltekopf, *J. Geophys. Res.*, 1987, **92**, 8329–8338.

74. R. L. de Zafra, M. Jaramillo, A. Parrish, P. Solomon, B. Connor and J. Barrett, *Nature*, 1987, **328**, 408–411.

75. A. F. Tuck, R. T. Watson, E. P. Condon, J. J. Margitan and O. B. Toon, *J. Geophys. Res.*, 1989, **94**, 11181–11222.

76. R. P. Turco, A. Plumb and E. Condon, *Geophys. Res. Lett.*, 1990, **17**, 313–316.

77. O. B. Toon, P. Hamill, R. P. Turco and J. Pinto, *Geophys. Res. Lett.*, 1986, **13**, 1284–1287.

78. P. J. Crutzen and F. Arnold, *Nature*, 1986, **342**, 651–655.

79. D. R. Hanson and K. Mauersberger, *Geophys. Res. Lett.*, 1988, **15**, 855–858, DOI: 10.1029/88GL00209.

80. C. Voigt, J. Schreiner, A. Kohlmann, P. Zink, K. Mauersberger, N. Larsen, T. Deshler, C. Kröger, J. Rosen, A. Adriani, F. Cairo, G. D. Donfrancesco, M. Viterbini, J. Ovarlez, H. Ovarlez, C. David and A. Dörnbrack, *Science*, 2000, **290**, 1756–1758.

81. D. J. Hofmann, T. L. Deshler, P. Aimedieu, W. A. Matthews, P. V. Johnston, Y. Kondo, W. R. S. G. J. Byrne and J. R. Benbrook, *Nature*, 1989, **340**, 117–121.

82. WMO, *Scientific assessment of ozone depletion: 1989, Report No. 20*, Geneva, Switzerland, 1990, p. 486.

83. D. W. Fahey, R. S. Gao, K. S. Carslaw, J. Kettleborough, P. J. Popp, M. J. Northway, J. C. Holecek, S. C. Ciciora, R. J. McLaughlin, T. L. Thompson, R. H. Winkler, D. G. Baumgardner, B. Gandrud, P. O. Wennberg, S. Dhaniyala, K. McKinley, T. Peter, R. J. Salawitch, T. P. Bui, J. W. Elkins, C. R. Webster, E. L. Atlas, H. Jost, J. C. Wilson, R. L. Herman, A. Kleinböhl and M. von König, *Science*, 2001, **291**, 1026–1031.

84. G. L. Manney, K. Krüger, J. L. Sabutis, S. Amina Sena and S. Pawson, *J. Geophys. Res.*, 2005, **110**, DOI: 10.1029/2004JD005367.

85. K. S. Carslaw, M. Wirth, A. Tsias, B. P. Luo, A. Dörnbrack, M. Leutbecher, H. Volkert, W. Renger, J. T. Bacmeister, E. Reimer and T. Peter, *Nature*, 1998, **391**, 675–678.

86. S. Fueglistaler, S. Buss, B. P. Luo, H. Wernli, H. Flentje, C. Hostetler, L. R. Poole, K. S. Carslaw and T. Peter, *Atmos. Chem. Phys.*, 2003, **3**, 697–712.

87. M. C. Pitts, L. R. Poole, A. Dörnbrack and L. W. Thomason, *Atmos. Chem. Phys.*, 2011, **11**, 2161–2177, DOI: 10.5194/acp-11-2161-2011.
88. J. M. Rodriguez, M. K. W. Ko and N. D. Sze, *Geophys. Res. Lett.*, 1988, **15**, 257–260.
89. D. J. Hofmann and S. Solomon, *J. Geophys. Res.*, 1989, **94**, 5029–5041.
90. G. Brasseur, C. Granier and S. Walters, *Nature*, 1990, **348**, 626–628.
91. M. A. Tolbert, M. J. Rossi and D. M. Golden, *Geophys. Res. Lett.*, 1988, **15**, 847–850.
92. D. R. Hanson and A. R. Ravishankara, *J. Geophys. Res.*, 1991, **96**, 17307–17314.
93. A. R. Ravishankara and D. R. Hanson, *J. Geophys. Res.*, 1996, **101**, 3885–3890.
94. E. W. Wolff and R. Mulvaney, *Geophys. Res. Lett.*, 1991, **18**, 1007–1010.
95. R. A. Cox, A. R. MacKenzie, R. H. Müller, T. Peter and P. J. Crutzen, *Geophys. Res. Lett.*, 1994, **21**, 1439–1442.
96. D. R. Hanson, A. R. Ravishankara and S. Solomon, *J. Geophys. Res.*, 1994, **99**, 3615–3629.
97. R. W. Portmann, S. Solomon, R. R. Garcia, L. W. Thomason, L. R. Poole and M. P. McCormick, *J. Geophys. Res.*, 1996, **101**, 22991–23006.
98. S. Tilmes, R. Müller and R. J. Salawitch, *Science*, 2008, **320**, 1201–1204, DOI: 10.1126/science.1153966.
99. Q. Shi, J. T. Jayne, C. E. Kolb, D. R. Worsnop and P. Davidovits, *J. Geophys. Res.*, 2001, **106**, 24259–24274, DOI: 10.1029/2000JD000181.
100. K. Drdla and R. Müller, *Atmos. Chem. Phys. Discuss.*, 2010, **10**, 28687–28720, DOI: 10.5194/acpd-10-28687-2010.
101. R. Müller, Th. Peter, P. J. Crutzen, H. Oelhaf, G. P. Adrian, T. v. Clarmann, A. Wegner, U. Schmidt and D. Lary, *Geophys. Res. Lett.*, 1994, **21**, 1427–1430.
102. A. R. Douglass, M. R. Schoeberl, R. S. Stolarski, J. W. Waters, J. M. Russell III, A. E. Roche and S. T. Massie, *J. Geophys. Res.*, 1995, **100**, 13967–13978.
103. L. T. Molina and M. J. Molina, *J. Phys. Chem.*, 1987, **91**, 433–436.
104. M. B. McElroy, R. J. Salawitch, S. C. Wofsy and J. A. Logan, *Nature*, 1986, **321**, 759–762.
105. M. P. Chipperfield and J. A. Pyle, *J. Geophys. Res.*, 1998, **103**, 28389–28403.
106. M. Rex, N. R. P. Harris, P. von der Gathen, R. Lehmann, G. O. Braathen, E. Reimer, A. Beck, M. Chipperfield, R. Alfier, M. Allaart, F. O'Connor, H. Dier, V. Dorokhov, H. Fast, M. Gil, E. Kyrö, Z. Litynska, I. S. Mikkelsen, M. Molyneux, H. Nakane, J. Notholt, M. Rummukainen, P. Viatte and J. Wenger, *Nature*, 1997, **389**, 835–838, DOI: 10.1038/39849.
107. G. Becker, R. Müller, D. S. McKenna, M. Rex and K. S. Carslaw, *Geophys. Res. Lett.*, 1998, **25**, 4325–4328.
108. G. Becker, R. Müller, D. S. McKenna, M. Rex, K. S. Carslaw and H. Oelhaf, *J. Geophys. Res.*, 2000, **105**, 15175–15184.

109. A. R. Ravishankara, J. S. Daniel and R. W. Portmann, *Science*, 2009, **326**, 123–125, DOI: 10.1126/science.1176985.
110. S. Tilmes, D. Kinnison, R. Müller, F. Sassi, D. Marsh, B. Boville and R. Garcia, *J. Geophys. Res.*, 2007, **112**, DOI: 10.1029/2006JD008334.
111. SPARC report on the evaluation of chemistry-climate models, ed. V. Eyring, T. G. Sheperd and D. W. Waugh, *World Meteorol. Organ.*, WMO/TD-No. 1526, Geneva, 2010, p. 434.

CHAPTER 2

Source Gases that Affect Stratospheric Ozone

STEPHEN A. MONTZKA

Earth System Research Laboratory, National Oceanic and Atmospheric
Administration, Boulder, Colorado, USA

2.1 Introduction

Human activities have altered the chemical composition of the global atmosphere. Changes are particularly noticeable for emitted chemicals degraded
slowly by natural processes. Emissions of such chemicals with lifetimes of a few
months or more, even those chemicals that are "heavier than air", are carried
by winds and the broad-scale general atmospheric circulation to become distributed throughout the troposphere and stratosphere. Chlorofluorocarbons
(CFCs) are human-made chemicals containing chlorine that persist in the
environment for decades to centuries. Continued emissions of these persistent
chemicals and others that contain chlorine and bromine led to increasing
concentrations during the second half of the 20th century and have caused the
depletion of stratospheric ozone.

  Long-lived, chlorine- and bromine-containing "source gases" were produced
by industry to fulfill societal needs. CFCs were particularly effective in applications for cooling and refrigeration, and they gained acceptance because they
were less toxic and flammable than the chemicals they replaced. CFCs were also
extensively used as blowing agents for foams and as aerosol propellants. Other
sources of ozone-depleting chlorine and bromine to the stratosphere include
methyl chloroform (CH_3CCl_3) and carbon tetrachloride (CCl_4), which are
chemicals used as solvents and chemical feedstocks; halon fire-extinguishing

Stratospheric Ozone Depletion and Climate Change
Edited by Rolf Müller
© Royal Society of Chemistry 2012
Published by the Royal Society of Chemistry, www.rsc.org

agents (halon-1211, CF_2ClBr; halon-1301, CF_3Br; and halon-2402, CF_2BrCF_2Br); and the fumigant methyl bromide (CH_3Br). Although hydro-chlorofluorocarbons (HCFCs) were designed as temporary substitutes for the main ozone-depleting substances (ODSs) mentioned above, they also contribute chlorine to the stratosphere and so are considered ODSs. The use of HCFCs as substitutes for the main ODSs was allowed temporarily because they are less efficient at delivering chlorine to the stratosphere and, therefore, less efficient at destroying ozone per amount emitted.

Human-produced CFCs were first detected in the atmosphere in air samples collected across the globe in the early 1970s.[1] The possibility that chlorine could catalyze ozone loss in the stratosphere was first discussed by Stolarski and Cicerone,[2] but only in a separate paper by Molina and Rowland were CFCs identified as potentially becoming a dominant source of ozone-destroying chlorine to the stratosphere.[3] Early model studies[4] predicted that enhanced levels of chlorine in the stratosphere would lead to a depletion of upper stratospheric ozone *via* a catalytic reaction cycle with chlorine. Additional studies[5] noted that bromine would also participate in catalytic reactions that destroy ozone. Some countries adopted policy measures on the basis of these early warnings. The United States phased out the use of CFCs as a propellant in spray cans as of 1 January 1979. This action was soon followed by similar bans in Canada, Sweden, and Norway. The use of CFCs as propellants accounted for only a fraction of all uses, however, and those other uses continued unabated for the next decade or longer. The urgency to restrict CFC production was muted in the early 1980s because theoretical studies indicated that ozone depletion would likely become severe only decades into the future.[4,6] This changed, however, in 1985 with the discovery, by Farman and coworkers,[7] of severe and worsening ozone depletion over Halley Station, Antarctica. Initially, a number of theories were proposed to explain the observed depletion, including enhanced concentrations of halogen from CFCs and related industrial gases.[8] Eventually, additional observational, laboratory, and theoretical work laid the blame squarely on chlorine and bromine from human-produced chemicals.[9–11] The influence of these chemicals was enhanced in the Antarctic by the unique atmospheric circulation patterns existing there (see Chapter 1). Within two years of the discovery of pronounced ozone depletion over Antarctica during Austral spring the first international agreement dealing with an environmental issue was signed, the "Montreal Protocol on Substances that Deplete the Ozone Layer".

As originally written and ratified, the Montreal Protocol provided only limited restrictions on the production and trade of some abundant ODSs. This was in part because significant uncertainties on the causes for the observed depletion were present at that time (the mid-1980s). The Protocol, however, included a mechanism for regularly revisiting these control measures as scientific uncertainties on the causes for depletion were reduced (see Box 2.1). As a result, in subsequent years, a number of amendments and adjustments were adopted by the Parties to the Protocol that strengthened controls so that at the present time (2011) anthropogenic production for dispersive uses (*i.e.*, those

BOX 2.1: The Success of the Montreal Protocol

The 1987 Montreal Protocol on Substances that Deplete the Ozone Layer has been very successful as a means of controlling the atmospheric abundances of ODSs. It did not specifically regulate or control atmospheric abundances or emissions of these gases. Instead, it targeted industrial production and international trade. This has proven to be a sufficient control mechanism because the fate of most ODS production is to eventually escape to the atmosphere.

Why was the Montreal Protocol successful?

The Montreal Protocol has been called one of the most successful international environmental agreements ever. Its success can be linked to a number of different features that are useful to review in part because not all international protocols (*e.g.*, the Kyoto Protocol) have been as successful in achieving their goals.

First, as initially written and ratified, the Montreal Protocol was not sufficiently restrictive to allow the ozone layer to heal. Instead, the 1987 Protocol represented a first step in which Parties to the Protocol acknowledged that there could be a problem and agreed to some relatively small limits on future production of some gases. The success of the Protocol ultimately stemmed from Policy-makers regularly revisiting existing controls on ODS production and, with updated input from scientific, technological, and economic assessment panels, deciding if tighter or looser controls on production were prudent. As a result of this process the Protocol was significantly adjusted and amended multiple times after 1987. The Protocol continues to be adjusted and amended; in 2007, for example, an accelerated schedule for the phase-out of hydrochlorofluorocarbons (HCFCs) was adopted because of the increasing threat their atmospheric increases posed to stratospheric ozone and because of the climate benefit such action would provide.[15] Additional amendments and revisions to the Montreal Protocol are even being considered today (2009–2010) to address environmental concerns related to the non-ozone depleting HFCs.

Second, the Montreal Protocol was also successful because its controls targeted an easily measureable and traceable quantity: the production and "consumption" of ODSs, where "consumption" refers to production plus net exchanges from international trade. This was a logical approach given that most production of ODSs ultimately escapes to the atmosphere, although, in the case of methyl bromide, it has some unique ramifications. Only a fraction (50–70%) of methyl bromide applied to soils for fumigation purposes is believed to ultimately reach the atmosphere.

Third, the Montreal Protocol set up schedules for cuts over time in the production of different groupings of chemicals (CFCs, for example) or individual compounds. It did not group all ODSs into a single "basket of gases" where cuts in any chemical could be offset by increases in any other compound. Instead, these compound groups were delineated by chemicals having similar atmospheric properties or uses so that it wasn't necessary to accurately assess the relative impacts of, for example, long-lived ODSs to those with shorter lifetimes. This effectively ensured production cuts in both long- and short-lived chemicals. Cuts in both long- and short-lived chemicals provided the best opportunity for a recovery that started early and was sustained over the long term,[52] whereas a focus on chemicals having only short or only long lifetimes would not ensure such an outcome (see text).

Fourth, the timescales for production and consumption phase-outs were designed based on numerous technological and political considerations. In particular, a ten-year grace period was provided for limiting and phasing down ODS production and consumption in developing countries. Allowing a delay in the phase-out of ODSs in developing countries recognized their lesser contribution to the initial stages of ozone depletion and ensured that future increases in ODS production and consumption of ODSs from these countries would not offset entirely the progress that had been made by developed countries to curb ODS emissions.

ultimately leading to emission of an ODS to the atmosphere) of the most potent ozone-depleting gases has practically ceased. Some production of ODSs continues, however, even today. For example, the production of ODS substitutes (*e.g.*, the HCFCs) continues and increased substantially during 2004–2010 because of rapid increased use in developing countries. Furthermore, production of ODSs for non-dispersive uses (*e.g.,* as chemical reagents or "feedstocks") continues and is not controlled by the Montreal Protocol as only a small fraction of this production is believed to leak into the atmosphere.

A critical step in assessing the progress of the Protocol includes tracking changes in the atmospheric abundance of ODSs and tracking the burden of ozone-depleting halogen in the stratosphere. Monitoring these changes and assessing their implications for stratospheric ozone and the success of the Montreal Protocol are discussed in this chapter. Although production of ODSs has been greatly reduced, atmospheric abundances and environmental impacts of ODSs respond to production decreases with delays that have different timescales for different ODSs. These delays arise from two considerations. First, although production may have stopped, halogen source gases are still present in existing or discarded equipment and chemical stockpiles. Emissions from these reservoirs or "banks" of chemical can sustain ODS emissions long after production has ceased. Banks represent the main source of emissions for some ODSs currently. The size and leak rate of ODSs from these banks are uncertain, making atmospheric data important for assessing the impact of these banks on the effectiveness of the Protocol. Second, ODSs are destroyed in the atmosphere by natural processes that typically remove only a few percent of an ODS each year from the atmosphere. Our understanding of the persistence of ODSs in the atmosphere, *i.e.* their "lifetimes", is limited, thus atmospheric measurements help verify our understanding of ODS destruction rates and refine our expectations for the timescales of ozone layer recovery.

Natural processes also contribute chlorine and bromine to the stratosphere, and atmospheric observations and modeling are critical for tracking changes in those contributions over time. Such changes are likely in a world modified by climate change because climate influences sources and sinks of halogenated trace gases. The most important natural contributor to chlorine in the stratosphere is methyl chloride, which accounts for about 16% of chlorine currently in the stratosphere. For bromine, naturally-emitted gases account for a much larger percentage of the amount of this halogen present in the stratosphere: about half of the bromine in the stratosphere today is believed to come from natural emissions of methyl bromide and very short-lived brominated gases.[12] While the contributions of long-lived ODSs to stratospheric chlorine and bromine are well estimated from ground-based observations at remote sites, the same is not true for the very short-lived halocarbons. This is because the atmospheric distributions of very short-lived halocarbons are highly variable in space and time. As a result, their contributions to stratospheric halogen burdens are more difficult to quantify and are discussed separately in Section 2.3.

Multiple metrics are used to describe how the threat to ozone from all long-lived ODSs is changing over time and are discussed in Section 2.4. The sum of the concentrations of tropospheric chlorine and bromine in source gases is useful to track, particularly once the enhanced efficiency for bromine to destroy ozone is included (the aggregated weighted sum has been called Equivalent Chlorine, or ECl). But source gases do not deplete ozone *per se*—halogen-catalyzed ozone depletion occurs after source-gases photochemically decompose in the stratosphere and liberate inorganic halogen (see Chapter 3). Equivalent Effective Chlorine (EECl) and Equivalent Effective Stratospheric Chlorine (EESC)[13] are metrics designed to estimate changes in stratospheric inorganic halogen from measured tropospheric changes in source gas concentrations and compound-dependent stratospheric decomposition rates. One additional metric, the Equivalent Stratospheric Chlorine (ESC), is a model-derived measure of equivalent inorganic stratospheric halogen (chlorine and bromine). Although these models are generally constrained by tropospheric source gas concentrations or emissions, ESC is the weighted sum of equivalent inorganic stratospheric halogen calculated based on the stratospheric degradation of source gases within these models.[14]

Human-influenced emissions of non-halogenated gases such as methane, nitrous oxide, and some sulfur-containing gases influence stratospheric ozone chemistry and so are discussed in Section 2.5. These chemicals affect the efficiency with which inorganic forms of chlorine and bromine deplete stratospheric ozone. In addition, HO_x from methane and NO_x from nitrous oxide catalyze the destruction of ozone in the upper stratosphere (where $HO_x = OH + HO_2$ and $NO_x = NO + NO_2$) (see also Chapter 1), although methane enhancements can lead to ozone increases in the lower stratosphere and troposphere. Sulfur-containing gases are also discussed because they are the primary component of stratospheric aerosol particles in the absence of significant volcanic contributions. This stratospheric aerosol affects the partitioning between reservoir forms of inorganic halogen and ozone-depleting forms in both the mid-latitude and polar stratospheres (see Chapter 4).

Ozone-depleting source gases, their substitutes (*e.g.*, HFCs), and the non-halogenated gases discussed in Section 2.5 are efficient greenhouse gases, though their current influence on climate is smaller than carbon dioxide (CO_2). Because climate change is expected to affect the abundance and distribution of stratospheric ozone (see Chapter 8), greenhouse gases have an indirect influence on stratospheric ozone. Although the Montreal Protocol was designed to address ozone depletion, it has provided a significant climate benefit as a result of its success at reducing emissions of the potent greenhouse gases that also deplete ozone.[15] The magnitude of these influences and their time-dependence are discussed briefly in Section 2.6.

2.2   Longer-lived Halogenated Source Gases

2.2.1   The Timescales and Processes that Remove Ozone-depleting Substances from the Atmosphere

Humans produce and release to the atmosphere a range of chemicals containing the halogens fluorine, chlorine, bromine, and iodine. The focus of this section is on those organic chlorinated or brominated source gases (*i.e.*, chemicals containing C–X bonds, where X = chlorine or bromine) that are decomposed on timescales of 0.5 years or longer by the natural oxidizing processes present in the atmosphere. We focus on these longer-lived compounds here because they account for the majority of ozone-depleting halogen reaching the stratosphere and because their contributions can be accurately estimated from tropospheric observations. Photolysis in the stratosphere is the most important process in removing the long-lived ODSs such as CFCs and CCl_4 from the atmosphere. Other processes that remove some ODSs from the atmosphere include reaction with atmospheric oxidants (OH, O_3, O atoms, *etc.*), hydrolysis, and for some chemicals, uptake by biota.[12] Terrestrial ecosystems act as sinks for a few long-lived halogenated trace gases (*e.g.*, methyl bromide), and in the marine environment irreversible loss can arise from hydrolysis or biological processing. Oceanic losses of the methyl halides, carbon tetrachloride, and methyl chloroform are well documented and must be included to derive accurate estimates of the overall lifetimes of these gases.[16,17] Most products of the atmospheric oxidation and photolysis of ODSs are water soluble and are subject to removal from the troposphere by wet and dry deposition on short timescales (< 1 year). Similarly, many other anthropogenically emitted halogenated chemicals used, for example, to clear roads of ice and disinfect pools do not significantly add to halogen in the stratosphere because they are highly soluble in water and are readily scrubbed from the atmosphere as they dissolve into aerosols and rain droplets that ultimately fall to the Earth's surface. The same is true for naturally emitted salts (sea salt, for example, and HCl emitted from volcanoes).

CFCs, however, are highly resistant to destructive processes in the lower atmosphere. CFCs are destroyed primarily by dissociative absorption of high-energy light that exists with substantial flux only above the low- to mid-stratosphere. In these stratospheric regions photolysis of CFCs is rapid, but because only a small fraction of the atmospheric burden of a CFC is transported through this stratospheric region each year, only a small amount of the global CFC burden becomes destroyed annually. Roughly 1% of the CFC-12 in the global atmosphere is transported each year to altitudes in the stratosphere where CFC-12 is readily photolyzed. Loss of $\sim 1\%$/yr of CFC-12 yields a lifetime of 100 years. CFC-11, on the other hand, is photolyzed by lower-energy radiation that penetrates to lower altitudes.[18] Because about 2% of the global CFC-11 burden is transported each year above an altitude where it is readily photolyzed, its lifetime is ~ 45 years.

The range of lifetimes for different CFCs (45–1000 years) reflects differences in their ability to dissociatively absorb UV light of different wavelengths. Light absorption cross-sections of CFCs are roughly proportional to the number of C–Cl bonds they contain. As a result, CFCs with multiple C–Cl bonds photolyze at longer wavelengths (less energy) and the lifetime for $CFCl_3$ (CFC-11; 45 years) is half that of CF_2Cl_2 (CFC-12; 100 yr) and less than one-tenth that of CF_3Cl (CFC-13; 640 years). This is also true for HCFCs; HCFC-141b (CH_3CFCl_2), given that two chlorine atoms are bonded to one carbon atom, is photolyzed more rapidly than HCFC-142b (CH_3CF_2Cl) (see Table 2.1).

Table 2.1 Properties and recent atmospheric abundances of the main ozone-depleting substances controlled by the Montreal Protocol.

Name	Formula	Lifetime (year)	ODP in the Montreal Protocol	Global mean during 2010[a] (ppt)	Peak mixing ratio[b] (ppt)
CFC-12	CCl_2F_2	100	1.0	531	543
CFC-11	CCl_3F	45	1.0	241	271
CFC-113	CCl_2FCClF_2	85	0.8	75	84
Halon-1211	$CBrClF_2$	16	3	3.9	4.2
Halon-1301	$CBrF_3$	65	10	3.1	3.1
Halon-2402	$CBrF_2CBrF_2$	20	6	0.5	0.5
Carbon tetrachloride	CCl_4	26	1.1	88	106
Methyl chloroform	CH_3CCl_3	5	0.1	7.6	135
Methyl bromide	CH_3Br	0.8	0.6	7.1	9.4
HCFC-22	$CHClF_2$	11.9	0.055	204	204
HCFC-141b	CH_3CCl_2F	9.2	0.11	20.4	20.4
HCFC-142b	CH_3CClF_2	17.2	0.065	20.1	20.1

Notes: Lifetimes at steady state given here are from ref. 12, although recent work[18] has suggested that the CFC-11 lifetime may be somewhat longer.

ODPs as listed in the Montreal Protocol (from ref. 12).

[a]Values represent mean global surface mixing ratios estimated from measurements made in the NOAA cooperative global sampling network. Units are pmol mol^{-1}, expressed as parts per trillion or ppt. Data available on line at ftp://ftp.cmdl.noaa.gov/hats/.

[b]Values represent the peak annual mean mixing ratio measured at any time in the past by the NOAA cooperative global sampling network. For compounds whose concentrations are still increasing or constant, the peak mixing ratio is the same as the 2010 global mean. Data available on line at ftp://ftp.cmdl.noaa.gov/hats/.

Long-lived halogenated source gases can be thought of as conduits for transporting chlorine and bromine from Earth's surface to the stratosphere. The portion of these chemicals that undergoes photochemical degradation in the stratosphere adds to the inorganic chlorine or bromine available for ozone destruction. Surface-based emissions of these chemicals have different efficiencies for contributing inorganic chlorine or bromine to the stratosphere based on the number of halogen atoms they contain and the fraction of emission that becomes destroyed in the stratosphere relative to the troposphere. This efficiency is described by a chemical's Ozone Depletion Potential (ODP), which is defined as the ozone depletion arising from a unit mass emission. ODPs are expressed on a scale relative to CFC-11, which is assigned a value of 1.0. The ODP of a chemical is also influenced greatly by the presence of bromine, as this halogen is about 60 times more potent at catalytically destroying ozone than chlorine. In contrast, HCFCs have much lower ODPs (<0.12) than CFCs because they are oxidized in part by OH in the troposphere and, therefore, the portion of HCFC emissions that are oxidized in the troposphere do not contribute chlorine to the stratosphere—essentially the stratospheric halogen conduit represented by HCFC is inefficient (see Table 2.1). Halons, on the other hand, are fully halogenated bromine-containing synthetic chemicals that have very high ODPs (3–10) because they are not destroyed appreciably in the troposphere and, more importantly, because they contain bromine.

2.2.2 The Relative Contribution of Human *vs.* Natural Sources to Ozone-depleting Halogen

The list of long-lived chemicals contributing chlorine and bromine to the stratosphere includes some that have only natural sources, some that originate entirely from human activities, and some that are produced from both natural and human activities. Identifying the relative importance of human *vs.* natural contributions to amounts observed in the modern atmosphere is critical for accurately attributing the cause of ozone depletion and also for understanding the magnitude of any potential benefit (*i.e.*, reduction in atmospheric concentration) from policy decisions affecting anthropogenic production rates.

Multiple pieces of evidence indicate that atmospheric increases observed for CFCs, halons, CH_3CCl_3, CCl_4, and HCFCs during the 19th century are entirely the result of human industrial activity (see below). Industry records indicate substantial increases in production of these chemicals after the mid 20th century for use in a variety of applications such as aerosol propellants, refrigeration, air-conditioning, foam-blowing, fire extinguishing, fumigation, and as solvents and cleaning agents. Originally, production magnitudes were compiled by most industrial producers and global emission estimates were derived based on time-scales for release of these chemicals in a range of different applications.[19] By the late 1990s, companies making up the Alternative Fluorocarbon Environmental Acceptability Study (AFEAS) no longer accounted for the majority of global industrial ODS production. Since that time, production data compiled by the

United Nations Environment Programme (UNEP) collected to monitor compliance with the Montreal Protocol have provided a more complete accounting of global production magnitudes.[20]

Magnitudes of industrial production reported by AFEAS or UNEP can be used to derive an emissions history for a source gas.[12,21,22] To derive an accurate emissions history, sales of source-gases to different applications (*e.g.*, foams blowing, refrigeration, air conditioning, spray cans, *etc.*) are combined with an understanding of how rapidly the source gases are emitted from these applications. Such an emissions history can be compared to emissions derived from atmospheric observations (see Box 2.2). For many ODSs, the emissions history based on production data matches fairly well emissions derived from observed changes in the background atmosphere (Figure 2.1).[12,21] This consistency demonstrates that the amount and changes in the atmospheric abundance of these ODSs are well described by anthropogenic production. Some discrepancies have been noted between measured and expected atmospheric abundances of ODSs. For example, CFC-11 emissions derived from atmospheric observations suggest slightly higher emissions than are derived from production data in recent years (Figure 2.1).[12,21] The source of this discrepancy may stem from uncertainties in the CFC-11 lifetime[18] or in the derivation of emissions from production data. Such explanations are more likely than there being significant natural sources of CFCs and most other industrially produced ODSs.

The conclusion that natural sources of most ODSs are insignificant is based in part on measurements of air trapped in unconsolidated snow above glacial ice, also called "firn air". Techniques for sampling firn air were developed in the 1980s,[23] but were not successfully applied to halocarbons until 1999, when sample collection methodologies were improved.[24–26] Measurements of CO_2 in firn air indicate, based on our understanding of the CO_2 atmospheric history, that firn air has been isolated from the ambient atmosphere for up to a century.[24,27] Given this, firn air data allow one to reconstruct the evolution of the atmospheric chemical composition before ongoing ODS measurements existed and before most ODSs were industrially produced in significant quantities.

In the oldest firn air samples collected during multiple sampling projects in Antarctica and Greenland (dated to the early 1900s) only very small amounts (or amounts indistinguishable from zero) of CFCs, halons, CH_3CCl_3, CCl_4, and HCFCs are measured (Figure 2.2). These results indicate that natural contributions of these industrially produced gases are insignificant. In contrast, concentrations of chemicals with known natural sources such as methyl bromide, methyl chloride, methane, and nitrous oxide are non-zero in the oldest firn air samples. Although conclusions derived from firn air data hinge on the assumption that any slow loss (or production) of these gases within the firn is negligible, such artifacts are unlikely given that consistent histories have been derived over time from air collected at multiple locations having quite a range of mean temperature and precipitation characteristics. One exception is methyl bromide in Greenland firn air and ice bubbles. Anomalously high and non-reproducible values for CH_3Br have been measured in air samples extracted at

BOX 2.2: Deriving Emissions from Observed Changes in Global Background Abundances

The abundance of a trace gas is determined by the balance of sources and loss processes (sinks). Given mass balance considerations, the measured global atmospheric abundance (G) of a trace gas and its change from year to year (dG/dt) can be used to derive a record of emissions, provided losses are well quantified. More specifically, for a long-lived trace gas:

$$dG/dt = E - k \times G \tag{2.1}$$

where E is the emission rate (mol yr^{-1}), k is the pseudo, first-order rate constant for loss, also known as a loss frequency (yr^{-1}), and G has units of moles. The inverse of k is the atmospheric lifetime (τ; units of years) of the trace gas and includes all loss processes affecting the global abundance of the trace gas:

$$k = 1/\tau \tag{2.2}$$

$$1/\tau = 1/\tau_{OH} + 1/\tau_{photolysis} + 1/\tau_{hydrolysis} + 1/\tau_{soils} + 1/\tau_{other}... \tag{2.3}$$

where τ_{OH}, $\tau_{photolysis}$, *etc.* are the partial lifetimes of the global trace gas burden with respect to those individual processes. When rearranged, eqn 2.1 provides the basis for deriving global annual emissions for a long-lived gas from a measure of its global abundance, changes in its global abundance over time, and knowledge of the trace gas loss frequency or lifetime. Global emissions derived in this way can have significant uncertainties because loss frequencies (lifetimes) have uncertainties (see below). Furthermore, this method requires that the product of a global mean loss frequency and the global mean distribution provides a reasonable representation of the mean of the sum of this product on local scales. As a result, global emissions derived in this way are accurate for gases whose concentrations are fairly uniformly distributed throughout the atmosphere, which is generally true for chemicals with lifetimes >0.5 yr. This method also has uncertainties related to estimating the total atmospheric burden of a gas and its change over time based on measurements at a limited number of sites. Long-term sampling networks typically acquire samples only at Earth's surface. This introduces errors if changes in the global mean mixing ratio at Earth's surface are not representative of the change in G throughout the entire atmosphere. Measurements of source gases from aircraft and balloons have improved our understanding of how long-lived gases become distributed throughout the atmosphere after being emitted, thus enabling more robust estimates of global mixing ratios derived from a network of surface sites.[37] For long-lived ODSs, uncertainties related to these issues are thought to be $<10\%$.[21]

Lifetimes for trace gases are derived from estimates of loss rates owing to the different processes removing or chemically transforming a trace gas in the atmosphere (see eqn 2.3). Stratospheric loss rates owing to photolysis are derived from laboratory measurements of a trace gas's ability to dissociatively absorb light as a function of wavelength (its absorption cross section). By incorporating these wavelength-dependent absorption cross sections into atmospheric general circulation models lifetimes associated with stratospheric photolysis are derived. Trace-gas measurements from balloons and aircraft in the stratosphere can also provide a measure of local chemical loss rates and have provided additional constraints to ODS lifetimes from stratospheric loss processes.[37] Recently it has been suggested that the lifetime of CFC-11 is 55–65 years (as opposed to 45 years), given an improved accuracy of models to calculate the residence time for air to travel through the stratosphere (or the "mean age" of stratospheric air).[18]

Rates of loss owing to oxidation by the hydroxyl radical and other oxidants (ozone, oxygen atoms) are derived from estimates of global oxidant concentrations and laboratory-measured reaction rate constants. Global hydroxyl radical concentrations are derived with the budget approach outlined in eqn 2.1 and measurements of a trace gas whose emissions are well known. In the past, the trace gas most often used for this purpose was methyl chloroform.[44,46] Lifetimes with respect to hydroxyl loss for other long-lived chemicals are fairly accurately derived with a global OH concentration estimated from the analysis of methyl chloroform observations and scaling the relative rate at which these other gases react with OH compared to CH_3CCl_3 at a 272 K.[96]

It should be noted that lifetimes are typically quoted for the fairly unique condition called steady-state, *i.e.*, when dG/dt is zero. Under these conditions, eqn 2.1 and 2.2 collapse to:

$$\tau = G/E \tag{2.4}$$

Under non-steady state conditions, however, lifetimes change because loss processes and ODS concentrations are not uniformly distributed throughout the atmosphere. As CFC emissions decline and concentrations decrease (no longer at steady-state), for example, the portion of global CFC molecules in the stratosphere where photolytic destruction occurs is larger than it is at steady state. With proportionally more molecules in the region where loss occurs, the trace-gas lifetime becomes shorter. Lifetime changes of a few percent are possible as trace gas distributions of ODSs respond to emission changes.[127]

Finally, global emissions derived from atmospheric measurements are sensitive to calibration uncertainty. Over the past 30 years advances in measurement and standardization technologies have significantly improved the consistency reported in measurements of ODSs in atmospheric samples. As a result, differences in global mean concentrations reported by different groups are typically <2%, although differences of 3–6% are still observed for some chemicals (CCl_4 and HCFC-142b).[12] Differences in reported global annual means also reflect the influence of instrumentation, site locations, and sampling frequency—and this good consistency suggests that these influences are fairly small.

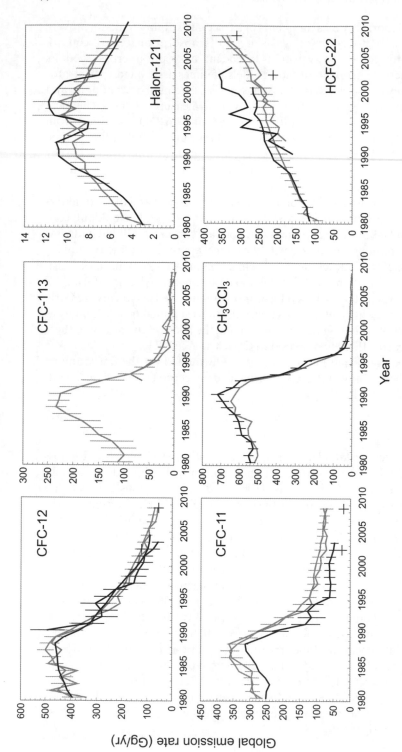

Figure 2.1 Total global emissions for selected ozone-depleting substances since 1980 (in Gg yr^{-1} or 1×10^9 g yr^{-1}). Global emissions were derived from independent surface-sampling networks (gray lines indicating NOAA and AGAGE-derived emissions)[36,52] with methods outlined in Box 2.2 or from inventories of industry production data (black lines). Plus symbols represent the estimated magnitude of emissions from banks in selected years (though for halon-1211, bank emissions in 2002 and 2008 are represented by the black line as 100% of emissions are from banks).[128,129] Inventory emissions are taken from sources relying primarily on AFEAS data (those during 1990–2004),[20,21,128,129] and an independent analysis of industrial production data (methyl chloroform).[131] (Adapted from ref. 12.)

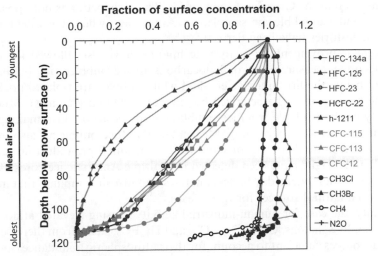

Figure 2.2 Firn-air concentrations of halocarbons measured at different depths in the unconsolidated snowpack at South Pole, Antarctica. Here, results are normalized by ambient air concentrations measured at the time air was sampled from the snow (for 2010 values see Table 2.1). While systematic changes in concentration during the past year including seasonal variations are reflected in results at 10–25 m depth, mean ages increase non-linearly with depth. The mean air age changes with depth most dramatically within the "lock-in zone" below ∼ 105 m, where vertical transport of firn air is reduced. The mean age of the deepest sample at this site is ∼ 1900 A.D.[24,27] Measured concentrations reflect past changes in atmospheric concentrations of these gases and are modulated by how the gases diffuse and mix in the snowpack. In the deepest, oldest samples, measured levels of industrially produced chemicals (CFCs, HCFCs, HFCs, halon-1211) are only a very small fraction of concentrations observed in the atmosphere today (in most cases they are at or below instrumental detection limits) suggesting no significant natural sources. Industrial chemicals produced and emitted earliest in the 20th century (CFCs) have penetrated more deeply into the firn than chemicals that have been produced and emitted for fewer years in the past (*e.g.*, halon-1211 and HCFCs); HFC-134a and HFC-125 have the shortest use history, and so have penetrated the least into the firn. HFC-23 results suggest a different emission history than other HFCs because its emissions are primarily associated with HCFC-22 production.[42] Chemicals with substantial natural sources (CH_3Cl, CH_3Br, N_2O, and CH_4) show smaller relative changes in the past. Results for CH_3Br are higher in mid-depths (40–100 m), suggesting a concentration decline in recent years, consistent with ongoing measurements (see also Figure 2.3).[51] These firn air samples were collected in December 2008–January 2009 by M. Aydin, (University of California, Irvine) and T. Sowers (Penn State University) and were analyzed in the NOAA-Boulder laboratories.

different times and locations from the Greenland ice sheet that are not likely representative of past atmospheric changes.[24,28] These findings suggest that this chemical is somehow produced in the low temperature and light conditions existing deep in the Greenland snow or ice. This production is not apparent in firn air and ice-bubble air sampled in Antarctica, where consistent methyl bromide histories have been derived.[24,29,30]

Recent improvements in analysis techniques have also allowed for precise and accurate measurements of halocarbons in ice bubbles (ice bubbles are located below the firn in a glacier—and while bubbles are isolated pockets of air, the firn region is characterized by interconnected air channels). Much less air is typically available from ice bubbles compared to the firn, making these measurements more challenging. Measurements of ice-bubble air also indicate insignificant natural sources of CFC-12.[31] Substantial pre-industrial levels of CH_3Cl and CH_3Br have been detected in multiple Antarctic ice cores at levels consistent with modern-day budget analyses and firn air samplings that indicate substantial natural sources for these gases.[29,32]

Finally, the case for human-industrial activity causing elevated atmospheric levels of CFCs, halons, CH_3CCl_3, CCl_4, and HCFCs, stems from there being no natural sources sufficient to account for their atmospheric abundances and how they have changed over time. Very small amounts of CFCs and other halogenated source gases may be produced from volcanic emissions,[33,34] although all available evidence indicates that this source cannot explain the increases observed during the 19th century in the atmospheric abundances of ozone-depleting source gases.

2.2.3 Measuring and Interpreting Modern-day Changes in the Atmospheric Abundance of Ozone-depleting Substances

Coordinated surface-based atmospheric measurements of ozone-depleting source gases began soon after concerns were raised about the potential for these chemicals to cause ozone destruction. By the late 1970s two programs were providing high-precision measurements of ODSs at multiple remote sites across the globe: the Atmospheric Lifetime Experiment (ALE, known today as the Advanced Global Atmospheric Gas Experiment or AGAGE) and the U.S. National Oceanic and Atmospheric Administration (NOAA). These programs, along with others, continue making measurements today.[12] Mean global ODS abundances at Earth's surface are derived in these programs from high-frequency ($\sim$hourly) measurements at four to five remote sites with *in situ* instrumentation and by collecting weekly flasks at five to ten remote sites that are analyzed subsequently at a central laboratory.[35,36] With so few observation sites, reliably estimating a global surface mean mixing ratio can be complicated by an uneven distribution of trace gas sources, sinks, and dynamics. To reliably estimate a global mean mixing ratio, sites are concentrated in regions where persistent mixing ratios gradients exist (*e.g.*, the northern hemisphere). Results from these two independent networks are generally quite consistent, once calibration differences are considered, suggesting that measurements at a fairly small number of remote sites can provide reliable estimates of global mean mixing ratios for long-lived gases (Figure 2.3).[12,21] Even the main features of

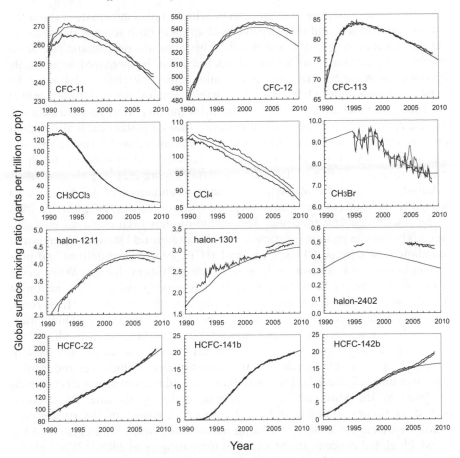

Figure 2.3 Mean global surface mixing ratios (expressed as dry air mole fractions in parts per trillion or ppt) of ozone-depleting substances over the past 18 years from independent sampling networks (colored lines) and from a scenario projection (black line)[132] made in 2006. Measured global surface monthly means are shown as red lines (NOAA data)[52] and blue lines (AGAGE data).[36] In years before 2006 the scenario mixing ratios were derived to match observations or to be consistent with industrial production data (for chemicals where observations were lacking). (Adapted from ref. 12.)

inferred global emissions, which are also sensitive to the rate of change in a global mean mixing ratio, are fairly independent of the sampling network or measurement technique for long-lived ODSs (lifetimes greater than 0.5 yr) (Figure 2.1).

Although measurements of ODSs are routinely made in these networks, the process of developing a long-term trace gas mixing ratio record is hardly routine. Given that time-dependent changes and spatial gradients in ODS mixing ratios can be quite small (<1%), only measurements made accurately, precisely, and consistently over time can provide detailed information about the location, magnitude, and time-dependent changes in both sources and sinks. In addition to requiring instrumental techniques that are precise and free from interferences, accurate calibration standards must be prepared in a highly reproducible manner, and consistent calibration reference scales must be maintained over time. Only then can reliable conclusions concerning source and sink magnitudes be derived on timescales ranging from years to decades. Firn air results provide independent evidence that consistent calibration scales have been successfully maintained for most ODSs over time because calibration scale consistency over multi-year periods is not critical to deriving an atmospheric mixing ratio history from firn air data (Figure 2.2).[24]

Atmospheric concentrations and column abundances of ODSs are also derived from Fourier transform absorption spectroscopy in the infrared region of the electromagnetic spectrum. This technique has been deployed on satellites, high-altitude balloons, the space shuttle, and ground stations. Trends and abundances of selected CFCs, CCl_4, and HCFCs have been derived and provide an independent measure of long-term concentration changes. When made from balloons or satellites, these data provide mixing ratios as a function of altitude and, therefore, estimates of stratospheric loss rates.[37–40] In general, these techniques report trends for ODS concentrations or column abundances over multi-annual periods that are comparable to surface-based results.[12]

The long-term measurement records of ODSs and chemicals used as substitutes show substantial changes in their atmospheric abundances and growth rates over time (Figure 2.3). The atmospheric abundances of nearly all chemicals regulated by the Montreal Protocol increased during the late 20th century, peaked at various times, and by 2010 were decreasing in the background atmosphere (see also Table 2.1). Exceptions include halon-1301 and the HCFCs, for which global concentrations were still increasing as of 2008.[12] While global production of halon-1301 for emissive uses was phased out in developed countries in 1994, emissions continue from stockpiles in applications where suitable replacements have not been found and where halon-1301 use is considered essential to human health and safety (*e.g.*, extinguishing fires without suffocating humans on aircraft and in military vehicles). Global atmospheric concentrations of HCFCs continue to increase because production for emissive uses continues at high levels.[12,41] Although HCFC production in developed nations for such uses has decreased in recent years, it has increased substantially in developing countries.[20,41] In 2004, HCFC production and consumption (where consumption = production + imports – exports) in developing countries for emissive uses surpassed developed country magnitudes.[20] Increases in HCFC production for these uses are expected to halt soon, as developing country production and consumption is capped beginning in 2013 by an amendment to the Montreal Protocol agreed to in 2007 (developed country consumption of HCFCs for emissive uses was capped in 1996). By 2030, global production and consumption of HCFCs for emissive uses will be effectively phased out.

It is important to recognize that only production of ODSs for uses that result in emission to the atmosphere is controlled by the Montreal Protocol.

Production for non-emissive uses, such as for reagent materials (chemical feedstocks), is not controlled or restricted by the Protocol, despite small fractions of this production leaking to the atmosphere during manufacture, transport, and use (thought to be $\sim 0.5\%$). Amounts produced as feedstock can be substantial: feedstock production of HCFC-22 (used to produce fluoropolymers) accounted for 37% of global HCFC-22 production in 2007 and was increasing rapidly then.[42,43] In the case of HCFC-22, continued production has adverse climate impacts associated with the co-production of HFC-23, a potent greenhouse gas.[42,43]

While changes in global mixing ratios of controlled ODSs are consistent with phased-in restrictions on production in the adjusted and amended Montreal Protocol, some observed trends are puzzling at first glance, given that the timing of production phase-outs can be very different than the timing of peak mixing ratios; for some controlled substances, observed atmospheric declines are much larger than for others. A closer analysis of the observations can provide insight into these differences, given our understanding of 1) emission rates implied from the observed changes, 2) knowledge of how different ODSs were used in the past, and 3) the fact that ODSs have different lifetimes.

As indicated earlier, global emission magnitudes can be derived for long-lived chemicals from changes in their global atmospheric abundance over time (Box 2.2). Such emission histories show substantial declines for many of the initially controlled ODSs (*i.e.*, not HCFCs) in recent years (Figure 2.1), but the relative changes in emissions are generally much different than the changes in global atmospheric abundances (Figure 2.3). For example, despite emissions of both CH_3CCl_3 and CFC-113 being nearly eliminated by the year 2000, their atmospheric abundances have responded with very different time constants. Since the early 1990s, the global abundance of CH_3CCl_3 has declined by about a factor of ten while the global abundance of CFC-113 has decreased by only 10%. This difference arises because these gases have very different lifetimes (Table 2.1). Methyl chloroform has a global lifetime of approximately five years, based on its reactivity with OH and considering mean global OH concentrations, while the absorption cross-section of CFC-113 suggests a lifetime of about 85 years from photolysis in the stratosphere. In the absence of emissions, global mixing ratios of a long-lived trace gas will decrease exponentially with a decay time constant that is approximately equal to its inverse lifetime (but not exactly its inverse lifetime, see Box 2.2). Observations of methyl chloroform have shown just this behavior, as they have decreased at $\sim 18.1\%$/yr over the past decade (1998–2007),[44] which is nearly the rate expected from the inverse of its nominal five-year lifetime. The CFC-113 global abundance has been decreasing at $\sim 0.8\%$/yr for a number of years,[12] which is slightly slower than the rate expected from its 85 year lifetime.

Emission declines have been less substantial for other CFCs and halons, despite similar schedules for production phase-outs in the Montreal Protocol as for CFC-113 and CH_3CCl_3. Emissions have decreased more slowly for these other gases because they were used in applications in which emissions lagged production by years or decades (*e.g.*, to blow foams, in refrigeration, and as fire extinguishing agents). As a result, reservoirs (or "banks") of these chemicals grew during the period when production exceeded emission and even today represent a substantial quantity of ODSs. Slow, continued leakage of ODSs from these banks has slowed the decline in emissions of these ODSs despite production for emissive uses having been nearly completely eliminated.

Emission magnitudes derived from observed changes in the remote atmosphere are sensitive to lifetime estimates (Box 2.2). As emissions become small, uncertainties in lifetimes make it difficult to assess if emissions are truly insignificant or if they continue at non-negligible rates. An additional independent approach can suggest qualitative emissions magnitudes from measured mean hemispheric differences.[12,45] This approach is useful for long-lived trace gases emitted primarily from the northern hemisphere and for which losses are distributed fairly evenly between the hemispheres. Purely anthropogenic ODSs fulfill these requirements as they are produced, sold, and emitted mostly in the northern hemisphere ($>85\%$ typically). This asymmetry in emissions leads to an asymmetry in atmospheric abundance for ODSs because there is a barrier for mixing of air between the hemispheres near the equator (the inter-tropical convergence zone). While air becomes mixed within a hemisphere on timescales of months, air mixes between the northern and southern hemispheres on a time scale of approximately one year. Observed, time-dependent changes in the N–S mean hemispheric mixing ratio difference for all controlled ODSs are qualitatively consistent with the emission changes implied from measured atmospheric changes. For example, during the 1980s and early 1990s when emissions were large for CFC-113 and CH_3CCl_3, measured N – S mixing ratio differences were also large for these gases.[46,47] But as emissions have declined to near zero and mixing ratios are decreasing at a rate close to their lifetime-limited rate, annual mean hemispheric differences for these gases are now much smaller. The hemispheric difference currently measured for CFC-113 (<0.2 ppt) suggests an upper limit on asymmetry in loss of a compound having a photolytic stratospheric sink. For CFC-11 and CFC-12, the mean hemispheric differences (North minus South) were 1.5 to 3 ppt in 2008, or significantly higher than expected from loss processes alone, consistent with continuing substantial emissions of these controlled ODSs, most likely from banks.[12]

2.2.4 Long-lived Halogen-containing Gases Emitted from Natural Processes

To assess the success or failure of the Montreal Protocol in reducing stratospheric levels of ozone-depleting chlorine and bromine, it is necessary to also quantify the contributions and systematic variations of natural processes to the atmospheric abundance of these halogens. Do systematic changes in natural sources of halogenated gases diminish the hard-won achievements of the Montreal Protocol or not? To be clear, the answer to this question is no, however, substantial amounts of chlorine and bromine are emitted from natural processes in the form of chemicals having lifetimes greater than 0.5 yr.

The main naturally emitted contributor to chlorine in the stratosphere is methyl chloride (CH_3Cl). At 550 ppt, CH_3Cl has a higher global mixing ratio than any other human-produced ODS although it contributes only ~16% to long-lived chlorine in the atmosphere, which is less than accounted for by the current abundances of CFC-12 or CFC-11 (given that these CFCs have 2 and 3 chlorine atoms per molecule, respectively; see Table 2.1). Methyl chloride is irreversibly removed from the atmosphere by hydroxyl radical oxidation, photolysis in the stratosphere, loss to polar ocean waters, and uptake by soils. Together these processes yield a mean lifetime for CH_3Cl emissions of ~1.0 yr. Most sources of methyl chloride are from natural processes, and the main sources are tropical plants, biomass burning, and the oceans.[48] The significance of biomass burning on the atmospheric abundance of CH_3Cl is reflected in the enhancements observed globally in its abundance during 1998, a year of enhanced biomass burning.[44] Some CH_3Cl emissions arise from human activities such as fossil fuel burning, waste incineration, and industrial processes.[45] Budget analyses suggest that this anthropogenic contribution amounts to 5 ± 4% of total emissions, slightly less than the atmospheric increase of ~10% derived for CH_3Cl during the 20th century from firn air samples (Figure 2.2).[30,49] Although this small discrepancy may suggest errors in our understanding of anthropogenic emission of CH_3Cl or changes in the balance of natural sources and sinks during the 20th century, both pieces of evidence suggest that anthropogenic emissions of CH_3Cl are relatively small compared to those of other ODSs. Observations showing similar CH_3Cl mean mixing ratios in both hemispheres (the NH annual mean is 2 ± 1% higher, on average than the SH mean; NOAA data) support this conclusion.

There are indications from the analysis of air trapped in glacial ice bubbles that the CH_3Cl abundance has varied periodically by ~10% in the past owing to cyclical changes in the balance of natural sources and sinks on times scales of ~100 yr.[49] Smaller variations are observed from year-to-year (±2%) and are thought to arise from annual variations in biomass burning and potentially changes in loss rates.[44] These smaller variations and those measured in ice are an important reminder that changes in the natural system could affect background levels of chlorine and bromine reaching the stratosphere, independent of control measures agreed upon in the Montreal Protocol. Ongoing global measurements of CH_3Cl suggest no significant systematic changes in its atmospheric abundance or in its influence of tropospheric chlorine abundance during the past 15 years.

Methyl bromide is unique among substances controlled by the Montreal Protocol because it has substantial natural and anthropogenic sources. It is both produced and destroyed in the ocean by natural processes, and its sinks are many: OH oxidation, photolysis in the stratosphere, hydrolysis in the ocean, and uptake by soils.[12,50] These factors and their relative importance for regulating atmospheric CH_3Br abundances were not well characterized in the late 1980s and early 1990s when controls on its industrial production were proposed. Much research during the 1990s and early 2000s was targeted towards improving our understanding of the contribution of human-produced

CH$_3$Br to ozone-depleting bromine in the stratosphere. These studies led to the conclusion that industrial production of CH$_3$Br in the mid-1990s accounted for 25–35% of the global atmospheric burden of CH$_3$Br.[12] Ongoing measurements show that global atmospheric mixing ratios and the mean hemispheric mixing ratio difference (NH–SH) both began to decrease in 1999 in response to reduced industrial production.[51] The atmospheric decline in global methyl bromide mixing ratios has continued each year since 1999 as production has been further reduced, confirming the significant role that human-derived emissions had in controlling the atmospheric abundance of this gas. By 2008, in response to industrial production and emissions reductions of 70% from peak levels, the methyl bromide global mean abundance had decreased by nearly 2 ppt (from a peak of just over 9 ppt), the NH-SH difference had decreased by about 50%, and the SH abundance had fallen about half of the way back to preindustrial values as estimated from the analysis of firn air and air trapped in glacial ice.[12,29,30,32,50]

2.2.5 Systematic Changes in Total Tropospheric Chlorine and Bromine from Long-lived ODSs

Deriving changes in the total tropospheric concentration of chlorine and bromine contained in long-lived substances controlled by the Montreal Protocol provides an assessment of the Protocol's progress in reducing the atmospheric abundance of ozone-depleting halogen. Global source-gas measurement networks show that total organic chlorine had been increasing in the troposphere up until the early 1990s (Figure 2.4).[12,35,36] Chlorine from source gases peaked in the troposphere between 1992 and 1994 at 3660 ± 23 ppt and, by 2008, had declined 8.4% below its peak.[12,35] Tropospheric bromine peaked at 16.7 ppt in 1998 and had declined by ~ 1 ppt to 15.4 ppt by 2008.[12,51] These results demonstrate that the Montreal Protocol has been successful in reducing the concentration of chlorine and bromine in the lower atmosphere.

How did the Montreal Protocol succeed in causing the global concentrations of chlorine and bromine to decline fairly quickly after controls were agreed to even when most chlorine and bromine existed in long-lived CFCs and halons in the early 1990s? The Protocol enabled a rapid turnaround in total tropospheric chlorine because it restricted production and consumption of relatively short-lived gases (lifetimes of 0.5–10 years) in addition to the substantial cuts mandated for long-lived ODSs. In the mid-1990s about 10% of tropospheric chlorine was accounted for by CH$_3$CCl$_3$, a chemical having a lifetime of only five years and that had been used primarily in applications where it was emitted soon after being sold. As mentioned above (Section 2.2.3), decreases in CH$_3$CCl$_3$ production led fairly quickly to declines in its global mixing ratio.[44,46] The rapid declines in CH$_3$CCl$_3$ mixing ratios enabled total tropospheric chlorine concentrations to begin decreasing by 1992–1994 despite continued increases in chlorine from the longer-lived CFCs and CCl$_4$.[52] Methyl chloroform accounted for most of the decline in tropospheric chlorine for over a

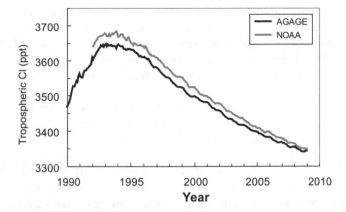

Figure 2.4 The tropospheric abundance (ppt) of organic chlorine (CCl_y) from the NOAA (grey) and AGAGE (black) global surface measurement networks (updates of refs. 51 and 133). Quantities are based upon independently measured mixing ratios of the source gases CFC-11, CFC-12, CFC-113, HCFC-22, HCFC-141b, HCFC-142b, methyl chloroform, carbon tetrachloride, and halon-1211. Updated results for CFC-114 and CFC-115 from AGAGE are used in both aggregations.[36] An additional constant 550 ppt was added for CH_3Cl and 80 ppt was added for short-lived gases such as CH_2Cl_2, $CHCl_3$, CCl_2CCl_2, and $COCl_2$. (Adapted from ref. 12.)

decade, but by 2008 the atmospheric abundance of methyl chloroform had decreased to 11 ppt (from a peak of over 130 ppt) and its contribution to declines in atmospheric chlorine had also become small ($\sim 1\%$).[12] Fortunately, emissions of CFCs also declined during the 1990s and early 2000s so that by 2008 mixing ratios of the three most abundant CFCs were all decreasing in the atmosphere (CFC-11, CFC-12, and CFC-113). When considered together, these CFCs are now the most important contributors to tropospheric chlorine declines.[12] So while a fairly rapid turnaround in total chlorine was possible because of cuts in production and emissions of a short-lived gas, a sustained decline in total chlorine was only ensured by deep, simultaneous cuts in longer-lived ODSs.

A similar story can be told about the turnaround in total tropospheric bromine concentrations; they did not begin to decrease in 1994 when production of halons was phased out in developed countries. This is due to a number of factors including: (1) the presence of large banks and stockpiles of halons that sustained emissions for years; (2) halons having long lifetimes (16 to 65 yr); (3) halons being produced and used after 1994 in developing countries; and (4) because a substantial amount of tropospheric bromine was supplied by methyl bromide. Total tropospheric bromine began to decline only in 1999 after a 25% cut in developed-country production of methyl bromide mandated by the Montreal Protocol. Methyl bromide concentrations declined quickly following this production cut because banks and stockpiles of methyl bromide were small and because methyl bromide is a fairly short-lived gas (lifetime ~ 0.8 yr).[12]

Total tropospheric bromine began declining in the same year methyl bromide production was reduced because the bromine declines from the short-lived methyl bromide outweighed bromine increases from the long-lived halons.[51] It took until 2008 for emissions of halons as a group to diminish to the point that bromine concentrations from halons alone had also stopped increasing in the troposphere.[12]

For both chlorine and bromine, a quick and sustained decline in total tropospheric concentrations of these halogens was achieved by implementing independent phase-out schedules of both short-lived and long-lived ODSs. Unlike the Kyoto Protocol in which a "basket of gases" approach was taken to facilitate trading and offsets of emissions for different gases, the Montreal Protocol addressed chemicals with different properties (*e.g.*, lifetimes) separately. This approach ensured that production of both long- and short-lived chemicals would be reduced concurrently, and this led to atmospheric chlorine and bromine reaching a peak quickly and declining in a sustained fashion thereafter.

2.3 Very Short-lived Substances: Accounting for All of the Chlorine and Bromine in the Stratosphere

Are CFCs, halons, and other chemicals controlled by the Montreal Protocol the dominant source of ozone-depleting halogens in the stratosphere? Or do other processes contribute a significant amount of ozone-depleting chlorine or bromine to the stratosphere? One way to address these questions is to make measurements of inorganic halogen compounds in the upper stratosphere where nearly all source gases have been oxidized or photolyzed and the available chlorine and bromine exist primarily as one or two inorganic chemicals. In the upper stratosphere at 55 km, for example, 95% of chlorine from all source gases is present as hydrogen chloride (HCl).[53,54] Measurements from satellites of HCl at this altitude, therefore, provide a measure of total stratospheric chlorine. Total column absorption measurements of HCl also provide a measure of total stratospheric chlorine and its change over time because HCl (and smaller contributions from $ClONO_2$) is only found in significant quantities in the stratosphere. For bromine, total atmospheric levels and trends are derived from stratospheric measurements of bromine monoxide (BrO) and models to account for the bromine not present as BrO (see also Chapter 3).[55] Measurements of these inorganic chlorinated and brominated chemicals are critical for discerning if stratospheric halogen abundances and their changes are accounted for by chemicals controlled by the Montreal Protocol and natural contributions of other gases such as methyl chloride.

These measurements show that the stratospheric chlorine content and its change over the past two decades are well described by tropospheric concentrations and changes of controlled ODSs and an additional constant contribution from CH_3Cl (Figure 2.5).[54,56,57] The results confirm that increased anthropogenic emissions of long-lived, synthetic chlorinated chemicals have

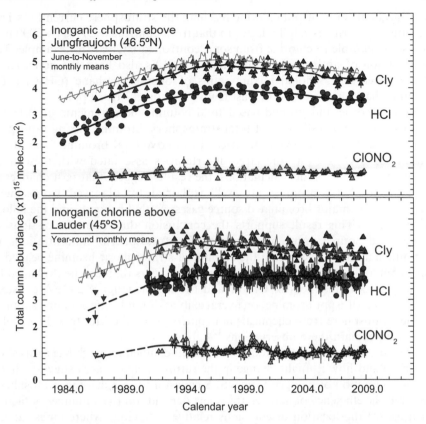

Figure 2.5 Time series of monthly-mean total column HCl (red circles) and ClONO$_2$ (green triangles) abundance as measured above the Jungfraujoch (46.5°N) and Lauder (45°S) (update of ref. 134 and 135). Total inorganic chlorine (Cl$_y$) column abundances (blue triangles) represent the sum of results for HCl and ClONO$_2$. Jungfraujoch data are shown only for June to November of each year; results from all months are displayed for Lauder. Fits are given by the black lines. Model-derived Cl$_y$ based on observed trends and abundances of source gases are calculated with the University of Leeds 2D model (orange lines and smoothed fit) and show good consistency with observed Cl$_y$.[136] (From ref. 12.)

caused stratospheric chlorine concentrations to substantially increase above their natural background.

Uncertainties in the spectroscopic measurements of HCl and ClONO$_2$ (0.1–0.3 ppb) do not preclude additional small contributions from very short-lived chlorinated substances (VSLS) emitted from natural processes or anthropogenic activities. Examples of chlorinated VSLS with anthropogenic sources that are not controlled by the Montreal Protocol include CH$_2$Cl$_2$, C$_2$Cl$_4$, C$_2$HCl$_3$, CHCl$_3$, and CH$_2$ClCH$_2$Cl. Although these chemicals likely have quite low ozone depletion potentials (ODPs), atmospheric measurements in the upper tropical tropopause suggest that anthropogenic emissions of these gases account for ~ 40 ppt chlorine reaching the stratosphere currently.[12] This amount of chlorine is small relative to that from controlled ODSs (> 3000 ppt) but is comparable to chlorine from some controlled HCFCs, for example. The contribution of anthropogenically emitted VSLSs also have the potential to change over time. Global mixing ratios of dichloromethane (CH_2Cl_2), for example, have increased in the past few years from 17 to 22 ppt.[12]

Balloon-borne and ground-based total column measurements of BrO are also used to derive estimates of total stratospheric bromine mixing ratios.[55,58] In this case, however, models are needed to derive total bromine from observations of BrO. Despite the large uncertainties associated with these measurements and the model calculations, the results suggest that past changes in total stratospheric bromine are well explained by observed tropospheric changes in the major brominated source gases: methyl bromide and the halons (Figure 2.6). This result supports the conclusion that human emissions of brominated source gases have augmented stratospheric concentrations of bromine but also suggests that the amount of stratospheric bromine peaked at 20–24 ppt and is substantially higher during all years than can be explained by tropospheric abundances of CH_3Br and the halons (Figure 2.6). A constant contribution of 6 ppt bromine, on average, in addition to the bromine supplied to the stratosphere from chemicals not controlled by the Montreal Protocol is needed to explain these observations.[12]

Since the discovery of this additional bromine, much research has focused on its origin and how it might change in the future. This research suggests that a non-zero fraction of halogen from VSLS emissions can reach the stratosphere, and that the efficiency of halogen transport depends on the a number of factors such as: (1) the location of emissions relative to regions where air is rapidly transported from the boundary layer to the upper troposphere; (2) the VSLS lifetime; and (3) the rate at which oxidation products become removed from the atmosphere.[59–62] Observations suggest that some VSLS (CH_2Br_2 and $CHBr_3$ for example) are emitted from natural sources in substantial quantities in the tropics where strong convective activity can transport them efficiently to the stratosphere.[12,59,63,64] Once a VSLS source gas or its inorganic oxidation products reach the upper tropical tropopause layer (above the 430 K level) and enter the stratosphere, the predominant air motions act to lift this air further into the mid-stratosphere so that this VSLS halogen adds to the stratospheric burden of ozone-depleting halogen.[62] Within the stratosphere, timescales for vertical transport and removal of inorganic halogen are much slower than in the troposphere. The mean residence times of molecules in the stratosphere above 430 K are determined by large-scale transport processes and are measured in years. At the altitudes where the ozone layer resides in mid-latitudes, stratospheric air was last in the troposphere an average of three years ago (its "mean stratospheric age" is three years). In polar regions, stratospheric air at ozone-layer altitudes has a mean age that is longer, typically five to six years.

Because of the spatial and temporal heterogeneity in VSLS emissions, in VSLS atmospheric concentrations, and in loss rates of VSLS and their oxidation products, accurately estimating the contribution of these short-lived

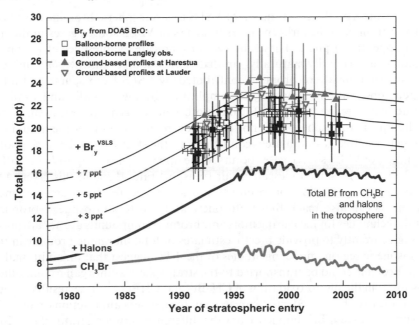

Figure 2.6 Changes in total stratospheric inorganic bromine (Br$_y$) derived from bal-
loonborne BrO observations (squares) (update of ref. 55) and annual
mean mixing ratios calculated from ground-based UV-vis measurements
of stratospheric BrO made at Harestua (60°N) and Lauder (45°S) (filled
and open orange triangles, respectively) (adapted from ref. 58). These
stratospheric trends are compared to trends in measured bromine at
Earth's surface (thick grey and purple lines)[24,51,137] with additional constant amounts of Br$_y$ added (thin lines). Squares show total inorganic
bromine derived from stratospheric measurements of BrO and photochemical modeling that accounts for BrO/Br$_y$ partitioning from slopes of
Langley BrO observations above balloon float altitude (filled squares) and
lowermost stratospheric BrO measurements (open squares). For the balloon-
borne observations, bold/faint error bars correspond to the precision/
accuracy of the estimates, respectively. For the ground-based Br$_y$ measurements (triangles), the error bars correspond to the total uncertainties. For
stratospheric data, the date corresponds to the time when the air was last in
the troposphere, *i.e.*, sampling date minus estimated mean age of the stratospheric air parcel. For tropospheric data, the date corresponds to the
sampling time, *i.e.*, no corrections are applied for the time required to
transport air from Earth's surface to the tropopause. Pre-industrial levels
were 5.8 ± 0.3 ppt for CH$_3$Br[32] and 0 ppt for the halons.[138] (Adapted from ref. 12.)

chemicals to stratospheric halogen is difficult. While much progress has been
made in understanding the spatial heterogeneity and time-varying nature of
VSLS emissions in recent years, basic questions remain unanswered, such as
whether or not the largest VSLS sources are from coastal regions or the open
ocean, and how important periodic vigorous convection through the tropopause is relative to large-scale transport circulation in delivering VSLS halogen
to the stratosphere.[12,64–66]

Constraints on the contribution of VSLS to stratospheric bromine have been derived from periodic airborne measurements of VSLS source gases in the tropical upper troposphere. These observations indicate the presence of 1–5 ppt of bromine in primarily naturally emitted VSLS source gases.[12] Obtaining a meaningful estimate of bromine contributed by VSLS from sparse and intermittent observations, however, is complicated by the periodic and localized nature of convective transport. Furthermore, source gas mixing ratios measured in the upper troposphere may not be a good representation of the amount of halogen a source gas ultimately contributes to the stratosphere because photo-oxidation of VSLS source gases occurs in the upper tropical troposphere (within the TTL) on the same timescale as air transport rates through this layer. As a result, source gas oxidation products not accounted for by source-gas measurements can also reach the stratosphere and add to stratospheric bromine. While techniques for measuring inorganic bromine compounds are not advanced enough currently to provide precise estimates of total inorganic bromine in this atmospheric region, modeling studies of the TTL suggest that an additional 0–4 ppt bromine could be transported to the stratosphere as inorganic compounds arising from the photo-oxidation of CH_2Br_2 and $CHBr_3$.[12] These results suggest that these two short-lived gases can account for a significant fraction or all of the ~6 ppt of "excess Br" implied from the difference between stratospheric BrO data and expectations from tropospheric halon and CH_3Br abundances. Other brominated source gases with natural sources may also contribute some additional bromine through these same mechanisms, although brominated VSLS emitted from anthropogenic activities (*e.g.*, n-propyl bromide) are thought to contribute only little to stratospheric bromine at the present time.[12,67]

Iodine is even more efficient at catalytically destroying ozone in the stratosphere than chlorine or bromine (see Chapter 3).[68] Recent atmospheric observations of iodinated source gases and of inorganic iodinated chemicals both suggest abundances of ≤ 0.1 ppt of iodine in air in the TTL (the source region of air to the stratosphere).[69,70] Models suggest that at such low concentrations, iodine plays only a minor role in influencing stratospheric ozone concentrations at the present time.

One other important point can be made with regard to very short-lived substances. A number of chlorine-, bromine-, or iodine-containing chemicals with very short lifetimes are being considered as substitutes for ODSs and HFCs. Methyl iodide (CH_3I), for example, has recently been approved for use as a substitute for methyl bromide in fumigation applications in California, U.S.A. Assessing the potential impact of these substitutes on the stratospheric ozone layer, climate, tropospheric ozone, and trifluoroacetic acid concentrations is an ongoing area of research. Some general relationships have been developed that provide rough estimates of mean annual ODPs for VSLS emissions from mid-latitudes[59] or from specific continental regions as a function of season.[62] Three-dimensional modeling work suggests that the ODP of CH_3I emitted from mid-latitudes is 0.017,[71] substantially less than 0.6 for methyl bromide. Despite iodine's high efficiency for depleting stratospheric ozone, CH_3I emitted from Earth's surface at mid-latitudes has a small annual mean ODP because only a very small fraction of the iodine from these emissions reaches the stratosphere. When emitted from other locations at certain times of the year, however, CH_3I can have a much larger ODP. Methyl iodide emitted from the Indian subcontinent during NH summer, for example, is estimated to have an ODP of ~ 1.0, or equivalent to CFC-11.[62] ODPs are also expected to be enhanced for emissions from aircraft when compared to emissions at Earth's surface.[72]

2.4 Past and Future Changes in Total Atmospheric Halogen Loading

As mentioned in the previous section, measured abundances of inorganic reservoir chemicals from total column absorbance data and satellites at 55 km provide useful information about trends and abundances of ozone-depleting halogen in the stratosphere. Changes in total inorganic stratospheric halogen can also be derived reliably from observed changes in the abundances of ODSs at Earth's surface.[13,35,73] While measuring ODSs in the troposphere to derive changes in inorganic stratospheric halogen may seem counterintuitive, aircraft measurement campaigns show that time-dependent changes in the amount of a long-lived source gas entering the stratosphere are well described by mean surface concentrations derived from measurements at remote surface sites.[74] This is true for long-lived source gases because vertical and horizontal mixing times in the troposphere are fast compared to tropospheric loss rates for most controlled ODSs and, as a result, these gases become fairly well mixed throughout the troposphere.

Similar airborne field experiments also provide a measure of the relative rates of destruction of different ODSs in the lower stratosphere.[37] These measurements show that the extent to which an ODS has decomposed in the stratosphere to liberate inorganic forms of halogen is a strong function of the mean time an air parcel has spent in the stratosphere (also called the "mean age" of that air parcel). With knowledge of: (1) tropospheric mixing ratios for ODSs; (2) ODS rates of decay in the stratosphere; and (3) the mean age of a particular air parcel, the effective concentration of inorganic halogen in that stratospheric air parcel can be calculated.[73,74]

Additionally, the total potential for an air parcel in the stratosphere to deplete ozone can be derived only if the enhanced efficiency of bromine to deplete ozone relative to chlorine is considered. ODSs containing bromine atoms are accounted for by multiplying their concentrations by an equivalency factor, called "alpha", to derive the "equivalent" chlorine burden from the presence of bromine-containing ODSs. Although alpha varies over time and space and is dependent on the relative abundance of chlorine and bromine,[9,75,76] a value of 60 is commonly used as an appropriate global mean.

With these considerations, different weighted sums of tropospheric mixing ratios can be derived. Equivalent Chlorine (ECl) is the sum of the concentration of tropospheric chlorine and bromine in source gases with the enhanced efficiency for bromine included as a multiplicative factor. It has been used in the past as a measure of inorganic halogen in regions of the atmospheric where nearly all source gases have photo-chemically decomposed (*e.g.*, above Antarctica during springtime).[35] Equivalent Effective Chlorine (EECl) is calculated from tropospheric abundances of chlorine and bromine in ODSs, consideration of source-gas degradation rates in the stratosphere (typically the mid-latitude stratosphere), and the enhanced efficiency for bromine to deplete ozone relative to chlorine. Because time lags associated with air being transported to the stratosphere are not included in EECl, this metric allows a discussion of the timing of tropospheric changes that are relevant for the future stratosphere. For example, tropospheric measurements of source-gas abundances indicate that EECl peaked in the troposphere during 1993–1994.[35] This important milestone was achieved because of the Montreal Protocol and represented an important first-step on the path towards ozone recovery. Ozone-depleting halogen concentrations in the stratosphere did not begin decreasing in 1993–1994, however, because it takes a few years for tropospheric trends to propagate to the stratosphere where the ozone layer is found.

Approximate changes in the summed stratospheric abundance of ozone-depleting inorganic halogen are derived from tropospheric data by accounting for these time lags and, in some formulations, stratospheric mixing processes. Equivalent Effective Stratospheric Chlorine (EESC) is such a metric and is calculated similarly to EECl but with transport-related time lags explicitly included. It can be calculated for different regions of the stratosphere based on the mean age of air in those regions and the extent of ODS decay associated with that mean age.[73] Air in the Antarctic stratosphere during springtime has a mean age of five to six years, while the mean "stratospheric age" of air in the mid-latitude stratosphere is less, about three years. As might be expected given the longer mean age and the associated additional source gas decomposition, much higher values of EESC are derived for the Antarctic stratosphere during spring than in the mid-latitude stratosphere (Figure 2.7, top panel). It is also apparent that the peak in EESC is delayed over Antarctica compared to mid-latitudes. This delay is primarily the result of three different factors. First, air in the Antarctic stratospheric vortex during Austral spring has a mean age of five to six years, hence the full extent of decline observed to date in the troposphere has yet to reach the Antarctic stratosphere. Second, declines in EESC in mid-latitudes are determined more by the subset of source gases that decompose fairly rapidly in the stratosphere. As a result, trends in mid-latitude EESC are more sensitive to changes in ODSs with shorter stratospheric lifetimes such as methyl chloroform. In the Antarctic vortex, trends in EESC reflect trends in a wider suite of ODSs, including the long-lived CFCs whose atmospheric concentrations have decreased relatively little from their peaks at the present time. Third, stratospheric mixing processes act to diffuse abrupt changes and spread them out over an extended period.[73,77] Air in the Antarctic stratosphere has a broader range of ages than air in mid-latitudes, and this has the effect of smoothing the peak in polar EESC and pushing it later in time.

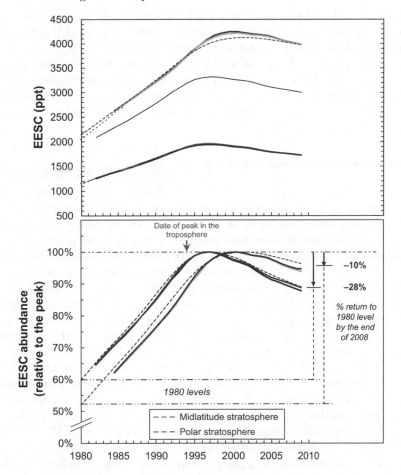

Figure 2.7 Top panel: Equivalent effective stratospheric chlorine (EESC) (ppt) calculated for the midlatitude stratosphere from surface measurements (see
Figure 2.3) and absolute fractional release values (black line)[45] or with
age-of-air-dependent fractional release values for the mid-latitude stratosphere (red lines; mean age = 3 years) and the Antarctic springtime stratosphere (blue lines; mean age = 5.5 years). EESC calculated with
stratospheric mixing processes included are shown as dashed colored
lines for the different stratospheric regions (*i.e.*, an air-age spectrum width
equal to one-half the mean age).[73] Different shades of the same colors
represent EESC calculated with tropospheric halocarbon data from the
different surface networks (NOAA and AGAGE, not always distin-
guishable from one another) without consideration of air-age spectra.
Bottom panel: EESC derived for the mid-latitude and Antarctic springtime stratospheric regions as a function of time plotted relative to peak
abundances (1950 ppt for the mid-latitude stratosphere and 4150 ppt for
the polar stratosphere). Percentages shown on right of this panel indicate
the observed change in EESC relative to the change needed for EESC to
return to its 1980 abundance (note that a significant portion of the 1980
EESC level is from natural emissions of CH_3Cl and CH_3Br). Given the
fully adjusted and amended Montreal Protocol, EESC is projected to
return to 1980s levels by ∼2050 in the mid-latitude stratosphere and by
∼2070 in the Antarctic springtime stratosphere.[78] Colors and line styles
represent the same quantities as in the top panel. (Adapted from ref 12.)

Secular trends in EESC show how the Montreal Protocol has been successful at reversing the long-term increases in stratospheric ozone-depleting halogen in both the mid-latitude and polar stratosphere (Figure 2.7, top panel). In both these stratospheric regions, ozone-depleting halogen has decreased in recent years as a result of declining tropospheric mixing ratios of chlorine- and bromine-containing source gases. Continued decreases are expected given adherence to the controls on production and consumption of ODSs in the adjusted and amended Montreal Protocol.

Projections of future EESC concentrations are made every four years as part of the WMO Scientific Assessment of Ozone Depletion.[78] The baseline scenario is initialized with updated surface observational records for ODSs, and provides the basis for a projection of ODS abundances and EESC into the future based on amounts of new production allowed in the existing Protocol (and associated emission), emissions of ODSs from existing banks, and ODS lifetimes. Different scenarios are also considered to gauge the potential benefits associated with additional controls on ODS production, consumption, or even emissions. The baseline scenario provides a timeline for ozone recovery when assessed by the return of EESC back to its 1980 level. The benchmark year for EESC of 1980 is used because ozone depletion became particularly noticeable after that time, although some analyses indicate that halogen-catalyzed ozone depletion may have been occurring even before 1980.[79,80] The most recent projections suggest that EESC will return to 1980 levels in the mid-latitude stratosphere in about 2050 and in the Antarctic stratosphere in about 2070.[78]

Measurements of stratospheric ozone suggest that ozone depletion has not worsened in recent years.[79,81] A clear signal of ozone increases is not expected for at least another decade,[82] although one report claims that small recent increases in Antarctic ozone can be attributed to declines in EESC.[83] Robustly detecting a turnaround in stratospheric ozone requires an analysis of many different factors including the natural variability of ozone abundance relative to an expected change, but also an assessment of the magnitude of the decline in EESC from its peak. The change in EESC normalized to its peak and to its level in 1980 can provide a measure of progress in returning EESC back to the benchmark 1980 level.[84,85] Normalizing the change in EESC based on these benchmark levels (the peak and 1980 levels) indicated that by 2008 the decline of ozone depleting halogen in mid-latitudes had been nearly one-third (28%) of that needed to return EESC back to the 1980 benchmark level; EESC declines in the Antarctic spring by 2008 were only 10% of that needed to return EESC back to 1980 values (Figure 2.7, bottom panel). Updated results suggest that EESC declines have continued since 2008.[85]

2.5 Non-halogenated Gases that Affect Stratospheric Ozone Chemistry

In addition to being affected by inorganic halogenated gases, stratospheric ozone chemistry is also influenced by the abundance of stratospheric aerosol particles, water vapor, methane, and nitrous oxide. Methane contributes HO_x and N_2O contributes NO_x to the stratosphere; these oxides are involved in catalytic reactions that destroy stratospheric ozone particularly in the middle and upper stratosphere (see Chapter 1). Methane also acts as a sink for stratospheric chlorine, and as methane abundance increases, the balance of inorganic stratospheric chlorine shifts from ozone-depleting forms to reservoir forms and chlorine-catalyzed ozone depletion is reduced. Methane also plays an important role in the formation of ozone in the troposphere and lower stratosphere, although in the troposphere this ozone production is strongly modulated by NO_x levels. As a result of these interactions, the overall influence of methane increases is to increase total column ozone, although the magnitude of this effect in models is influenced by atmospheric chlorine loadings.[86,87]

2.5.1 Methane (CH_4)

Methane is naturally produced in many different ecosystems by anaerobic degradation of organic compounds and other biological processes. Most naturally emitted methane is from tropical wetlands; smaller contributions stem from extra-tropical wetlands, the oceans, and termites.[88,89] These natural sources sustain a background methane mixing ratio of about 0.7 ppm in the global atmosphere, based on the analysis of preindustrial air contained in ice bubbles.[90] This background has varied during the past 800 000 years between 0.35 and 0.8 ppm owing to past changes in climate and temperature.[91]

Measurements of air extracted from ice bubbles also show that methane levels began increasing above the 0.8 ppm natural background by about 1850.[90] The largest increases are apparent during the 1900s, and during this latter period additional measurements of firn air confirm the substantial increases in methane over time (Figure 2.2).[25] Regular ongoing measurements of methane by two independent globally distributed sampling networks began in the mid-1980s when the global methane mixing ratio was 1.65 ppm, which is more than two times higher than the natural background level (Figure 2.8).[92,93] By 2009 the global methane mixing ratio had reached nearly 1.79 ppm.[94,95]

The increases in atmospheric methane concentrations since the 1800s are primarily the result of increases in anthropogenic emissions, which are thought to currently amount to 340 Tg CH_4/yr, or about two-thirds of total global methane emissions.[88,89] Methane is the predominant component of the fossil fuel known as "natural gas". This methane was created over millennia from the sub-surface anaerobic degradation of hydrocarbon molecules. Exploration, recovery, transport, and use of this fossil fuel and others account for 110 Tg/yr of CH_4. Human-derived methane emissions also arise from rice agriculture, from digestive processes in ruminant animals, and from agriculture-related waste. Emissions from these agricultural activities are estimated to account for 120 Tg CH_4/yr. Additional anthropogenic emissions of approximately 105 Tg CH_4/yr stem from landfills and biomass burning.

Methane is removed from the atmosphere by reaction with the hydroxyl radical. The rate of this reaction combined with our understanding of the

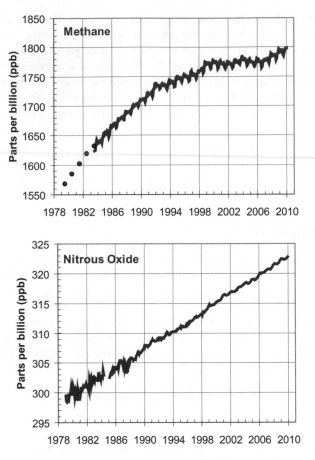

Figure 2.8   Mean global surface mixing ratios (expressed as dry air mole fractions in parts per billion or ppb) of methane (CH_4) and nitrous oxide (N_2O) from the NOAA cooperative air sampling network.[92,94,106] Annual mean estimates for CH_4 derived from firn and ice cores in years before ongoing measurements began are shown as blue points.[90] Blue lines represent global monthly means. For methane, monthly means (blue line) have been smoothed to show running annual changes (red line).

concentration of hydroxyl radicals results in methane emissions having an atmospheric lifetime of $\sim$9 years.[46] This oxidation process links atmospheric methane levels to global photochemical processes. As a result, when projecting future methane concentrations, it is not entirely sufficient to consider only how emissions might change, an understanding of global photochemistry and hydroxyl radical concentrations on global scales is also required. Modeling and indirect measurements suggest, however, that global mean hydroxyl radical concentrations are fairly well buffered on annual timescales.[44,89,96] The story is a bit more complex for methane, however, because methane is not a benign player in atmospheric chemistry. With a global mean concentration of

~ 1.8 ppm, methane acts as a significant sink for OH and affects the balance of OH concentrations globally. This results in a positive feedback between methane emissions and its atmospheric abundance.[97–99] In other words, theory suggests that a pulse of methane emission depletes hydroxyl concentrations so that the lifetime of methane in the atmosphere (and other chemicals oxidized by OH) is longer than it was before the emission perturbation.

Ongoing measurements of methane from global sampling networks during the past 25 years show increasing concentrations over much of that time but with a rate of increase that has gradually become smaller. For the eight years following 1998 methane's global background concentration remained nearly constant.[92,93] This stabilization was the result of total sources nearly balancing sinks—the system was close to steady state. But despite the near-constant total emission rate implied from these observations, inventory analyses of methane sources suggested that human-derived emissions had increased substantially (~ 10%) during the 2000s.[100] Such a trend, if true, would require concurrent decreases in natural methane emissions or a systematic trend in loss rates (*i.e.*, primarily increasing hydroxyl radical concentrations) to explain the fairly constant global background levels. Such a large systematic trend in loss (*i.e.* OH concentrations) would not be consistent with observations of other trace gases for which total emissions are better constrained (*e.g.*, CH_3CCl_3).[44,92,93] Unfortunately, current observational networks for discerning changes in methane fluxes from different sources are not dense enough, nor are inverse modeling tools sophisticated enough to derive regional emissions independently of inventory estimates or determine changes in specific natural fluxes.

Since 2007, global methane concentrations have increased again (Figure 2.8). An analysis of those changes suggests that they stem in part from increased emissions in the Arctic and the tropics, owing to anomalously warm and wet years in those regions.[92,93] Such changes are an important reminder that natural methane emissions are not likely to remain constant in the future as climate changes. Climate changes leading to warmer and wetter ecosystems will likely lead to increased methane emissions. A significant positive feedback would mean that larger reductions in human-derived methane emissions will be required to reduce the climate impact of methane in the future than estimated from analyses not considering these feedbacks.

2.5.2 Nitrous Oxide (N_2O)

Nitrogen is a critical nutrient for life in many ecosystems and is a limiting reagent to biosphere global primary production. Most of the nitrogen in the atmosphere is present as N_2 and is unavailable for use by most organisms. Bio-available nitrogen is created from N_2 in the natural system primarily by nitrogen-fixing bacteria.[101] Smaller amounts of bio-available or reactive nitrogen are created naturally from lightning and other processes. During the microbially-mediated cycling of nitrogen between oxidized and reduced forms in aerobic and anaerobic environments, a small fraction is converted to N_2O. In marine environments emissions of N_2O are enhanced in equatorial upwelling zones and coastal zones where nutrients and biological activity are enhanced. In terrestrial systems N_2O sources are also enhanced in nutrient-rich ecosystems. These sources have sustained a natural background level of N_2O in the global atmosphere of between 200 and 300 ppb in data available from the past 800 000 years based on the analysis of pre-industrial air contained in ice bubbles.[102–104]

Humans have nearly doubled the amount of reactive nitrogen naturally cycled in the environment through the combustion of fossil fuels, the production and application of nitrogen fertilizer, the cultivation of nitrifying legumes, and other industrial processes. The increase in N_2O emissions resulting from these activities caused the global background level for N_2O to increase beginning in the 19th and 20th century.[25,105] The most rapid increases have been observed in the latter half of the 20th century and are well documented by firn air measurements (Figure 2.2) and ongoing ambient-air measurements (Figure 2.8).[25,106,107] These ongoing measurements show that global N_2O mixing ratios have increased fairly linearly at a rate of 0.8 ppb/yr since the late 1970s. In 2010 the global mean N_2O mixing ratio was 323 ppb.[106]

The removal rate of N_2O from the atmosphere is determined primarily by the flux of high-energy UV light in the stratosphere. Because only a small fraction ($\leq 1\%$) of the total mass of the atmosphere is transported through this stratospheric region the lifetime of N_2O is 114 years. The photochemical destruction of N_2O in the stratosphere produces NO_x, which participates in the catalytic destruction of ozone (See Chapter 1). As a result, emissions of N_2O cause ozone destruction. Unlike methane, the lifetime for N_2O is less strongly coupled to tropospheric chemistry. However, the N_2O lifetime is influenced by stratospheric chemical processes because N_2O loss is determined by UV light flux in the stratosphere, which is affected by stratospheric ozone. As a result, perturbations to ODSs, methane, and N_2O itself will affect stratospheric ozone and, therefore, the N_2O loss rate.[108]

Human-caused N_2O emissions currently account for 6.7 Tg N/yr, or about 40% of all N_2O emissions.[88] Agricultural emissions account for about 4.1 Tg N/yr, while industrial processes and fossil fuel combustion combined account for ~ 2 Tg N/yr. Considering that N_2O produces NO_x in the stratosphere and given the substantial decrease in emissions of ODSs since the 1990s, it has been estimated that ODP-weighted anthropogenic emissions of N_2O are now larger than ODP-weighted emissions of ODSs.[109] This implies that current emissions of N_2O will cause more ozone depletion integrated over their lifetime than will current emissions of ODSs. To be clear, this does not imply that most of the current ozone depletion is caused by N_2O, despite some confusion on this point in the literature; Current rates of anthropogenic ozone depletion are determined primarily by current atmospheric mixing ratios of ODSs.

2.5.3 Sulfur Compounds

Stratospheric aerosol particles affect ozone abundances because they influence the proportion of inorganic chlorine present in reactive forms (*e.g.*, ClO) relative to non-reactive forms (HCl, $ClONO_2$) (see Chapter 4). These aerosols also affect the partitioning of inorganic brominated chemicals, but their influence on bromine-catalyzed ozone loss is reduced because the partitioning of inorganic bromine chemicals is shifted more in favor of reactive forms (*e.g.*, BrO) even in the absence of high aerosol loadings.[110] The Junge aerosol layer in the stratosphere consists primarily of sulfuric acid aerosol.[111] This sulfur is accounted for in part from explosive volcanic eruptions, which rapidly transport sulfur-containing gases directly into the stratosphere. The large variations in stratospheric aerosol mass in the past reflect the occurrence of explosive volcanic eruptions.[112] Under high aerosol loadings after powerful eruptions, ozone depletion can be enhanced.[113] Very little of non-volcanic H_2SO_4 from surface emissions or that is produced in the lower troposphere becomes transported to the stratosphere, however, owing to its high solubility and rapid loss from the troposphere by dry and wet deposition. The Junge layer is also sustained by less soluble forms of sulfur that are transported from the surface and become oxidized within the stratosphere or in the upper troposphere as they are transported into the stratosphere. In the absence of substantial volcanic input, a substantial fraction of non-volcanic sulfur in the stratosphere is believed to be transported there in the form of carbonyl sulfide (COS).[112,114] It is also likely that some of the sulfur reaching the stratosphere is from emissions of sulfur dioxide (SO_2) and carbon disulfide (CS_2).[112] Sulfur dioxide is highly soluble in water and has a short lifetime against oxidation in the troposphere. As is true for VSLS (Section 2.3), because timescales for SO_2 removal in the troposphere are comparable to the timescales for its transport to the stratosphere, it is difficult to accurately quantify the amount of sulfur reaching the stratosphere from SO_2 emissions given our current understanding. Observations of aerosol backscatter from ground-based lidars suggest large recent increases in aerosol mass.[115] Observed changes in the atmospheric abundance of COS are insufficient to account for these aerosol increases; surface-based monitoring of the COS abundance suggests only very small relative changes from 2000–2010.[116,117] Changes in SO_2 abundances are much more difficult to capture with global measurement networks, given the high temporal and spatial variability of this short-lived gas. Inventory estimates of sulfur emissions (and SO_2) suggest a significant increase in recent years owing to the dramatic increase in fossil-fuel coal combustion in China, for example, and this may contribute to the recent aerosol mass increase.[115,118]

2.6 The Contributions of Ozone-depleting Gases to Changes in Climate

The long-lived substances discussed in this chapter that affect the chemistry of stratospheric ozone also directly influence climate through their absorption of available electromagnetic radiation. They indirectly influence climate because they catalyze the destruction of stratospheric ozone, which is itself a greenhouse gas. The contributions of long-lived gases to current climate forcing are described in this section to provide background for Chapter 8, which focuses on how climate change is expected to affect the future abundance of stratospheric ozone.

Climate forcing (or radiative forcing) is the change in the balance of radiative energy at the top of the troposphere, expressed in W/m^2, relative to some earlier period. Direct climate forcing is usually calculated for a change in concentration of a trace gas relative to its concentration in 1750, and it is a measure of the heat imbalance arising from this change based upon the trace gas absorption spectrum relative to other absorbing species. It does not account for the persistence of a trace gas in the atmosphere (it is not a time-integrated measure of climate effects); instead, it is a measure of the forcing on the climate system arising from a greenhouse gas given its atmospheric abundance at some point in time relative to a different time.

Of all the long-lived gases present in the atmosphere today, carbon dioxide is currently by far the largest contributor to direct radiative forcing.[119] Based on measured global tropospheric abundances in 2009, the direct climate forcing from all long-lived gases amounted to about 2.8 W/m^2. Carbon dioxide contributed 1.77 W/m^2, methane 0.5 W/m^2, ODSs (including HCFCs) accounted for 0.32 W/m^2, and N_2O contributed 0.17 W/m^2.[120,121] Other contributors include the HFCs, PFCs, SF_6, and NF_3, which together amounted to about 0.03 W/m^2 in 2009.[12,121] The largest increase in direct climate forcing over the past decade has been from CO_2 (0.27 W/m^2/decade); smaller decadal increases arose from N_2O (0.025 W/m^2/decade), HCFCs (0.015 W/m^2/decade), and HFCs (0.011 W/m^2/decade). PFCs and SF_6 have each accounted for ~ 0.001 W/m^2/decade to direct forcing. The direct radiative forcing from these long-lived trace gases (including CO_2) in 2009 was 27.5% higher than it was in 1990, the reference year for the Kyoto Protocol.[121] Climate forcing from decreases in stratospheric ozone is a cooling influence; this forcing is estimated at $-0.05 \pm$ 0.1 W/m^2 and is primarily an indirect forcing associated with ODSs.[119]

Future changes in climate forcing will be determined by the lifetimes and also by the magnitude of current and future emissions of these gases and others that affect climate. A measure of the time-integrated climate forcing influence of a trace gas emission can be derived with different metrics, although one often used and that is also included in the Kyoto Protocol is the Global Warming Potential (GWP). GWPs represent the time-integrated change in climate forcing associated with emission of a trace gas relative to the time-integrated change in climate forcing from an equivalent emission of CO_2 over some time horizon. When different time horizons are considered they can yield quite different GWPs depending on the lifetime of a trace gas relative to the time-dependent atmospheric response to a pulse emission of CO_2.[122] GWPs derived for a 100-year integration period are typically used as weighting factors on emissions of non-CO_2 gases to derive so-called "CO_2-equivalent" emissions (often abbreviated as CO_2-eq emissions). In this way the time-integrated climate effect of a trace-gas emission can be estimated and compared to other trace-gas emissions. CO_2-eq emissions from all long-lived trace gases totaled approximately 50 GtCO_2-eq/yr in 2009; most of this emission is from CO_2 (33 Gt/yr).[123] Equivalent emissions of long-lived non-CO_2 greenhouse gas were less than half as large (~ 15 GtCO$_2$-eq/yr) as that of CO$_2$ in 2009 and are accounted for as follows: 9 GtCO$_2$-eq/yr from CH$_4$, 3 GtCO$_2$-eq/yr from N$_2$O, 1.7 GtCO$_2$-eq/yr from ODS (including HCFCs), ~ 0.5 GtCO$_2$-eq/yr from HFCs, ~ 0.17 GtCO$_2$-eq/yr from SF$_6$, and ~ 0.12 GtCO$_2$-eq/yr from PFCs.[41,78,124–126] Total emissions from these gases (including CO$_2$) have generally increased over time, although they were fairly constant during the 1990s when emissions of ODSs decreased from 9 GtCO$_2$-eq/yr down to ~ 2 GtCO$_2$/yr as a result of the Montreal Protocol.[15,78] The climate protection provided by these emission declines is substantially greater than the reduction target of the first commitment period of the Kyoto Protocol (~ 2 GtCO$_2$-eq/yr).[15] By controlling emissions of ODSs and thus being responsible for these emission reductions, the Montreal Protocol has contributed substantially to protecting climate.

Acknowledgements

I appreciate the thoughtful reviews of earlier versions of this chapter provided by R. Müller, S. Reimann, J. Daniel, B. Hall, M. von Hobe, and F. Stroh. I owe a debt of gratitude to colleagues I worked closely with in preparing Chapter 1 of the 2010 Scientific Assessment of Ozone Depletion, and to E.J. Dlugokencky and J.H. Butler for many recent conversations related to non-CO$_2$ greenhouse gases. For updated NOAA data presented here I owe thanks to G. Dutton, C. Siso, B. Miller, B. Hall, T. Conway, and E.J. Dlugokencky. I also thank J. Elkins for his ongoing support, and scientists within the AGAGE community for their generosity and collegiality in our common pursuit to better understand trace gases in the atmosphere.

References

1. J. E. Lovelock, R. J. Maggs and R. J. Wade, *Nature*, 1973, **241**, 194.
2. R. S. Stolarski and R. J. Cicerone, *Can. J. Chem.*, 1974, **52**, 1610.
3. M. J. Molina and F. S. Rowland, *Nature*, 1974, **249**, 820.
4. P. J. Crutzen, *Geophys. Res. Lett.*, 1974, **1**, 205.
5. S. C. Wofsy, M. B. McElroy and Y. L. Yung, *Geophys. Res. Lett.*, 1975, **2**, 215.
6. WMO, 1986: *Scientific assessment of ozone depletion: 1985– Report No. 16,* Geneva, Switzerland, 1986.
7. J. C. Farman, B. G. Gardiner and J. D. Shanklin, *Nature*, 1985, **315**, 207.
8. S. Solomon, *Revs. Geophys.*, 1999, **37**, 275.
9. S. Solomon, R. R. Garcia, F. S. Rowland and D. J. Wuebbles, *Nature*, 1986, **321**, 755.
10. J. G. Anderson, W. H. Brune and M. H. Proffitt, *J. Geophys. Res.*, 1989, **94**, 11465.
11. R. L. deZafra, M. Jaramillo, A. Parrish, P. Solomon and J. Barrett, *Nature*, 1987, **328**, 408.

12. S. A. Montzka, S. Reimann, S. O'Doherty, A. Engel, K. Krüger, W. T. Sturges, D. Blake, M. Dorf, P. Fraser, L. Froidevaux, K. Jucks, K. Kreher, M. J. Kurylo, W. Mellouki, J. Miller, O.-J. Nielsen, V. L. Orkin, R. Prinn, R. Rhew, M. L. Santee, A. Stohl and D. Verdonik, Chapter 1 in *Scientific Assessment of Ozone Depletion: 2010, Global Ozone Research and Monitoring Project—Report No. 52*, World Meteorological Organization, Geneva, 2011.

13. J. S. Daniel, S. Solomon and D. L. Albritton, *J. Geophys. Res.*, 1995, **100**, 1271.

14. V. Eyring, D. W. Waugh, G. E. Bodeker, E. Cordero, H. Akiyoshi, J. Austin, S. R. Beagley, B. A. Boville, P. Braesicke, C. Brühl, N. Butchart, M. P. Chipperfield, M. Dameris, R. Deckert, M. Deushi, S. M. Frith, R. R. Garcia, A. Gettelman, M. A. Giorgetta, D. E. Kinnison, E. Mancini, E. Manzini, D. R. Marsh, S. Matthes, T. Nagashima, P. A. Newman, J. E. Nielsen, S. Pawson, G. Pitari, D. A. Plummer, E. Rozanov, M. Schraner, J. F. Scinocca, K. Semeniuk, T. G. Shepherd, K. Shibata, B. Steil, R. S. Stolarski, W. Tian and M. Yoshiki, *J. Geophys. Res.*, 2007, **112**, D16303.

15. G. J. M. Velders, S. O. Andersen, J. S. Daniel, D. W. Fahey and M. McFarland, *U.S. Proc. Natl. Acad. Sci.*, 2007, **104**, 4814.

16. J. H. Butler, J. W. Elkins, T. M. Thompson and B. D. Hall, *J. Geophys. Res.*, 1991, **96**(D12), 22347.

17. S. A. Yvon-Lewis and J. H. Butler, *J. Geophys. Res.*, 2002, **107**, 4414.

18. A. R. Douglass, R. S. Stolarski, M. R. Schoeberl, C. H. Jackman, M. L. Gupta, P. A. Newman, J. E. Nielsen and E. L. Fleming, *J. Geophys. Res.*, 2008, **113**, D14309.

19. AFEAS data available at: http://www.afeas.org/data.php.

20. UNEP data available at: http://ozone.unep.org/Data_Reporting/Data_Access/.

21. L. Kuijpers, P. Ashford, S. A. Montzka, N. Campbell, D. Clodic, J. S. Daniel, P. Midgley, I. D. Rae, G. Velders and D. Verdonik, *Task Force on Emissions Discrepancies Report*, UNEP, October, 2006.

22. A. McCulloch, P. M. Midgley and P. Ashford, *Atmos. Environ.*, 2003, **37**, 889.

23. J. Schwander, J.-M. Barnola, C. Andrié, M. Leuenberger, A. Ludin, D. Raynaud and B. Stauffer, *J. Geophys. Res.*, 1993, **98**(D2), 2831.

24. J. H. Butler, M. Battle, M. Bender, S. A. Montzka, A. D. Clarke, E. S. Saltzman, C. Sucher, J. Severinghaus and J. W. Elkins, *Nature*, 1999, **399**, 749.

25. M. Battle, M. Bender, T. Sowers, P. P. Tans, J. H. Butler, J. W. Elkins, J. T. Ellis, T. Conway, N. Zhang, P. Lang and A. D. Clarke, *Nature*, 1996, **383**, 231.

26. M. L. Bender, T. Sowers, J.-M. Barnola and J. Chappellaz, *Geophys. Res. Lett.*, 1994, **21**, 189.

27. J. P. Severinghaus, M. R. Albert, Z. R. Courville, M. A. Fahnestock, K. Kawamura, S. A. Montzka, J. Muhle, T. A. Scambos, E. Shields,

C. A. Shuman, M. Suwa, P. Tans and R. F. Weiss, *Earth and Planet. Sci. Lett.*, 2010, **293**, 359.

28. M. Aydin, M. B. Williams and E. S. Saltzman, *J. Geophys. Res.*, 2007, **112**, D07312.

29. E. S. Saltzman, M. Aydin, C. Tatum and M. B. Williams, *J. Geophys. Res.*, 2008, **113**, D05304.

30. C. M. Trudinger, D. M. Etheridge, G. A. Sturrock, P. J. Fraser, P. B. Krummel and A. McCulloch, *J. Geophys. Res.*, 2004, **109**, D22310.

31. M. Aydin, S. A. Montzka, M. O. Battle, M. B. Williams, W. J. DeBruyn, J. H. Butler, K. R. Verhulst, C. Tatum, B. K. Gun, D. A. Plotkin, B. D. Hall and E. S. Saltzman, *Atmos. Chem. Phys.*, 2010, **10**, 5135.

32. E. S. Saltzman, M. Aydin, W. J. De Bruyn, D. B. King and S. A. Yvon-Lewis, *J. Geophys. Res.*, 2004, **109**, D05301.

33. M. Frische, K. Garofalo, T. H. Hansteen, R. Borchers and J. Harnisch, *Environ. Sci. Pollut. Res.*, 2006, **13**(6), 406.

34. F. M. Schwandner, T. M. Seward, A. P. Gize, P. A. Hall and V. J. Dietrich, *J. Geophys. Res.*, 2004, **109**, D04301.

35. S. A. Montzka, J. H. Butler, R. C. Myers, T. M. Thompson, T. H. Swanson, A. D. Clarke, L. T. Lock and J. W. Elkins, *Science*, 1996, **272**, 1318.

36. R. G. Prinn, R. F. Weiss, P. J. Fraser, P. G. Simmonds, D. M. Cunnold, F. N. Alyea, S. O'Doherty, P. Salameh, B. R. Miller, J. Huang, R. H. J. Wang, D. E. Hartley, C. Harth, L. P. Steele, G. Sturrock, P. M. Midgley and A. McCulloch, *J. Geophys. Res.*, 2000, **105**(D14), 17751.

37. C. M. Volk, J. W. Elkins, D. W. Fahey, G. S. Dutton, J. M. Gilligan, M. Loewenstein, J. R. Podolske, K. R. Chan and M. R. Gunson, Evaluation of source gas lifetimes from stratospheric observations, *J. Geophys. Res.*, 1997, **102**, 25,543.

38. R. Zander, E. Mahieu, P. Demoulin, P. Duchatelet, G. Roland, C. Servais, M. De Mazière and C. P. Rinsland, *Environ. Sci.*, 2005, **2**(2–3), 295.

39. C. P. Rinsland, C. Boone, R. Nassar, K. Walker, P. Bernath, E. Mahieu, R. Zander, J. C. McConnell and L. Chiou, *Geophys. Res. Lett.*, 2005, **32**, L16S03.

40. N. D. C. Allen, P. F. Bernath, C. D. Boone, M. P. Chipperfield, D. Fu, G. L. Manney, D. E. Oram, G. C. Toon and D. K. Weisenstein, *Atmos. Chem. Phys.*, 2009, **9**(19), 7449.

41. S. A. Montzka, B. D. Hall and J. W. Elkins, *Geophys. Res. Lett.*, 2009, **36**, L03804.

42. S. A. Montzka, L. Kuijpers, M. O. Battle, M. Aydin, K. R. Verhulst, E. S. Saltzman and D. W. Fahey, *Geophys. Res. Lett.*, 2010, **37**, L02808.

43. B. R. Miller, M. Rigby, L. J. M. Kuijpers, P. B. Krummel, L. P. Steele, M. Leist, P. J. Fraser, A. McCulloch, C. Harth, P. Salameh, J. Mühle, R. F. Weiss, R. G. Prinn, R. H. J. Wang, S. O'Doherty, B. R. Greally and P. G. Simmonds, *Atmos. Chem. Phys.*, 2010, **10**, 7875.

44. S. A. Montzka, M. Krol, E. Dlugokencky, B. Hall, P. Jöckel and J. Lelieveld, *Science*, 2011, **331**, 67.

45. C. Clerbaux, D. Cunnold, J. Anderson, P. Bernath, A. Engel, P. J. Fraser, E. Mahieu, A. Manning, J. Miller, S. A. Montzka, R. Prinn, S. Reimann, C. P. Rinsland, P. Simmonds, D. Verdonik, D. Wuebbles and Y. Yokouchi, *Scientific Assessment of Ozone Depletion: 2006, Global Ozone Research and Monitoring Project—Report No. 50*, World Meteorological Organization, Geneva, 2007.

46. R. G. Prinn, J. Huang, R. F. Weiss, D. M. Cunnold, P. J. Fraser, P. G. Simmonds, A. McCulloch, C. Harth, S. Reimann, P. Salameh, S. O'Doherty, R. H. J. Wang, L. W. Porter, B. R. Miller and P. B. Krummel, *Geophys. Res. Lett.*, 2005, **32**, L07809.

47. P. Fraser, D. Cunnold, F. Alyea, R. Weiss, R. Prinn, P. Simmonds, B. Miller and R. Langenfelds (1996), Lifetime and emission estimates of 1,1,2-trichlorotrifluorethane (CFC-113) from daily global background observations June 1982–June 1994, *J. Geophys. Res.*, 1996, **101**(D7), 12,585.

48. X. Xiao, R. G. Prinn, P. J. Fraser, P. G. Simmonds, R. F. Weiss, S. O'Doherty, B. R. Miller, P. K. Salameh, C. M. Harth, P. B. Krummel, L. W. Porter, J. Mühle, B. R. Greally, D. Cunnold, R. Wang, S. A. Montzka, J. W. Elkins, G. S. Dutton, T. M. Thompson, J. H. Butler, B. D. Hall, S. Reimann, M. K. Vollmer, F. Stordal, C. Lunder, M. Maione, J. Arduini and Y. Yokouchi, *Atmos. Chem. Phys.*, 2010, **10**, 5515.

49. M. Aydin, E. S. Saltzman, W. J. De Bruyn, S. A. Montzka, J. H. Butler and M. Battle, *Geophys. Res. Lett.*, 2004, **31**, L02109.

50. S. A. Yvon-Lewis, E. S. Saltzman and S. A. Montzka, *Atmos. Chem. Phys.*, 2009, **9**, 5963.

51. S. A. Montzka, J. H. Butler, B. D. Hall, J. W. Elkins and D. J. Mondeel, *Geophy. Res. Lett.*, 2003, **30**(15), 1826.

52. S. A. Montzka, J. H. Butler, J. W. Elkins, T. M. Thompson, A. D. Clarke and L. T. Lock, *Nature*, 1999, **398**, 690.

53. R. Zander, E. Mahieu, M. R. Gunson, M. C. Abrams, A. Y. Chang, M. Abbas, C. Aellig, A. Engel, A. Goldman, F. W. Irion, N. Kämpfer, H. A. Michelson, M. J. Newchurch, C. P. Rinsland, R. J. Salawitch, G. P. Stiller and G. C. Toon, *Geophys. Res. Lett.*, 1996, **23**, 2357.

54. L. Froidevaux, N. J. Livesey, W. G. Read, R. J. Salawitch, J. W. Waters, B. Drouin, I. A. MacKenzie, H. C. Pumphrey, P. Bernath, C. Boone, R. Nassar, S. Montzka, J. Elkins, D. Cunnold and D. Waugh, *Geophys. Res. Lett.*, 2006, **33**, L23812.

55. M. Dorf, J. H. Butler, A. Butz, C. Camy-Peyret, M. P. Chipperfield, L. Kritten, S. A. Montzka, B. Simmes, F. Weidner and K. Pfeilsticker, *Geophys. Res. Lett.*, 2006, **33**, L24803.

56. E. Mahieu, P. Duchatelet, R. Zander, P. Demoulin, C. Servais, C. P. Rinsland, M. P. Chipperfield and M. De Mazière, in Ozone Vol. II, Proceedings of the XX Quadrennial Ozone Symposium, 1–8 June 2004, Kos, Greece, edited by C. S. Zerefos, *Int. Ozone Comm.*, Athens, Greece, 2004, 997–998.

57. C. P. Rinsland, E. Mahieu, R. Zander, N. B. Jones, M. P. Chipperfield, A. Goldman, J. Anderson, J. M. Russell III, P. Demoulin, J. Notholt, G. C.

Toon, J. F. Blavier, B. Sen, R. Sussmann, S. W. Wood, A. Meier, D. W. T. Griffith, L. S. Chiou, F. J. Murcray, T. M. Stephen, F. Hase, S. Mikuteit, A. Schulz and T. Blumenstock, *J. Geophys. Res.*, 2003, **108**(D8), 4252.

58. F. Hendrick, P. V. Johnston, M. De Mazière, C. Fayt, C. Hermans, K. Kreher, N. Theys, A. Thomas and M. Van Roozendael, *Geophys. Res. Lett.*, 2008, **35**, L14801.

59. K. S. Law, W. T. Sturges, D. R. Blake, N. J. Blake, J. B. Burkholder, J. H. Butler, R. A. Cox, P. H. Haynes, M. K. W. Ko, K. Kreher, C. Mari, K. Pfeilsticker, J. M. C. Plane, R. J. Salawitch, C. Schiller, B.-M. Sinnhuber, R. von Glasow, N. J. Warwick, D. J. Wuebbles and S. A. Yvon-Lewis, *Scientific Assessment of Ozone Depletion: 2006, Global Ozone Research and Monitoring Project—Report No. 50*, World Meteorological Organization, Geneva, Switzerland, 2007.

60. R. Hossaini, M. P. Chipperfield, B. M. Monge-Sanz, N. A. D. Richards, E. Atlas and D. R. Blake, *Atmos. Chem. Phys.*, 2010, **10**, 719.

61. A. Gettelman, P. H. Lauritzen, M. Park and J. E. Kay, Processes regulating short-lived species in the tropical tropopause layer, *J. Geophys. Res.*, 2009, **114**, D13303.

62. J. Brioude, R. W. Portmann, J. S. Daniel, O. R. Cooper, G. J. Frost, K. H. Rosenlof, C. Granier, A. R. Ravishankara, S. A. Montzka and A. Stohl, *Geophys. Res. Lett.*, 2010, **37**, L19804.

63. Y. Yokouchi, F. Hasebe, M. Fujiwara, H. Takashima, M. Shiotani, N. Nishi, Y. Kanaya, S. Hashimoto, P. Fraser, D. Toom-Sauntry, H. Mukai and Y. Nojiri, *J. Geophys. Res.*, 2005, **110**, D23309.

64. J. H. Butler, D. B. King, J. M. Lobert, S. A. Montzka, S. A. Yvon-Lewis, B. D. Hall, N. J. Warwick, D. J. Mondeel, M. Aydin and J. W. Elkins, *Global Biogeochem. Cycles*, 2007, **21**, GB1023.

65. P. Konopka, G. Günther, R. Müller, F. H. S. dos Santos, C. Schiller, F. Ravegnani, A. Ulanovsky, H. Schlager, C. M. Volk, S. Viciani, L. L. Pan, D.-S. McKenna and M. Riese, *Atmos. Chem. Phys.*, 2007, **7**, 3285.

66. P. Ricaud, B. Barret, J.-L. Attié, E. Motte, E. Le Flochmoën, H. Teyssèdre, V.-H. Peuch, N. Livesey, A. Lambert and J.-P. Pommereau, *Atmos. Chem. Phys.*, 2007, **7**, 5639.

67. R. Hossaini, M. P. Chipperfield, W. Feng, T. J. Breider, E. Atlas, S. A. Montzka, B. R. Miller, F. Moore and J. Elkins, 2011, submitted.

68. S. Solomon, R. R Garcia and A. R. Ravishakara, *J. Geophys. Res.*, 1994, **99**(D10), 20491.

69. J. Aschmann, B.-M. Sinnhuber, E. L. Atlas and S. M. Schauffler, *Atmos. Chem. Phys.*, 2009, **9**, 9237.

70. A. Butz, H. Bösch, C. Camy-Peyret, M. P. Chipperfield, M. Dorf, S. Kreycy, L. Kritten, C. Prados-Román, J. Schwärzle and K. Pfeilsticker, *Atmos. Chem. Phys.*, 2009, **9**(18), 7229.

71. D. Youn, K. O. Patten, D. J. Wuebbles, J. Lee and C.-W. So, *Atmos. Chem. Phys.*, 2010, **10**, 10129.

72. Y. Li, K. O. Patten, D. Youn and D. J. Wuebbles, *Atmos. Chem. Phys.*, 2006, **6**, 4559.

73. P. A. Newman, J. S. Daniel, D. W. Waugh and E. R. Nash, *Atmos., Chem. Phys.*, 2007, **7**, 4537.
74. S. M. Schauffler, E. L. Atlas, S. G. Donnelly, A. Andrews, S. A. Montzka, J. W. Elkins, D. F. Hurst, P. A. Romashkin and V. Stroud, *J. Geophys. Res.*, 2003, **108**(D5), 4173.
75. M. Y. Danilin, N.-D. Sze, M. K. W. Ko, J. M. Rodriguez and M. J. Prather, *Geophys. Res. Lett.*, 1996, **23**, 153.
76. J. S. Daniel, S. Solomon, R. W. Portmann and R. R. Garcia, *J. Geophys. Res.*, 1999, **104**(D19), 23871.
77. T. M. Hall and R. A. Plumb, *J. Geophys. Res.*, 1994, **99**, 1059.
78. J. S. Daniel. G. J. M. Velders, O. Morgenstern, D. W. Toohey, T. J. Wallington, D. J. Wuebbles, H. Akiyoshi, A. F. Bais, E. L. Fleming, C. H. Jackman, L. J. M. Kuijpers, M. McFarland, S. A. Montzka, M. N. Ross, S. Tilmes and M. B. Tully, A Focus on Information and Options for Policymakers, *Scientific Assessment of Ozone Depletion: 2010, Global Ozone Research and Monitoring Project—Report No. 52*, World Meteorological Organization, Geneva, 2011.
79. A. Douglass, V. Fioletov, S. Godin-Beekmann, R. Müller, R. S. Stolarski, A. Webb, A. Arola, J. B. Burkholder, J. P. Burrows, M. P. Chipperfield, R. Cordero, C. David, P. N. den Outer, S. B. Diaz, L. E. Flynn, M. Hegglin, J. R. Herman, P. Huck, S. Janjai, I. M. Jánosi, J. W. Krzyścin, Y. Liu, J. Logan, K. Matthes, R. L. McKenzie, N. J. Muthama, I. Petropavlovskikh, M. Pitts, S. Ramachandran, M. Rex, R. J. Salawitch, B.-M. Sinnhuber, J. Staehelin, S. Strahan, K. Tourpali, J. Valverde-Canossa and C. Vigouroux, *Scientific Assessment of Ozone Depletion: 2010, Global Ozone Research and Monitoring Project—Report No. 52*, World Meteorological Organization, Geneva, 2011.
80. C. H. Jackman, E. L. Fleming, S. Chandra, D. B. Considine and J. E. Rosenfield, *J. Geophys. Res.*, 1996, **101**(D22), 28753.
81. J. A. Mäder, J. Staehelin, T. Peter, D. Brunner, H. E. Rieder and W. A. Stahel, *Atmos. Chem. Phys.*, 2010, **10**, 12161.
82. P. A. Newman, E. R. Nash, S. R. Kawa, S. A. Montzka and S. M. Schauffler, *Geophys. Res. Lett.*, 2006, **33**, L12814.
83. Salby, M., E. Titova and L. Deschamps, *Geophys. Res. Lett.*, 2011, **38**, L09702.
84. D. J. Hofmann and S. A. Montzka, *EOS Trans.*, American Geophysical Union, 2009, **90**, 1.
85. http://www.esrl.noaa.gov/gmd/odgi/ accessed April, 2011.
86. R. W. Portmann and S. Solomon, *Geophys. Res. Lett.*, 2007, **34**, L02813.
87. J.-F. Lamarque, D. E. Kinnison, P. G. Hess and F. M. Vitt, *J. Geophys. Res.*, 2008, **113**, D12301.
88. K. L. Denman, G. Brasseur, A. Chidthaisong, P. Ciais, P. M. Cox, R. E. Dickinson, D. Hauglustaine, C. Heinze, E. Holland, D. Jacob, U. Lohmann, S. Ramachandran, P. L. da Silva Dias, S. C. Wofsy and X. Zhang, in *Climate Change 2007: The Physical Science Basis*, (Solomon, S., D. Qin, M. Manning, Z. Chen, M. Marquis, K. B. Averyt, M. Tignor

and H. L. Miller (eds.)) Ch. 7 (Cambridge Univ. Press, Cambridge, U.K. 2007).

89. P. Bousquet, P. Ciais, J. B. Miller, E. J. Dlugokencky, D. A. Hauglustaine, C. Prigent, G. R. Van der Werf, P. Peylin, E.-G. Brunke, C. Carouge, R. L. Langenfelds, J. Lathière, F. Papa, M. Ramonet, M. Schmidt, L. P. Steele, S. C. Tyler and J. White, *Nature*, 2006, **443**, 439.

90. D. M. Etheridge, L. P. Steele, R. J. Francey and R. L. Langenfelds, *J. Geophys. Res.*, 1998, **103**, 15979.

91. L. Loulergue, A. Schilt, R. Spahni, V. Masson-Delmotte, T. Blunier, B. Lemieux, J.-M. Barnola, D. Raynaud, T. F. Stocker and J. Chappellaz, *Nature*, 2008, **453**, 383.

92. E. J. Dlugokencky, L. Bruhwiler, J. W. C. White, L. K. Emmons, P. C. Novelli, S. A. Montzka, K. A. Masarie, P. M. Lang, A. M. Crotwell, J. B. Miller and L. V. Gatti, *Geophys. Res. Lett.*, 2009, **36**, L18803.

93. M. Rigby, R. G. Prinn, P. J. Fraser, P. G. Simmonds, R. L. Langenfelds, J. Huang, D. M. Cunnold, L. P. Steele, P. B. Krummel, R. F. Weiss, S. O'Doherty, P. K. Salameh, H. J. Wang, C. M. Harth, J. Mühle and L. W. Porter, *Geophys. Res. Lett.*, 2008, **35**, 446, L22805.

94. NOAA CH_4 data available at: http://www.esrl.noaa.gov/gmd/dv/ftpdata.html.

95. AGAGE CH_4 data available at: http://agage.eas.gatech.edu/data.htm.

96. C. M. Spivakovsky, J. A. Logan, S. A. Montzka, Y. J. Balkanski, M. Foreman-Fowler, D. B. A. Jones, L. W. Horowitz, A. C. Fusco, C. A. M. Brenninkmeijer, M. J. Prather, S. C. Wofsy and M. B. McElroy, *J. Geophys. Res.*, 2000, **105**, 8931.

97. J. Lelieveld, F. J. Dentener, W. Peters and M. C. Krol, *Atmos. Chem. Phys.*, 2004, **4**, 2337.

98. M. J. Prather, *Geophys. Res. Lett.*, 1996, **23**(19), 2597.

99. F. Dentener, W. Peters, M. Krol, M. van Weele, P. Bergamaschi and J. Lelieveld, *J. Geophys. Res.*, 2003, **108**, 4442.

100. EDGAR Emission Source: EC-JRC/PBL. EDGAR version 4.1. http://edgar.jrc.ec.europa.eu/, 2009.

101. N. Gruber and J. N. Galloway, *Nature*, 2008, **451**, 293.

102. R. Spahni, R. Spahni, J. Chappellaz, T. F. Stocker, L. Loulergue, G. Hausammann, K. Kawamura, J. Flückiger, J. Schwander, D. Raynaud, V. Masson-Delmotte and J. Jouzel, *Science*, 2005, **310**, 1317.

103. J. Flückiger, T. Blunier, B. Stauffer, J. Chappellaz, R. Spahni, K. Kawamura, J. Schwander, T. F. Stocker and D. Dahl-Jensen, *Global Biogeochem. Cycles*, 2004, **18**, GB1020.

104. A. Schilt, M. Baumgartner, T. Blunier, J. Schwander, R. Spahni, H. Fischer and T. F. Stocker, *Quatern. Sci. Revs.*, 2010, **29**, 82.

105. J. Flückiger, A. Dällenback, T. Blunier, B. Stauffer, T. F. Stocker, D. Raynaud and J.-M. Barnola, *Science*, 1999, **285**, 227.

106. NOAA N_2O global surface means available at: ftp://ftp.cmdl.noaa.gov/hats/n2o/insituGCs/RITS/.

107. AGAGE N_2O data available at: http://agage.eas.gatech.edu/data.htm.

108. M. J. Prather and J. Hsu, *Science*, 2010, **330**, 952.
109. A. R. Ravishankara, J. S. Daniel and R. W. Portmann, *Science*, 2009, **326**, 123.
110. N. Theys, M. Van Roozendael, Q. Errera, F. Hendrick, F. Daerden, S. Chabrillat, M. Dorf, K. Pfeilsticker, A. Rozanov, W. Lotz, J. P. Burrows, J.-C. Lambert, F. Goutail, H. K. Roscoe and M. De Mazière, *Atmos. Chem. Phys.*, 2009, **9**, 831.
111. C. E. Junge and J. E. Manson, *J. Geophys. Res.*, 1961, **66**(7), 2163.
112. *SPARC Assessment of Stratospheric Aerosol Properties*, ed. L. Thomason and Th. Peter, WCRP-124, WMO/TD No. 1295, 2006, p. 322.
113. D. J. Hofmann and S. Solomon, *J. Geophys. Res.*, 1989, **94**(D4), 5029.
114. P. J. Crutzen, *Geophys. Res. Lett.*, 1976, **3**, 73.
115. D. Hofmann, J. Barnes, M. O'Neill, M. Trudeau and R. Neely, *Geophys. Res. Lett.*, 2009, **36**, L15808.
116. S. A. Montzka, P. Calvert, B. D. Hall, J. W. Elkins, T. J. Conway, P. P. Tans and C. Sweeney, *J. Geophys. Res.*, 2007, **112**, D09302.
117. NOAA surface measurement data available at ftp://ftp.cmdl.noaa.gov/hats/carbonyl sulfide/.
118. S. J. Smith, J. van Aardenne, Z. Klimont, R. J. Andres, A. Volke and S. Delgado Arias, *Atmos. Chem. Phys.*, 2011, **11**, 1101.
119. P. Forster, V. Ramaswamy, P. Artaxo, T. Berntsen, R. Betts, D. W. Fahey, J. Haywood, J. Lean, D. C. Lowe, G. Myhre, J. Nganga, R. Prinn, G. Raga, M. Schulz, and R. Van Dorland, *Climate Change 2007: The Physical Science Basis*. Contribution of Working Group I to the Fourth Assessment Report of the Intergovernmental Panel on Climate Change, ed. S. Solomon, D. Qin, M. Manning, Z. Chen, M. Marquis, K. B. Averyt, M. Tignor, and H. L. Miller, Cambridge University Press, Cambridge, U.K., and New York, NY, U.S.A., 2007, p. 996.
120. D. J. Hofmann, J. H. Butler, E. J. Dlugokencky, J. W. Elkins, K. Masarie, S. A. Montzka and P. Tans, *Tellus*, 2006, **58B**, 614.
121. http://www.esrl.noaa.gov/gmd/aggi/
122. D. Archer and V. Brovkin, *Climatic Change*, 2008, **90**, 283.
123. P. Friedlingstein, R. A. Houghton, G. Marland, J. Hackler, T. A. Boden, T. J. Conway, J. G. Canadell, M. R. Raupach, P. Ciais and C. Le Quéré, *Nature Geosci.*, 2010, **3**, 811.
124. J. Mühle, A. L. Ganesan, B. R. Miller, P. K. Salameh, C. M. Harth, B. R. Greally, M. Rigby, L. W. Porter, L. P. Steele, C. M. Trudinger, P. B. Krummel, S. O'Doherty, P. J. Fraser, P. G. Simmonds, R. G. Prinn and R. F. Weiss, *Atmos. Chem. Phys.*, 2010, **10**, 6485.
125. M. Rigby, J. Mühle, B. R. Miller, R. G. Prinn, P. B. Krummel, L. P. Steele, P. J. Fraser, P. K. Salameh, C. M. Harth, R. F. Weiss, B. R. Greally, S. O'Doherty, P. G. Simmonds, M. K. Vollmer, S. Reimann, J. Kim, K.-R. Kim, H. J. Wang, J. G. J. Olivier, E. J. Dlugokencky, G. S. Dutton, B. D. Hall and J. W. Elkins, *Atmos. Chem. Phys.*, 2010, **10**, 10305.
126. I. Levin, T. Naegler, R. Heinz, D. Osusko, E. Cuevas, A. Engel, J. Ilmberger, R. L. Langenfelds, B. Neininger, C. v. Rohden, L. P. Steele, R. Weller, D. E. Worthy and S. A. Zimov, *Atmos. Chem. Phys.*, 2010, **10**, 2655.

127. M. J. Prather, *Global Biogeochem. Cycles*, 1997, **11**(3), 393.
128. IPCC/TEAP (Intergovernmental Panel on Climate Change/Technology and Economic Assessment Panel), *IPCC/TEAP Special Report on Safeguarding the Ozone Layer and the Global Climate System: Issues Related to Hydrofluorocarbons and Perfluorocarbons*, prepared by Working Groups I and III of the Intergovernmental Panel on Climate Change, and the Technical and Economic Assessment Panel, Cambridge University Press, Cambridge, UK and New York, NY, USA, 2005.
129. UNEP (United Nations Environment Programme), *2006 Report of the Halons Technical Options Committee 2006 Assessment*, edited by D. Catchpole, and D. Verdonik, Nariobi, Kenya, 2007.
130. A. McCulloch, P. Ashford and P. M. Midgley, *Atmos. Environ.*, 2001, **35**, 4387.
131. A. McCulloch and P. M. Midgley, *Atmos. Environ.*, 2001, **35**, 5311.
132. J. S. Daniel and G. J. M. Velders, A. R. Douglass, P. M. D. Forster, D. A. Hauglustaine, I. S. A. Isaaksen, L. J. M. Kuijpers, A. McCulloch and T. J. Wallington, Halocarbon scenarios, ozone depletion potentials, and global warming potentials, *Scientific Assessment of Ozone Depletion: 2006, Global Ozone Research and Monitoring Project—Report No. 50*, World Meteorological Organization, Geneva, 2007.
133. S. O'Doherty, D. M. Cunnold, A. Manning, B. R. Miller, R. H. J. Wang, P. B. Krummel, P. J. Fraser, P. G. Simmonds, A. McCulloch, R. F. Weiss, P. Salameh, L. W. Porter, R. G. Prinn, J. Huang, G. Sturrock, D. Ryall, R. G. Derwent and S. A. Montzka, *J. Geophys. Res.*, 2004, **109**, D06310.
134. E. Mahieu, P. Duchatelet, R. Zander, P. Demoulin, C. Servais, C. P. Rinsland, M. P. Chipperfield and M. De Mazière, The evolution of inorganic chlorine above the Jungfraujoch station: An update, in *Ozone Vol. II, Proceedings of the XX Quadrennial Ozone Symposium*, 1–8 June 2004, Kos, Greece, edited by C. S. Zerefos, *Int. Ozone Comm.*, Athens, Greece, 2004, 997–998.
135. C. P. Rinsland, E. Mahieu, R. Zander, N. B. Jones, M. P. Chipperfield, A. Goldman, J. Anderson, J. M. Russell III, P. Demoulin, J. Notholt, G. C. Toon, J. F. Blavier, B. Sen, R. Sussmann, S. W. Wood, A. Meier, D. W. T. Griffith, L. S. Chiou, F. J. Murcray, T. M. Stephen, F. Hase, S. Mikuteit, A. Schulz and T. Blumenstock, *J. Geophys. Res.*, 2003, **108**(D8), 4252.
136. M. P. Chipperfield, M. Burton, W. Bell, C. P. Walsh, T. Blumenstock, M. T. Coffey, J. W. Hannigan, W. G. Mankin, B. Galle, J. Mellqvist, E. Mahieu, R. Zander, J. Notholt, B. Sen and G. C. Toon, *J. Geophys. Res.*, 1997, **102**(D11), 12901.
137. P. J. Fraser, D. E. Oram, C. E. Reeves, S. A. Penkett and A. McCulloch, *J. Geophys. Res.*, 1999, **104**(D13), 15985.
138. C. E. Reeves, W. T. Sturges, G. A. Sturrock, K. Preston, D. E. Oram, J. Schwander, R. Mulvaney, J.-M. Barnola and J. Chappellaz, *Atmos. Chem. Phys.*, 2005, **5**(8), 2055.

CHAPTER 3

Stratospheric Halogen Chemistry

MARC VON HOBE AND FRED STROH

Forschungszentrum Jülich, Institute for Energy and Climate Research
(IEK-7), 52425 Jülich, Germany

3.1 Introduction

The members of the seventh main group of the periodic system of the elements (fluorine, chlorine, bromine, and iodine) are commonly referred to as the halogen group (the Greek term "halogen" points to the fact that the elements like to form salts when in contact with metals). We will ignore here the higher members of the halogen group, *i.e.* the radioactive astatine, which is of extremely low abundance in the Earth system, and ununseptium (the preliminary name of element 117), which was recently generated in amounts of a few atoms and seems to support the long sought-for island of enhanced stability for superheavy nuclei.[1] Of the halogen family, chlorine and bromine in particular play an important role in stratospheric as well as tropospheric ozone chemistry. In both regimes halogen reactions can cause almost complete destruction of ambient ozone concentrations under specific atmospheric conditions.[2–4] Fluorine and iodine have much less pronounced effects on the ozone budget (see section 3.3 for a brief discussion) and will therefore not be dealt with in detail here. We are focusing on the upper tropospheric and stratospheric issues and give an account of halogen abundance, chemistry, interconversion, and partitioning. Furthermore, a critical discussion on the

Stratospheric Ozone Depletion and Climate Change
Edited by Rolf Müller
© Royal Society of Chemistry 2012
Published by the Royal Society of Chemistry, www.rsc.org

major measurement techniques for halogen compounds regularly applied in the atmosphere is given (see Box 3.1). We then describe how chlorine and bromine radical species destroy ozone in catalytic cycles under different atmospheric conditions. Topics with major open questions and where progress has been made lately are discussed in special subsections, as are the photolysis of chlorine peroxide, ClOOCl, halogen abundance and chemistry in the tropopause region.

Throughout the chapter a capital "X" in chemical formulas denotes a halogen atom (F, Cl, Br, or I). Any mention of the term "stratosphere" will also imply the upper troposphere and tropopause region, as generally the same chemical processes dominate the halogen chemistry in these atmospheric regions as opposed to the planetary boundary layer where halogens also play an important role. However, the chemical reactions driving halogen partitioning and influencing ozone and other trace gases in the boundary layer are different from those in the stratosphere. For a description of these reactions see ref. 5 and 6.

When studying the literature we became aware of several fascinating aspects in the history of the chemistry of relevant halogen compounds, which have often been right at the center of interest when major progress has been made in physical chemistry. Therefore, we will open this chapter with a brief historical account.

3.2 A Brief History of Halogen Chemistry

Free chlorine was probably prepared by alchemists from the 13th century on.[7] The first reported notice, however, occurred in 1626 when Johann Rudolph Glauber, an early German-Dutch chemist, treated a solution of hydrochloric acid with metal oxides (probably containing manganese dioxide) freeing up elementary chlorine. The yellow distillate of the solution (chlorine dissolved in water) he described as a "fire spirit" that was able to dissolve some metals and most minerals.[8]

But Glauber did not categorize his unique observations. Fast progress was made at the end of the 18th and beginning of the 19th centuries when chemists began analyzing all kinds of natural compounds employing more or less standardized and proven experimental techniques. In 1792, Scheele described the preparation of chlorine and characterized some of its properties, not being aware that he had discovered a new element.[8] This was systematically shown by Humphrey Davy by demonstrating that the so-called oxymuratic acid did not contain oxygen, as had been believed by the community for many years.[9] Still this result was not accepted by famous chemists such as Berzelius.[7]

A French manufacturer, Bernard Courtois first isolated iodine when treating ashes of seaweed with sulfuric acid.[10] Again, it was up to Davy to show that iodine was a new element.[11] Around 1824, both the German chemist Carl Löwig and the French chemist Antoine-Jerome Balard produced bromine from mineral water and seaweed, respectively. Balard, however, was the one to publish his results first.[12,13] Due to its high reactivity, fluorine was isolated electrolytically much later by H. Moissan,[14] leading to the award of the Nobel Prize in chemistry in 1906. For a more complete description of the history of chlorine and bromine see Wisniak[7,15] or Chabot.[16]

Once chlorine was available in the laboratories, the first halogen oxides were synthesized and described. The symmetric chlorine dioxide OClO was identified in 1815 by Davy, who named the gas "euchlorine" (*i.e.* "very green").[17] Cl_2O was identified somewhat later by Balard.[18] Also, the higher oxides of chlorine up to Cl_2O_7 were synthesized and characterized. The interesting early overview by Millon[8] shows that, due to the lack of knowledge of the correct elemental composition ratio, OClO was still termed chlorine oxide "ClO" and accordingly for the higher oxides. For this reason the dichlorine oxide, Cl_2O, identified by Balard was at that time questioned by colleagues such as Gay-Lussac and Millon.

The fact that even small amounts of chlorine were able to destroy ozone was reported by Hautefeuille and Chappuis already around 1880[19] and in more detail in 1884,[20] where they speculated that an unstable chlorine-oxygen compound was formed intermediately. In the following years, in an effort to understand such photochemical reactions, studies on halogens played an important role in the development of basic kinetic theory. The chemists then were puzzled by the fact that a few quanta of light were able to cause the reaction of millions of molecules, and the reactions of chlorine, bromine, and iodine with hydrogen and also ozone were in the center of interest.[21,22] In 1929, Bodenstein postulated ClO as an intermediate in a catalytic reaction cycle in the decomposition of ozone by chlorine, however, still missing the fact that halogen atoms were involved.[23] Similarly, BrO, which had been described by Lewis and Schumacher,[24] was postulated as an intermediate in the analog ozone bromine reaction by Spinks.[25]

Rapid progress in the understanding of reactive intermediates was made with major advances with respect to analytical tools—in particular spectroscopic techniques—in the late 1940s and early 1950s. The unambiguous identification of ClO by its emission spectrum in flames was reported in 1948 by Pannetier.[26] In the early 1950s ClO was one of the first radicals to be observed by the newly developed flash photolysis technique by Porter[27] who has termed chlorine as the "photochemists element".[28] For their follow-on research on fast elementary reactions of Cl, ClO and other reactive species, Porter and Norrish were awarded the Nobel Prize in chemistry in 1967 together with Eigen.

However, it was not until the 1970s that it was recognized that halogens could play a role in stratospheric chemistry.[29–31] Molina and Rowland[32] pointed out the possible threat of emissions of chlorofluorocarbon (CFC) compounds leading to an increased stratospheric chlorine loading that could thin the ozone shield protecting the Earth from harmful ultraviolet radiation via light-driven catalytic cycles.

BOX 3.1: Measurement of Stratospheric Halogen Species

The proper understanding of stratospheric halogen chemistry and ozone loss depends critically on accurate atmospheric measurements. Furthermore, laboratory measurements of all relevant reaction rates and their temperature dependence as well as absorption cross sections and quantum yields are a prerequisite for good model performance. The ultimate validation of state-of-the-art models however, can only be provided by critical comparisons to actual measurement data collected within different atmospheric domains under a realistic range of atmospheric conditions (temperature, pressure, radiation, aerosols/clouds). Absence of good measurement data on key species prevents proper understanding of chemical and dynamical processes. For the halogen species low to very low atmospheric abundances down to the ppt range as well as the reactive nature of many species require most sophisticated measurement techniques in order to provide data of suitable quality.

Table 3.1 gives an overview of *in situ* (IS) and remote sensing (RS) measurement techniques and their key specifications currently available for the measurement of inorganic halogen species at atmospheric concentrations. While IS measurements are usually better suited for process studies due to a much better characterisation of the measurement volume and better accuracy, the RS measurements provide a much greater temporal and regional (up to global) coverage yielding data more suitable for the study of large scale dynamical processes and trends. Due to more or less complicated algorithms needed to retrieve trace gas profiles, the RS measurements need validation which should primarily be supplied by independent measurement techniques. Therefore, intercomparisons through parallel deployment of different instrumental techniques are worthwhile in order to corroborate the atmospheric observations.

Table 3.1 shows that only for the stable chlorine reservoir species HCl and $ClONO_2$ good accuracies of better than 10% can be achieved. The more reactive chlorine species directly or indirectly (OClO) involved in the ozone destruction cycles can hardly be quantified to better than 20% accuracy, which sets a natural limit to our ability to judge the quality of our current chemical models. However, except for Cl_2 (the major primary product of heterogeneous activation) and BrCl (one product branch of the ClO + BrO reaction), all chlorine species with an appreciable stratospheric concentration have been measured and currently no major inconsistencies between measured and modeled concentrations have been identified.

The situation is much worse for the inorganic bromine species where only BrO concentrations can be measured to somewhat better than 20% accuracy. For all other Br_y species only very sparse RS measurements with quite large error bars exist. Due to their extremely low ambient concentrations, all measurement techniques rely on averaging hundreds of spectra taken over quite large temporal and spatial intervals, limiting their use for detailed

Table 3.1 Overview of selected examples of remote-sensing (RS) and in situ (IS) measurement techniques employed for atmospheric profile measurements of halogen species. Generally, the best reported measurement quality specifications from the most up-to-date publications are given. Values given as "ca." are visually extracted from figures or similar. Generally 1σ accuracy is stated.

Species	Measurement	Technique	Altitude range/km	Det. limit	Resolution[a]	Accuracy	Selected references
Chlorine species							
HCl	IS	Chemical ionization mass spectrometry	5–20	5 ppt	1 s	20 %	125, 126
HCl	IS	Tunable diode laser absorption	19–31	40 ppt	1 s	30–6 %	127, 128
HCl	RS	FIR emission MLS-Aura[b]	11–55	ca. 50 ppt	3–6 km	15–5 %	129
HCl	RS	IR occultation ACE-FTS[c]	10–50	ca. 20 ppt	3–6 km	10 %	122
ClONO$_2$	IS	Thermal disso-ciation CCRF[d]	15–20	10 ppt	35 s	21 %	98, 130
ClONO$_2$	RS	IR emission MIPAS-E[e]	18–27	50 ppt*	3.5 km	6–11 %	131
ClO	IS	CCRF[d]	10–30	3 ppt	10 s	17 %	79, 130, 132, 133
ClO	RS	IR emission MIPAS-E	12–25	200 ppt	3–4.5 km	30 %	134
ClO	RS	FIR emission MLS-Aura	10–45	~50 ppt	3 km	50–300 ppt	78, 135
ClOOCl	IS	Thermal disso-ciation CCRF	15–20	10 ppt	1–3 min	25 %	79, 98
ClOOCl	RS	IR emission MIPAS-B	15–30	500 ppt	2 km	50 %	136
OClO	RS	DOAS/ SCIAMACHY[g]	15–40	~20 ppt	2.5 km	50–100 %	137
HOCl	RS	FIR emission	15–40	~10 ppt	3 km	20–30 %	58
HOCl	RS	IR emission MIPAS-E	20–50	~30 ppt	9 km	30–80 ppt	138
Cl	IS	CCRF	30–40		1 km	50 %	139
Bromine species							
BrO	IS	CCRF	15–30	2–3 ppt	1 min	25 %	140
BrO	RS	DOAS Balloon	5–30	1–2 ppt	2 km	18 %	141
BrO	RS	DOAS SCIAMACHY	5–30	~2–3 ppt	2.5 km	20–50 %	137
BrO	RS	FIR emission MLS-Aura	30–40	2–3 ppt	5.5 km	25 %	135
BrONO$_2$	RS	IR emission MIPAS-E	20–35	~2 ppt	3–10 km	25–50 %	121
HBr	RS	FIR emission Balloon	20–35	~1 ppt	6 km	35 %	142
HOBr	RS	FIR emission Balloon	22–34	<2 ppt	12 km	upper limit	57

[a] For in situ measurements the temporal resolution needed to achieve the stated precision is given. For remote-sensing measurements the vertical resolution is given.
[b] MLS-Aura is the Microwave Limb Sounder onboard the NASA Earth Observing System Aura satellite.
[c] ACE-FTS is the Atmospheric Chemistry Experiment-Fourier Transform Spectrometer onboard the NASA Earth Observing System satellites.
[d] CCRF stands for Chemical Conversion Resonance Fluorescence.
[e] MIPAS is the Michelson Interferometer for Passive Atmospheric Sounding, E refers to the ENVISAT satellite instrument, B to the balloon-borne version.
[f] DOAS refers to Differential Optical Absorption Spectroscopy technique (in the UV spectral region).
[g] SCIAMACHY is the Scanning Imaging Absorption Spectrometer for Atmospheric ChartographY onboard the ENVISAT satellite.

process studies. However, the results from those measurements are consistent with our current understanding of stratospheric inorganic bromine chemistry. First BrONO$_2$ measurements have only recently been reported from MIPAS-ENVISAT[121] and have to be consolidated by thorough checks of the spectroscopic parameters. Due to the regular global coverage of the satellite, these measurements will hopefully provide a very valuable data set in the future.

Table 3.1 does not contain measurement techniques supplying column amounts of trace gases since these data are hard to directly compare to concentration data as supplied by the techniques listed. However, column amounts have been very valuable in various aspects. Ground based as well as satellite based measurements of infrared absorption spectra have supplied important data on stratospheric trace gas trends for species such as HF and HCl which are not abundant in the troposphere.[122,123] In the case of the major chlorine reservoirs, both HCl and ClONO$_2$ could be monitored in this way to supply a good measure of the evolution of total stratospheric inorganic chlorine[124] (*cf.* Figure 2.4). Some aspects of BrO column measurements are discussed in Section 3.6.1.

Due to their stability, bromine as well as chlorine source gases including VSLS can be measured with better accuracy than most of the respective inorganic family members (up to the 1–2% range for Cl SGs, 5–10% for Br SGs, and around 15% for Br VSLS) by gas chromatographic techniques employed *in situ* or after sampling air from the atmosphere.[44] This fact helps to constrain stratospheric Cl$_y$ and Br$_y$ (see Chapter 2 and Section 3.6.1).

3.3 Overview of Stratospheric Halogen Chemistry, Abundances and Partitioning

All-important members of the stratospheric inorganic chlorine and bromine families, termed Cl$_y$ and Br$_y$, respectively, and their various chemical interconversions are shown in Figures 3.1 and 3.2. For more detailed discussions of the kinetics and properties of the species, we recommend Bedjanian and Poulet, 2003,[33] Wayne *et al.*, 1995,[34] Lary *et al.*, 1996,[35] and Lary, 1996.[36] Here we give a wrap-up of the main features of chlorine and bromine abundance, partitioning, and interconversion. Our knowledge today is based on the results of atmospheric measurements as summarized in Box 3.1 combined with laboratory studies of rate constants of the relevant halogen reactions and the photochemical absorption cross sections and quantum yields of the halogen species.[37,38] Intercomparison of species concentrations from model simulations employing the laboratory chemical kinetics data with abundances observed in the atmosphere are the standard tool for testing and improving our models of atmospheric halogen chemistry.

Fluorine and iodine do not significantly destroy ozone in the stratosphere. Sizeable amounts of fluorine are released by photolysis or chemical reactions from CFCs, but the catalytic efficiency for ozone destruction under stratospheric conditions is more than 100 times lower than for chlorine.[39] Iodine, on the other hand, is a very efficient catalyst, but stratospheric iodine levels are low

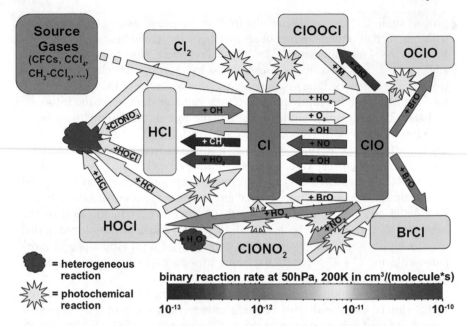

Figure 3.1 Overview of stratospheric chlorine chemistry. The relevant reactive and reservoir species in the stratospheric inorganic chlorine family are shown with their major chemical interconversions. Arrows indicating gas phase reactions are colored according to the binary reaction rate calculated for a temperature of 200 K and pressure of 50 hPa. The reaction of Cl and CH$_4$ has a rate constant of 1.2×10^{-14} cm^3/(molecule $\times$ s) only, *i.e.* lower than the range of the color scale. Generally only the reaction partners are indicated on the arrows. Photolysis and heterogeneous reactions are marked by sun and cloud symbols, respectively. All given reaction rates and branching ratios are taken from JPL06, JPL09.[37,38] Uncertainties of the reaction rates are generally in the 20–60% range.

so that at present iodine compounds do not make a significant contribution to stratospheric ozone depletion.[40–42] Fluorine and iodine will be included in the general discussions but no detailed representation of their chemistry will be given here.

3.3.1 Abundances

The sum of all chlorine compounds shown in Figure 3.1—the total available stratospheric chlorine or Cl$_{tot}$—is determined by the loading of chlorine-containing gases in the troposphere when the air entered the stratosphere and is conserved during the gradual conversion from mostly organic source gases (SGs) to inorganic forms. The same is true for the respective bromine compounds shown in Figure 3.2. However, there a non-negligible amount seems to be injected into the stratosphere in inorganic form (product gas injection, PGI).[43,44] The nature and atmospheric abundance of the SGs and their fate during transport into the stratosphere as well as estimates of PGI have been described in Chapter 2. We will discuss some general chemical aspects of the

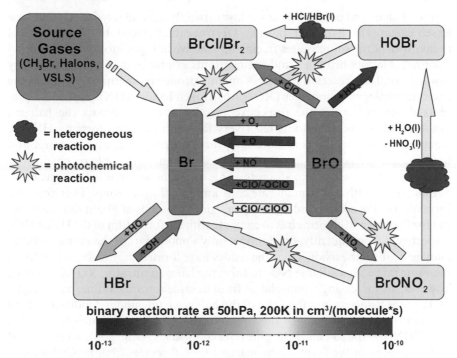

Figure 3.2 Overview of stratospheric bromine chemistry. The relevant reactive and reservoir species in the stratospheric inorganic bromine family are shown with their major chemical interconversions. Same coding as in Figure 3.1. Uncertainties of the reaction rates are generally in the 30–70% range.

involved processes below and in Section 3.6.1 for the very short-lived species (VSLS) contributing to Br$_y$.

A mixing ratio of 3.65 ppb (+/− 0.13 ppb) total available chlorine in the stratosphere has been determined for 2004 in a recent inventory published by Nassar *et al.*[45] that is based on a comprehensive set of satellite observations of atmospheric chlorine compounds. In the lower stratosphere (∼15 km), measured total chlorine was approximately 3.5 ppb, fairly consistent, given uncertainties, with the 3.4 ppb observed in the troposphere then (see Figure 2.4).

While total available stratospheric chlorine is constrained quite well,[45] the uncertainty in the estimates of total available stratospheric bromine is much larger (*cf.* Chapter 2) mainly due to the fact that VSLS and PGI contribute a much higher fraction as compared to Cl$_y$. In the latest WMO assessment, a median of 22.5 ppt of total bromine[44] has been derived with an uncertainty range of 19.5–24.5 ppt (see Section 3.6.1 for a critical discussion).

3.3.2 Partitioning

Once halogen source gases (see Chapter 2) ascend through the troposphere and lower stratosphere, increasing UV radiation fields and radical concentrations below, within, and above the ozone layer (mainly atomic oxygen, O, and to a lesser extent, the hydroxyl radical, OH) eventually lead to photochemical breakdown of the source gases freeing up reactive halogen atoms. The resulting fluorine, chlorine, bromine, and iodine atoms exhibit quite different reactivity towards available reaction partners—mainly ozone and methane—forming the respective halogen monoxides (XO) or hydrogen halides (HX). Figure 3.3 presents a comparison of the respective reaction rates[37,38] across the halogen families. While fluorine atoms react very fast with methane forming HF, the reaction of chlorine atoms is more than three orders of magnitude slower and bromine and iodine virtually do not react at all under atmospheric conditions. The reactivity of the halogen atoms towards ozone, however, is in the same order of magnitude, with a maximum rate constant for chlorine atoms. Therefore, the formation of the corresponding halogen oxides proceeds at about the same rate but for fluorine and chlorine has to compete with the formation of the HF or HCl, respectively. Consequently, for fluorine, only a minor part of the atoms will ever manage to form the oxide. Once the oxides have formed they will react at about the same rate with available NO_2 to form the halogen nitrates, $XONO_2$. These, however, are increasingly photolabile from fluorine towards iodine, partitioning more of the halogens into the form of the halogen oxide under conditions of the mid-latitude stratosphere. For these reasons, Cl_y is partitioned mainly into HCl and $ClONO_2$—the so-called chlorine reservoir species—while only a few percent will be present in the form of the reactive radical species (mainly ClO and, to much lesser extent, Cl) in the mid-latitude sunlit stratosphere as obvious from

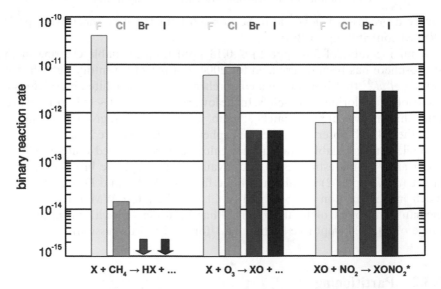

Figure 3.3   Comparison of reaction rates in cm^3/(molecule × s) of basic atmospheric reactions across the halogen family. Reaction rates were calculated for 50 hPa and 200 K employing the currently recommended rate constants.[37,38]

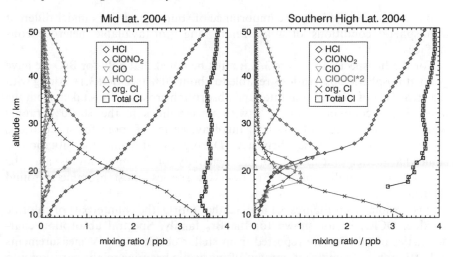

Figure 3.4 Stratospheric chlorine inventory for 2004: averages for northern and southern mid latitudes, 30–60° (left panel), and for southern high latitudes, 60–85° (right panel), as derived from ACE-FTS local sunset measurements, MIPAS measurements (HOCl) and model data (ClOOCl)[45] for the time interval Feb. 2004 until Jan. 2005. The mixing ratios of the most important Cl_y species (accounted for the number of Cl atoms per molecule) within each domain as well as the sum of the organic Cl species and total Cl, respectively, are plotted for the altitude range 10–50 km. Data taken from Nassar *et al.* 2006[45] (auxiliary data).

Figure 3.4 (left panel). Only at altitudes in excess of 30 km, $ClONO_2$ becomes more short-lived due to faster photolysis. Consequently, somewhat higher partitioning into ClO is found (see Figure 3.4). For Br_y, the fast reaction of HBr with OH and the fast photolysis of $BrONO_2$ leads to a much higher partitioning into BrO under sunlit conditions of the mid-latitude stratosphere. At night almost all BrO will be converted into the reservoir forms $BrONO_2$, HOBr, and HBr or mainly BrCl under conditions of chlorine activation.

The conversion to the inorganic forms occurs at different rates for the different primary source gases. Therefore, the total abundance of Cl_y and Br_y increases with the so-called age-of-air, *i.e.* the time that an air mass has spent in the stratosphere. Consequently, the abundance and partitioning of these compounds strongly depends on altitude and, to some extent, on latitude, as shown in Figure 3.4 for Cl_y.

3.3.3 Chlorine *versus* Bromine Chemistry

Figures 3.1 and 3.2 present schemes of the stratospheric Cl_y and Br_y chemistry including gas phase as well as heterogeneous chemical reactions and photolysis processes. The reaction rates given for the gas phase reactions span a quite large range of three orders of magnitude. However, the final turnover within a reaction depends a lot on the species concentrations, which should be taken into account when judging the importance of sources and sinks under different atmospheric conditions of solar zenith angle or available heterogeneous surfaces.

For the Br_y species in Figure 3.2, it must be noted that only for BrO we have good atmospheric observations reliable to about 20% (see Box 3.1). Along with the error bars of the reaction rates and absorption cross sections taken from the latest JPL recommendations[37,38] (with uncertainties in the 30–70% range) which are all in excess of 30–50%, this leaves a lot of room for speculations especially into night-time chemistry. During the daytime the situation is somewhat better since, due to the reactivity of bromine atoms and the photochemical lability of the Br_y reservoir species, most bromine is found as BrO.

It is obvious from Figures 3.1 and 3.2 that OBrO, the counterpart of chlorine dioxide, OClO, is not shown for the Br_y family. Spectral absorption characteristics of OBrO were reported from stellar occultation UV measurements indicating that a significant amount of inorganic bromine might partition into this species at night.[46,65] However, this observation could not be reproduced by other UV absorption measurements,[47] nor are there efficient reactions that could generate even a few ppt of OBrO.[48] Therefore it is not currently regarded as an important or abundant Br_y species.

The gas phase reactions of ClO and BrO are among the more complex ones in chemical kinetics: association reactions like the formation of the halogen nitrates (from $XO + NO_2$) or dichlorine peroxide, ClOOCl, (from $ClO + ClO$, R13 in Section 3.4) are challenging in terms of the proper description of their rate constants since they do not only depend on the concentrations of the reactants and temperature but also on the concentration of the so-called bath gas, *i.e.* they depend on atmospheric pressure (*e.g.* Troe, 2003[49]). On the other hand multi-channel reactions like the reaction of ClO and BrO (R16a-c in Section 3.4) pose an experimental and a theoretical problem. It is as important to know the proper branching ratios as well as the overall reaction rate (*e.g.* Seakins, 2007[50]) and even nowadays, with the evolution of fast and sensitive detection methods (*e.g.* mass spectrometry), it remains extremely challenging to properly sort out minor but still very important branching ratios.

The heterogeneous reaction of HCl and $ClONO_2$ on sulfate aerosol or on liquid and solid PSC surfaces (see Chapter 4 for details) producing Cl_2 (followed by ClO and ClOOCl formation upon photolysis and reaction with ozone) can dramatically change the partitioning within the Cl_y family from reservoirs to reactive species. This is the central chlorine activation process. Similar heterogeneous reactions occur in the Br_y family, as there are the hydrolysis of $BrONO_2$ producing HOBr[51] and the reactions of the halides (HBr/HCl) with HOCl/HOBr and $ClONO_2$/$BrONO_2$, respectively, both producing BrCl.[36] For clarity, the halogen nitrate reactions have been neglected in Figures 3.1 and 3.2. While the $BrONO_2$ hydrolysis proceeds on background sulfate aerosol independent of temperature, the bulk-phase reactions of the halides need temperatures below 210 K to proceed on sulfate aerosol or PSC surfaces. Both reaction products, HOBr and BrCl, are readily photolyzed but the processes do only slightly enhance the BrO partitioning. This was shown for the polar vortices[52,53] as well as for $BrONO_2$ hydrolysis on cold sulfate aerosol around the vortex[54] and under midsummer conditions when stratospheric aerosol peaked after the Mount Pinatubo eruption in the 1990's.[55] However, strong signals resulting from the heterogeneous bromine chemistry can be seen in other parameters. Erle *et al.*[54] report elevated amounts of OClO generated *via* two sequences: HOBr photolysis generates elevated HO_x that will convert HCl into ClO_x. HO_x produced during sunrise has been observed earlier,[56] which can at least partly be attributed to HOBr photolysis. On the other hand, HOBr can also react heterogeneously with HCl producing BrCl, which will also enhance ClO_x and thereby OClO production. Finally, Slusser *et al.*[55] report sizeable reductions in NO_2 that can only be explained when including $BrONO_2$ hydrolysis into the models.

3.4 Halogen Catalyzed Ozone Loss Cycles in the Stratosphere

The simplest ozone destroying catalytic cycle is

$$
\begin{array}{llll}
\text{Cycle 1:} & O_3 + h\nu & \rightarrow & O_2 + O & \text{R1} \\
& XO + O & \rightarrow & X + O_2 & \text{R2} \\
& \underline{X + O_3} & \rightarrow & \underline{XO + O_2} & \text{R3} \\
\text{Net:} & 2O_3 & \rightarrow & 3O_2 &
\end{array}
$$

X can be any halogen atom (in fact, X can also represent non-halogen radical species such as OH and NO, but they are not discussed here). Fluorine, as mentioned above, is very inefficient in destroying ozone due to the fast loss reactions of fluorine atoms with CH_4 and the stability of the HF reservoir gas. The concentration of iodine in the stratosphere is far too low to have any significant effect. Therefore, it is mainly chlorine and bromine destroying ozone by this catalytic cycle.

At mid-latitudes, the relatively small mixing ratios of the ozone destroying species ClO and Cl (*cf.* above) normally limit the amount of ozone destruction. Furthermore, Cycle 1 proceeds only in the presence of atomic oxygen. In the atmosphere, oxygen atoms are almost exclusively produced by the UV-photolysis of either O_2 or O_3 molecules. Consequently, the abundance of O atoms increases with the intensity and photon energy of UV radiation and therefore with altitude. It reaches a maximum at around 40 km, where cycle 1 is the most important catalytic ozone destroying cycle.

Besides this simplest catalytic ozone destruction cycle, Cl and Br are also involved in mixed catalytic cycles, where radicals from different "chemical families" (*i.e.* HO_x, NO_x, ClO_x) interact. Both, chlorine and bromine monoxide can react with the HO_2 radical, leading to ozone destruction *via* the HOCl cycle[57,58] and the equivalent HOBr cycle[59]:

Cycle 2:

$O_3 + OH$	$\rightarrow$	$HO_2 + O_2$	R4
$XO + HO_2$	$\rightarrow$	$HOX + O_2$	R5
$HOX + hv$	$\rightarrow$	$X + OH$	R6
$X + O_3$	$\rightarrow$	$XO + O_2$	R3
Net: $2O_3$	$\rightarrow$	$3O_2$	

The production of a halogen atom upon photolysis of the reservoir gases $ClONO_2$[60] and $BrONO_2$[36,61] also leads to ozone destruction *via* the following cycle:

Cycle 3:

$NO + O_3$	$\rightarrow$	$NO_2 + O_2$	R7
$XO + NO_2 + M$	$\rightarrow$	$XONO_2 + M$	R8
$XONO_2 + hv$	$\rightarrow$	$X + NO_3$	R9
$NO_3 + hv$	$\rightarrow$	$NO + O2$	R10
$X + O_3$	$\rightarrow$	$XO + O_2$	R3
Net : $2O_3$	$\rightarrow$	$3O_2$	

In a similar manner to Cycle 3 $BrONO_2$ can heterogeneously hydrolyse producing HOBr (see Section 3.3.3). HOBr will then either photolyse directly or—at cold temperatures—react heterogeneously with HCl to form $BrCl$[36] that photolyses subsequently. These heterogeneous ozone destruction cycles will of course only work in the presence of sufficient aerosol surface area but have the potential to simultaneously generate HO_x from H_2O and/or ClO_x from HCl. Because it is temperature-independent, the $BrONO_2$ hydrolysis cycle will work year-round at all latitudes. Under background aerosol conditions this cycle may contribute by around 20% to mid-latitude ozone loss.[36]

ClO_x and NO_x interact in yet another gas-phase catalytic cycle:

Cycle 4:

$O_3 + hv$	$\rightarrow$	$O_2 + O$	R1
$O + NO_2$	$\rightarrow$	$NO + O_2$	R11
$NO + ClO$	$\rightarrow$	$NO_2 + Cl$	R12
$Cl + O_3$	$\rightarrow$	$ClO + O_2$	R3
Net : $2O_3$	$\rightarrow$	$3O_2$	

The relative contribution to ozone destruction of these and other catalytic cycles in different regions and under different conditions has been assessed in much detail by Grenfell *et al.*[62]

In the lower stratosphere and in particular at high latitudes, the relatively low abundance of atomic oxygen limits the rate of ozone loss by cycle 1. The discovery of the Antarctic ozone hole in the mid-1980s[63] was thus completely unexpected. The almost quantitative removal of ozone over the Antarctic continent in polar spring could not be explained by the catalytic ozone cycles known at that time to destroy ozone in the atmosphere and thus presented a puzzle to atmospheric scientists.

Even though the isolation of air within the polar vortex would prevent the replenishment of ozone by dynamic processes over the polar winter and spring, the concentrations of ClO, BrO and atomic oxygen—or any other known catalysts destroying ozone—were expected to be far too low for ozone to be removed at a rate even close to that needed to explain the ozone hole.

One piece of the puzzle was the fast heterogeneous activation of chlorine from its reservoir species HCl and $ClONO_2$ on aerosol particles at cold temperatures,[3] forming Cl_2, which is readily photolysed as soon as light is available. Throughout most of the stratosphere, the total aerosol surface area is too low and temperatures are too high for this process to be significant. However, in the polar stratosphere in winter, the formation of polar stratospheric clouds leads to significantly enhanced ClO concentrations. PSC formation and heterogeneous chlorine activation are discussed in great depth in Chapter 4.

Molina and Molina[64] and McElroy *et al.*[65] recognised that, due to the high ClO concentrations resulting from heterogeneous activation, two additional catalytic cycles could effectively remove ozone without involving atomic oxygen (Figure 3.5)

Cycle 5:

$$
\begin{array}{llll}
ClO + ClO + M & \leftrightarrow & ClOOCl + M & R13 \\
ClOOCl + hv & \rightarrow & Cl + ClOO & R14 \\
ClOO + M & \rightarrow & Cl + O_2 + M & R15 \\
2[Cl + O_3 & \rightarrow & ClO + O_2] & R2a \\
\hline
\text{Net:} \quad 2O_3 & \rightarrow & 3O_2 &
\end{array}
$$

Cycle 6:

$$
\begin{array}{llll}
ClO + BrO & \rightarrow & Cl + Br + O_2 & R16a \\
& \rightarrow & BrCl + O_2 & R16b \\
& \rightarrow & OClO + Br & R16c \\
BrCl + hv & \rightarrow & Br + Cl & R17 \\
Cl + O_3 & \rightarrow & ClO + O_2 & R2a \\
Br + O_3 & \rightarrow & BrO + O_2 & R2b \\
\hline
\text{Net:} \quad 2O_3 & \rightarrow & 3O_2 &
\end{array}
$$

These two cycles have been studied extensively in the past two decades and are discussed in detail below. A third cycle making a minor contribution to ozone loss in the polar stratosphere in winter is the HOCl cycle (Cycle 2 with $X = Cl$).

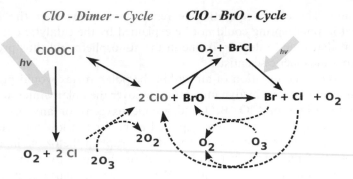

Figure 3.5 Schematic of the two most important "polar" catalytic cycles and their interaction.

3.4.1 The ClO Dimer Cycle

The ClO dimer cycle (cycle 5) is initiated by the recombination of two ClO radicals (R13), which is favored by high ClO abundances, cold temperatures and high pressures. The ClOOCl formation rate constant k_{13} has been determined in a number of laboratory studies employing flash photolysis with time resolved UV absorption spectroscopy.[66–70] Except for the most recent investigation,[67] the values agree well above 220 K, but at stratospheric temperatures between 180 and 220 K there is some discrepancy. The same is true for the equilibrium constant $K_{EQ} = k_{13}/k_{-13}$ between ClO and ClOOCl. Here k_{-13} determines the thermal dissociation of ClOOCl back into ClO. Parameterisations from different laboratory studies[68,71–75] and inferred from field observations[76–79],[*] result in K_{EQ} values at stratospheric temperatures that are different by up to an order of magnitude. Both k_{13} and K_{EQ} have a significant influence on the partitioning of active chlorine. However, within the uncertainties set by the various laboratory experiments, the impact on calculated ozone loss rates is small, and a detailed discussion of these parameters is beyond the scope of this chapter. An extensive review of the kinetic parameters governing the ClO-dimer-cycle is found in von Hobe et al.[80]

Reactions R15 and R2a of cycles 5 and 6 proceed essentially instantaneously and thus do not limit the overall rate of the catalytic cycle. Under twilight conditions—prevailing in the polar stratosphere in winter and spring—the rate-limiting step of ozone loss by cycle 5 is the UV-photolysis of the ClO dimer (R14). Until very recently, the rate of this reaction has been one of the biggest uncertainties in our understanding of polar ozone depletion, and considerable progress has been made since the publication of the overview by von Hobe et al.[80] So it is worth taking a more detailed look at the photolysis of ClOOCl, which will be done in Section 3.5 below.

---

[*]Avallone and von Hobe give parameterisations for the temperature dependence of K_{EQ} calculated from observations, while Berthet et al. and Santee et al. test the consistency of existing parameterisations with observed night-time ClO from satellite observations.

3.4.2 The ClO/BrO Cycle

In the ClO/BrO cycle (Cycle 6) ClO and BrO recombine to form Cl and Br atoms, BrCl or OClO and Br, respectively *via* reactions 16a-c. No mixed dimer (*e.g.* ClOOBr) formation occurs. Only Reactions 16a and b lead to catalytic ozone removal, because the OClO produced in Reaction 16c is subsequently photolysed, yielding an oxygen atom, which can react with O_2 to produce an O_3 molecule. Therefore, the branching of Reaction 16 into the different channels a, b and c is one of the key parameters determining O_3 loss by this catalytic cycle. Laboratory studies on the branching ratios as well as the individual rate constants at stratospheric temperatures are in good agreement[38] and are generally consistent with field observation.[81] At 200K branching into the OClO (R16c) and BrCl (R16b) channels is around 60% and 8%[38], respectively. Still, Kawa *et al.*[82] identified the ClO + BrO reaction and its branching ratios as one of the larger uncertainties in a sensitivity study on polar stratospheric ozone loss.

The overall rate of Reaction 16 obviously depends on the amount of ClO present and the ClO/BrO cycle runs faster when chlorine activation is high. Besides the partitioning of chlorine between ClO_x and the reservoir species, the partitioning of ClO_x into ClO and its dimer is also important: any ClO present as dimer does not participate in the ClO/BrO cycle. This means, that the dimer photolysis (R14) not only limits the rate of the ClO dimer cycle, but by governing the ClO abundance also regulates the rate of the ClO/ BrO cycle. As a consequence, the overall O_3 loss rate by the two cycles depends on the ClO dimer photolysis rate more than on any other kinetic parameter.

3.5 ClOOCl Photolysis

In the atmosphere, the photolysis rate is represented by the product of the actinic flux and the photolysis cross section, both integrated over the relevant wavelength range:

$$J = \int_{\lambda} I(\lambda) \cdot \phi(\lambda) \cdot \sigma(\lambda) \cdot d\lambda$$

The actinic flux $I(\lambda)$ depends on a number of atmospheric parameters including solar zenith angle and altitude as well as the presence and quantity of other absorbing species and aerosol particles. Nevertheless, it can be reasonably well constrained in models under most conditions. The photolysis cross section is given by the absorption cross section $\sigma(\lambda)$ multiplied by the photolysis quantum yield $\phi(\lambda)$, both of which are traditionally determined in laboratory experiments. For ClOOCl, virtually all studies addressing the photolysis quantum yield came to the conclusion that it is close to unity over the relevant wavelength range.[83–85] However, significant differences exist in the various published ClOOCl absorption cross sections[72,86–92] or UV-vis absorption spectra.[†,93–96] Until recently, there was some consensus on the absorption cross section at 246 nm, where the maximum absorption is observed and which is generally used as a calibration point to scale relative UV-vis absorption spectra into absolute cross sections. However, the shape of the spectrum and the absorption cross sections in the atmospherically relevant wavelength region above 310 nm varied by up to a factor of ten. A comparison of the most important published cross sections and spectra is shown in Figure 3.6. The earlier studies carried out in the 1980s and 90s are more[72,86,88] or less[93] consistent with observed atmospheric ClO$_x$ partitioning[80,97,98] and ozone loss.[80,99]

In 2007, Pope *et al.*[95] published a ClOOCl absorption spectrum that would result in ClOOCl photolysis rates so low that the ClO-dimer and ClO/BrO catalytic cycles would not explain the observed ozone loss in most Antarctic and Arctic winters. Nevertheless, this study represented a step forward in determining ClOOCl absorption cross sections because Pope *et al.*[95] developed a method for preparing almost pure ClOOCl. The only impurity that could not be entirely removed was Cl$_2$. However, in the analysis of their results, Pope *et al.*[95] over corrected for this Cl$_2$ impurity and consequently proposed very low ClOOCl absorption above 300 nm. Both the absence of impurities other than Cl$_2$ as well as the over correction have been demonstrated by von Hobe *et al.*[96] who prepared and purified ClOOCl using the procedure described by Pope *et al.* but measured the spectrum in a Ne-Matrix. Von Hobe *et al.*[96] also made efforts to quantitatively explain the observed differences to previously published spectra, and even though they were not able to measure absolute cross sections, they presented evidence that the earlier measurements of the ClOOCl absorption at 246 nm may be inaccurate. Further evidence that the widely accepted absolute absorption cross section at the maximum are indeed questionable comes from other new experiments. Using diode array spectroscopy, Papanastasiou *et al.*[91] reported ClOOCl cross sections for the wavelength range 200–420 nm that in the atmospherically relevant region are in good agreement with those reported previously by the same laboratory.[86] However, they find roughly 20% higher cross sections at the maximum[‡] as compared to that earlier study. Lien *et al.*[90] presented an even higher maximum cross section from an experiment where ClOOCl in a molecular beam was photolyzed by laser light while monitoring the total amount by mass spectroscopy. On the other hand, Wilmouth *et al.*[92] carried out experiments measuring the production of Cl atoms from ClOOCl photolysis at three distinct wavelengths and reported the product of the ClOOCl cross section and the quantum yield of Cl atoms at 248 nm to be 6.6×10^{-18} cm^2, in good agreement with the values determined in the late 1980s and early 90s if the quantum yield is assumed to be 1.

---

† To obtain *absolute* absorption *cross sections*, it is necessary to determine the amount of ClOOCl present during the absorption measurement, *e.g.* by mass balance calculations. All other experiments yield only *relative* absorption *spectra*.

‡ Note that only ClOOCl absorption above 310 nm is relevant for ozone loss, which is not influenced by the higher cross sections in the peak region if the concurrent changes in the spectral shape compared to the older studies are also taken into account.

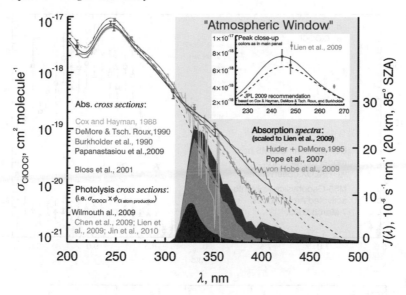

Figure 3.6 ClOOCl absorption/photolysis cross sections (solid lines and symbols) and spectra (long dashed lines) measured in different laboratories. Short dashed lines indicate values obtained by a spectral fit or extrapolation. In the case of Pope *et al.*[95] the short dashed line represents the published "method 1" spectrum obtained by their Gaussian correction for Cl_2 while the long dashed line shows the "method 2" spectrum corrected for all impurities other than Cl_2. Also shown are typical atmospheric photolysis rates resulting from some of these studies ("method 1" is shown for Pope).[95] The overall *J* effective in the atmosphere is obtained by integrating the areas under the curves.

Clearly the absolute ClOOCl cross sections are still poorly established. The extremely low values published by Pope *et al.*[95] that would question even our qualitative understanding of the catalytic chemistry leading to polar ozone depletion have been dismissed on the basis of new laboratory work employing several independent techniques. But the *overall uncertainty range* of laboratory derived ClOOCl absorption spectra has not been significantly reduced yet. The recently questioned absolute cross section of the absorption maximum calls for more work to be done.

It is noted that if the spectrum from von Hobe *et al.*[96] or the Cl_2-uncorrected gas phase spectrum from Pope *et al.*[95] described as "method 2 spectrum" in Figure 3.6 are scaled to the new calibration point given by Lien *et al.*[90] good agreement with molecular beam cross sections at other wavelengths[87] (Figure 3.6) and with observed atmospheric ClO_x partitioning and O_3 loss is reached (Figure 3.7).

3.6 Halogen Chemistry in the Upper Troposphere and Lowermost Stratosphere (UTLS)

Ozone trends in the tropopause region play an important role in the radiative forcing of the Earth's climate system.[100] Over the past two decades, evidence

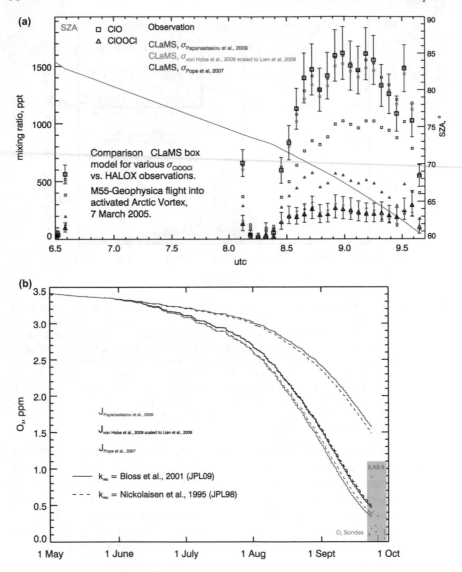

Figure 3.7 (a) ClO$_x$ partitioning and (b) O$_3$ loss for different kinetic parameters updated from von Hobe *et al.*[80]

has emerged for the presence of active halogen species in this region that could lead to significant chemical ozone loss.[101,102]

3.6.1   The Inorganic Bromine Budget

Since inorganic bromine is much more efficient in destroying stratospheric ozone than chlorine, the budget of Br$_y$ in the stratosphere is a very important quantity, which at the same time is not easily accessible. This subject is discussed in much detail also in Chapter 2 and the latest WMO Assessment[44] where the 2008 stratospheric Br_y is given as 22.5 ppt with a range of 19.5–24.5 ppt. Here, we present a critical discussion of the uncertainties on the determinations of stratospheric Br_y levels and argue that uncertainties may in fact be somewhat bigger while we agree on the general statements and conclusions of the above references.

The major uncertainty in the inorganic bromine budget of the stratosphere is a result of the extend of the contribution of very short-lived species (VSLS) to the budget, that is hard to quantify. These species are partly introduced into the stratosphere as source gases (SGI, for source gas injection) but due to their relatively fast photolytic and/or chemical breakdown also feed the tropospheric inorganic bromine that can be directly introduced into the stratosphere (so-called product gas injection, PGI).[43] The VSLS concentrations can be quantified quite well with accuracies generally better than 15% through gas chromatographic measurements of air samples collected in the tropical tropopause layer (TTL, see section 1.1.3). However, these measurements are made infrequently and therefore problems of spacial homogeneity can occur since injections may locally vary a lot due to source distributions connected closely with convective events (*e.g.* Gettelman *et al.*, 2009,[103] Aschmann *et al.*, 2011).[104] A synthesis of available measurements is given by Montzka and Reiman[44] (Table 1.7), resulting in 2.7 ppt of bromine contained in VSLS in the TTL at 15 km altitude and 1.5 ppt at the cold point tropopause at *ca.* 17 km which probably is a good measure of the minimum SGI by VSLS. These data can be compared within models of VSLS transport and chemistry throughout the troposphere and TTL in order to check the consistency (*e.g.* Sinnhuber and Folkins, 2006,[105] Aschmann *et al.*, 2009)[106] of VSLS vertical evolution and constrains the resulting PGI. Models can also be used for global upscaling provided source distributions and convective events are well constrained which also is a quite optimistic assumption. Latest results of Hossaini *et al.*[107] and Aschmann[104] simulate average total inputs (SGI + PGI) from VSLS of 2.4 ppt and 3.4 ppt, respectively. A major uncertainty is the amount of PGs scavenged and washed out upon transport. A maximum total input of 5 ppt is given by Aschmann *et al.*[104] for a scenario with zero PG removal.

PGI can only be quantified by the determination of actual Br_y in the TTL and lowest stratosphere regions preferably along with parallel source gas measurements. This as a minimum requires the measurement of BrO which is the main daytime Br_y species and the only one that can be quantified to better than 20% accuracy (see Box 3.1 and Section 3.3.3). A summary of available BrO measurements has been compiled by Montzka and Reiman[44] (Table 1.14). Almost all measurements listed are based on the DOAS technique, mostly measuring vertical column densities from ground stations or satellites. Vertical profiles are retrieved from measured columns employing photochemical models thereby introducing additional uncertainties.[108,109]

For all column measurements the separation of the tropospheric contribution to the column measurement poses a serious problem and recent results are not well consistent. Richter *et al.*[110] report average free tropospheric mixing ratios of 0.5–2 ppt from BrO column measurements of GOME while

Schofield *et al.*[109] find average free tropospheric BrO mixing ratios of 0.2 ppt with a maximum of 0.9 ppt retrieved from ground based measurements at Lauder, New Zealand. Even a small difference will naturally have a major impact on the resulting stratospheric columns (*e.g.* Salawitch *et al.*, 2010[111]). Based on comparisons of column observations of the OMI instrument onboard the NASA AURA satellite and tropospheric BrO in-situ measurements emplyoing the CIMS (chemical ionization mass spectrometry) technique Salawitch *et al.*[111] show that extremely careful consideration of local meteorological conditions, especially tropopause heights, is needed when dealing with column measurements. This is shown for special situations where also high boundary layer BrO values were expected but also has general relevance. However, some direct profile measurements are available[112] and agreement with the column measurements is reasonable.

Given the uncertainties of the chemical kinetics parameters alone (see section 3.3) models can hardly constrain the Br_y/BrO ratio to better than 18%[113] in order to calculate Br_y from BrO. Combined with the measurement uncertainty for BrO of also 18% for the very best reported measurement accuracy[112] this leaves a combined error of around 6 ppt (or *ca.* 26%) for an assumed 22 ppt of Br_y (with 70–80% residing in BrO). This results in a an uncertainty range of 16–28 ppt encompassing any VSLS and PG contribution from close to zero to 12 ppt based on the reported and reliable budget from long-lived SGs of 15.7 ppt (+/- 0.2 ppt) for 2008.[44] The minimum contribution of VSLS and PG injection is probably at least 1–2 ppt as discussed above and in Chapter 2 and ref. 44, however from a single stratospheric BrO measurement alone zero VSLS contribution can hardly be reliably ruled out. There are studies reporting somewhat better accuracies for determining Br_y[112,113] however, regarding the limits of the accuracy of BrO measurements and the strong involvement of models with inherent error sources that are hard to quantify, great care must be taken with the accuracy of stratospheric Br_y.

Montzka and Reiman[44] (Table 1.14) show that stratospheric Br_y derived from different column measurements is quite consistent with retrievals from balloon-borne DOAS measurements[112] and also *in situ* measurements.[101] The average taken from these column and profile measurements suggests a total VSL SGI and PGI of 3–8 ppt[44] (range of mean values from different methods) with a mean of 6 ppt. We suggest that these uncertainties are somewhat optimistic and would not completely rule out a range from 1–3 ppt.

3.6.2 Chlorine Chemistry in the Tropopause Region

The amount of inorganic chlorine present in the UTLS region depends on the age of the air mass, *i.e.* the time since it has entered the stratosphere. But even when significant amounts of Cl_y are present, they must first be converted into active forms (*i.e.* ClO_x) in order to stimulate chemical ozone loss. Under cloud-free conditions, the amount of ClO_x does not exceed a few ppt even in air masses that have undergone some photochemical processing (Figure 3.8). However, heterogeneous chlorine activation has been observed on ice particles

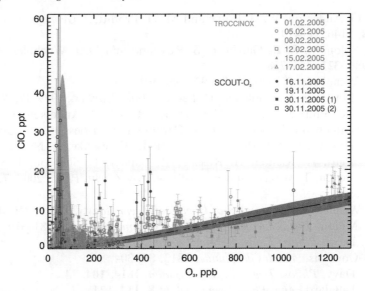

Figure 3.8 Relation between ClO and ozone in the tropical UTLS from airborne measurements made during TROCCINOX and SCOUT-O₃ field campaigns. The black line and grey shadings (uncertainty) represent ClO mixing ratios expected at local noon from model simulations without heterogeneous reactions included. Blue shading indicates the range of ClO predicted with heterogeneous chemistry included in the model. From von Hobe *et al.*[120]

in high altitude cirrus clouds.[114–120] The extremely cold temperatures at the tropical tropopause could foster not only heterogeneous activation but also fast catalytic loss cycles such as the ClO dimer cycle under certain conditions. This possibility has been demonstrated in box-model simulations.[120]

Acknowledgements

We thank Thorsten Benter and Steve Montzka for reviewing this chapter and providing valuable comments.

References

1. Y. T. Oganessian, F. S. Abdullin, P. D. Bailey, D. E. Benker, M. E. Bennett, S. N. Dmitriev, J. G. Ezold, J. H. Hamilton, R. A. Henderson, M. G. Itkis, Y. V. Lobanov, A. N. Mezentsev, K. J. Moody, S. L. Nelson, A. N. Polyakov, C. E. Porter, A. V. Ramayya, F. D. Riley, J. B. Roberto, M. A. Ryabinin, K. P. Rykaczewski, R. N. Sagaidak, D. A. Shaughnessy, I. V. Shirokovsky, M. A. Stoyer, V. G. Subbotin, R. Sudowe, A. M. Sukhov, Y. S. Tsyganov, V. K. Utyonkov, A. A. Voinov, G. K. Vostokin and P. A. Wilk, *Phys. Rev. Lett.*, 2010, **104**, 14502, DOI:10.1103/PhysRevLett.104.14502.

2. J. G. Anderson, W. H. Brune and M. H. Proffitt, *J. Geophys. Res.*, 1989, **94**, 11465–11479.
3. S. Solomon, R. R. Garcia, F. S. Rowland and D. J. Wuebbles, *Nature*, 1986, **321**, 755–758.
4. P. Wennberg, *Nature*, 1999, **397**, 299–301.
5. M. Martinez, T. Arnold and D. Perner, *Ann. Geophys.*, 1999, **17**, 941–956.
6. W. R. Simpson, R. von Glasow, K. Riedel, P. Anderson, P. Ariya, J. Bottenheim, J. Burrows, L. J. Carpenter, U. Friess, M. E. Goodsite, D. Heard, M. Hutterli, H. W. Jacobi, L. Kaleschke, B. Neff, J. Plane, U. Platt, A. Richter, H. Roscoe, R. Sander, P. Shepson, J. Sodeau, A. Steffen, T. Wagner and E. Wolff, *Atmos. Chem. Phys.*, 2007, **7**, 4375–4418.
7. J. Wisniak, *Indian Journal of Chemical Technology*, 2002, **9**, 450–463.
8. E. Millon, *Justus Liebigs Annalen der Chemie*, 1843, **46**, 281–319.
9. H. Davy, *Philos. Trans. R. Soc. London*, 1810, **100**, 231.
10. B. Courtois, *Annales de Chimie*, 1813, **88**, 304.
11. H. Davy, *Philos. Trans. R. Soc. London*, 1814, **104**, 74.
12. A. J. Balard, *Ann. Phys. Chem.*, 1826, **8**, 114–124.
13. C. Löwig, *Magazin für Pharmazie*, 1828, **21**, 31–36.
14. H. Moissan, *Comptes rendus*, 1886, **102**, 1543–1544.
15. J. Wisniak, *Indian Journal of Chemical Technology*, 2002, **9**, 263–271.
16. H. Chabot, *Actualite Chimique*, 2008, 41–45.
17. H. Davy, *Philos. Trans. R. Soc. London.*, 1815, **105**, 214–219.
18. A. J. Balard, *Ann. Chim. Phys.*, 1834, **57**, 225–304.
19. P. Hautefeuille and J. Chappuis, *Comptes rendus de l'Académie des Sciences*, 1880, **91**, 762–765.
20. J. Chappuis and P. Hautefeuille, *Annales scientifiques de l'Ecole Normale Superieure*, 1884, **1**, 55–84.
21. M. Bodenstein, *Zeitschrift für Physikalische Chemie–stochiometrie und Verwandtschaftslehre*, 1926, **120**, 129–143.
22. M. Bodenstein and G. Kistiakowski, *Zeitschrift für Physikalische Chemie–stochiometrie und Verwandtschaftslehre*, 1925, **116**, 371–390.
23. M. Bodenstein, *Sitzungsberichte der Preußischen Akademie der Wissenschaften Physikalisch-Mathematische Klasse*, 1929, 367–369.
24. B. Lewis and H. J. Schumacher, *Zeitschrift für Physikalische Chemie–stochiometrie und Verwandtschaftslehre*, 1928, **138**, 462–462.
25. J. W. T. Spinks, *Nature*, 1931, **128**, 548–548.
26. G. Pannetier and A. G. Gaydon, *Nature*, 1948, **161**, 242–243.
27. G. Porter, *Proc. R. Soc. London Ser. A*, 1950, **200**, 284–300.
28. G. Porter, *Pure Appl. Chem.*, 1996, **68**, 1683–1687.
29. R. J. Cicerone, R. S. Stolarski and S. Walters, *Science*, 1974, **185**, 1165–1167.
30. R. J. Cicerone and R. S. Stolarski, *Transactions-American Geophysical Union*, 1974, **55**, 275–275.
31. R. S. Stolarski and R. J. Cicerone, *Canadian J. Chem.*, 1974, **52**, 1610–1615.

32. M. J. Molina and F. S. Rowland, *Nature*, 1974, **249**, 810–812.
33. Y. Bedjanian and G. Poulet, *Chemical Reviews*, 2003, **103**, 4639–4655.
34. R. P. Wayne, G. Poulet, P. Biggs, J. P. Burrows, R. A. Cox, P. J. Crutzen, G. D. Hayman, M. E. Jenkin, G. Lebras, G. K. Moortgate, U. Platt and R. N. Schindler, *Atmos. Environ.*, 1995, **29**, 2677–2881.
35. D. J. Lary, *J. Geophys. Res.*, 1996, **101**, 1505–1516.
36. D. J. Lary, M. P. Chipperfield, R. Toumi and T. Lenton, *J. Geophys. Res.*, 1996, **101**, 1489–1504.
37. S. P. Sander, B. J. Finlayson-Pitts, R. R. Friedl, D. M. Golden, R. E. Huie, H. Keller-Rudek, C. E. Kolb, M. J. Kurylo, M. J. Molina, G. K. Moortgat, V. L. Orkin, A. R. Ravishankara and P. H. Wine, *Chemical Kinetics and Photochemical Data for Use in Atmospheric Studies,* JPL Publication 06-2, Jet Propulsion Laboratory, Pasadena, 2006.
38. S. P. Sander, J. Abbatt, J. R. Barker, J. B. Burkholder, R. R. Friedl, D. M. Golden, R. E. Huie, C. E. Kolb, M. J. Kurylo, G. K. Moortgat, V. L. Orkin and P. H. Wine, *Chemical Kinetics and Photochemical Data for Use in Atmospheric Studies* JPL Publication 09-31, Jet Propulsion Laboratory, Pasadena, 2009.
39. R. S. Stolarski and R. D. Rundel, *Geophys. Res. Lett.*, 1975, **2**, 443–444.
40. H. Bosch, C. Camy-Peyret, M. P. Chipperfield, R. Fitzenberger, H. Harder, U. Platt and K. Pfeilsticker, *J. Geophys. Res.*, 2003, 108.
41. A. Butz, H. Bosch, C. Camy-Peyret, M. P. Chipperfield, M. Dorf, S. Kreycy, L. Kritten, C. Prados-Roman, J. Schwarzle and K. Pfeilsticker, *Atmos. Chem. Phys.*, 2009, **9**, 7229–7242.
42. I. Pundt, J. P. Pommereau, C. Phillips and E. Lateltin, *J. Atmos. Chem.*, 1998, **30**, 173–185.
43. M. K. W. Ko, N. D. Sze, C. J. Scott and D. K. Weisenstein, *J. Geophys. Res.*, 1997, **102**, 25507–25517.
44. WMO, *Scientific assessment of ozone depletion* 52, World Meteorological Organization, Global Ozone Research and Monitoring Project, Geneva, Switzerland, 2010.
45. R. Nassar, P. F. Bernath, C. D. Boone, C. Clerbaux, P. F. Coheur, G. Dufour, L. Froidevaux, E. Mahieu, J. C. McConnell, S. D. McLeod, D. P. Murtagh, C. P. Rinsland, K. Semeniuk, R. Skelton, K. A. Walker and R. Zander, *J. Geophys. Res.*, 2006, **111**, D22312.
46. J.-B. Renard, M. Pirre, C. Robert and D. Huguenin, *J. Geophys. Res.*, 1998, **103**, 25383–25395.
47. F. Erle, U. Platt and K. Pfeilsticker, *Geophys. Res. Lett.*, 2000, **27**, 2217–2220.
48. M. P. Chipperfield, T. Glassup, I. Pundt and O. V. Rattigan, *Geophys. Res. Lett.*, 1998, **25**, 3575–3578.
49. J. Troe, *Chemical Reviews*, 2003, **103**, 4565–4576.
50. P. W. Seakins, *Annu. Rep. Prog. Chem., Sect. C*, 2007, **103**, 173–222.
51. D. R. Hanson, A. R. Ravishankara and E. R. Lovejoy, *J. Geophys. Res.*, 1996, **101**, 9063–9069.

52. W. H. Brune, J. G. Anderson and K. R. Chan, *J. Geophys. Res.*, 1989, **94**, 16639–16647.
53. D. W. Toohey, J. G. Anderson, W. H. Brune and K. R. Chan, *Geophys. Res. Lett.*, 1990, **17**, 513–516.
54. F. Erle, A. Grendel, D. Perner, U. Platt and K. Pfeilsticker, *Geophys. Res. Lett.*, 1998, **25**, 4329–4332.
55. J. R. Slusser, D. J. Fish, E. K. Strong, R. L. Jones, H. K. Roscoe and A. Sarkissian, *J. Geophys. Res.*, 1997, **102**, 12987–12993.
56. R. J. Salawitch, S. C. Wofsy, P. O. Wennberg, R. C. Cohen, J. G. Anderson, D. W. Fahey, R. S. Gao, E. R. Keim, E. L. Woodbridge, R. M. Stimpfle, J. P. Koplow, D. W. Kohn, C. R. Webster, R. D. May, L. Pfister, E. W. Gottlieb, H. A. Michelsen, G. K. Yue, J. C. Wilson, C. A. Brock, H. H. Jonsson, J. E. Dye, D. Baumgardner, M. H. Proffitt, M. Loewenstein, J. R. Podolske, J. W. Elkins, G. S. Dutton, E. J. Hintsa, A. E. Dessler, E. M. Weinstock, K. K. Kelly, K. A. Boering, B. C. Daube, K. R. Chan and S. W. Bowen, *Geophys. Res. Lett.*, 1994, **21**, 2547–2550.
57. D. G. Johnson, W. A. Traub, K. V. Chance and K. W. Jucks, *Geophys. Res. Lett.*, 1995, **22**, 1373–1376.
58. L. J. Kovalenko, K. W. Jucks, R. J. Salawitch, G. C. Toon, J. F. Blavier, D. G. Johnson, A. Kleinbohl, N. J. Livesey, J. J. Margitan, H. M. Pickett, M. L. Santee, B. Sen, R. A. Stachnik and J. W. Waters, *Geophys. Res. Lett.*, 2007, **34**.
59. G. Poulet, M. Pirre, F. Maguin, R. Ramaroson and G. Lebras, *Geophys. Res. Lett.*, 1992, **19**, 2305–2308.
60. R. Toumi, R. L. Jones and J. A. Pyle, *Nature*, 1993, **365**, 37–39.
61. J. B. Burkholder, A. R. Ravishankara and S. Solomon, *J. Geophys. Res.*, 1995, **100**, 16793–16800.
62. J. L. Grenfell, R. Lehmann, P. Mieth, U. Langematz and B. Steil, *J. Geophys. Res.*, 2006, **111**, D17311.
63. J. C. Farman, B. G. Gardiner and J. D. Shanklin, *Nature*, 1985, **315**, 207–210.
64. L. T. Molina and M. J. Molina, *J. Phys. Chem.*, 1987, **91**, 433–436.
65. M. B. McElroy, R. J. Salawitch, S. C. Wofsy and J. A. Logan, *Nature*, 1986, **321**, 759–762.
66. W. J. Bloss, S. L. Nickolaisen, R. J. Salawitch, R. R. Friedl and S. P. Sander, *J. Phys. Chem. A*, 2001, **105**, 11226–11239.
67. G. Boakes, W. H. H. Mok and D. M. Rowley, *Phys. Chem. Chem. Phys.*, 2005, **7**, 4102–4113.
68. S. L. Nickolaisen, R. R. Friedl and S. P. Sander, *J. Phys. Chem.*, 1994, **98**, 155–169.
69. S. P. Sander, R. R. Friedl and Y. L. Yung, *Science*, 1989, **245**, 1095–1098.
70. M. Trolier, R. L. Mauldin and A. R. Ravishankara, *J. Phys. Chem.*, 1990, **94**, 4896–4907.
71. N. Basco and J. E. Hunt, *Int. J. Chem. Kinet.*, 1979, **11**, 649–664.
72. R. A. Cox and G. D. Hayman, *Nature*, 1988, **332**, 796–800.

73. T. Ellermann, K. Johnsson, A. Lund and P. Pagsberg, *Acta Chem. Scand.*, 1995, **49**, 28–35.
74. V. Ferracci and D. M. Rowley, *Phys. Chem. Chem. Phys.*, 2010, **12**, 11596–11608.
75. J. Plenge, S. Kühl, B. Vogel, R. Müller, F. Stroh, M. von Hobe, R. Flesch and E. Rühl, *J. Phys. Chem. A*, 2005, **109**, 6730–6734.
76. L. M. Avallone and D. W. Toohey, *J. Geophys. Res. [Atmos.]*, 2001, **106**, 10411–10421.
77. G. Berthet, P. Ricaud, F. Lefèvre, E. Le Flochmoen, J. Urban, B. Barret, N. Lautie, E. Dupuy, J. De la Noe and D. Murtagh, *Geophys. Res. Lett.*, 2005, **32**.
78. M. L. Santee, S. P. Sander, N. J. Livesey and L. Froidevaux, *Proc. Natl. Acad. Sci. U. S. A.*, 2010, **107**, 6588–6593.
79. M. von Hobe, J. U. Grooß, R. Müller, S. Hrechanyy, U. Winkler and F. Stroh, *Atmos. Chem. Phys.*, 2005, **5**, 693–702.
80. M. von Hobe, R. J. Salawitch, T. Canty, H. Keller-Rudek, G. K. Moortgat, J.-U. Grooß, R. Müller and F. Stroh, *Atmos. Chem. Phys.*, 2007, **7**, 3055–3069.
81. A. Butz, H. Bösch, C. Camy-Peyret, M. Dorf, A. Engel, S. Payan and K. Pfeilsticker, *Geophys. Res. Lett.*, 2007, **34**, L05801.
82. S. R. Kawa, R. S. Stolarski, P. A. Newman, A. R. Douglass, M. Rex, D. J. Hofmann, M. L. Santee and K. Frieler, *Atmos. Chem. Phys.*, 2009, **9**, 8651–8660.
83. M. J. Molina, A. J. Colussi, L. T. Molina, R. N. Schindler and T. L. Tso, *Chem. Phys. Lett.*, 1990, **173**, 310–315.
84. T. A. Moore, M. Okumura, J. W. Seale and T. K. Minton, *J. Phys. Chem. A*, 1999, **103**, 1691–1695.
85. J. Plenge, R. Flesch, S. Kühl, B. Vogel, R. Müller, F. Stroh and E. Rühl, *J. Phys. Chem. A*, 2004, **108**, 4859–4863.
86. J. B. Burkholder, J. J. Orlando and C. J. Howard, *J. Phys. Chem. A*, 1990, **94**, 687–695.
87. H. Y. Chen, C. Y. Lien, W. Y. Lin, Y. T. Lee and J. J. Lin, *Science*, 2009, **324**, 781–784.
88. W. B. DeMore and E. TschuikowRoux, *J. Phys. Chem.*, 1990, **94**, 5856–5860.
89. B. Jin, I. C. Chen, W. T. Huang, C. Y. Lien, N. Guchhait and J. J. Lin, *J. Phys. Chem. A*, 2010, **114**, 4791–4797.
90. C. Y. Lien, W. Y. Lin, H. Y. Chen, W. T. Huang, B. Jin, I. C. Chen and J. J. Lin, *J. Chem. Phys.*, 2009, **131**.
91. D. K. Papanastasiou, V. C. Papadimitriou, D. W. Fahey and J. B. Burkholder, *J. Phys. Chem. A*, 2009, **113**, 13711–13726.
92. D. M. Wilmouth, T. F. Hanisco, R. M. Stimpfle and J. G. Anderson, *J. Phys. Chem. A*, 2009, **113**, 14099–14108.
93. K. J. Huder and W. B. DeMore, *J. Phys. Chem.*, 1995, **99**, 3905–3908.
94. J. R. McKeachie, M. F. Appel, U. Kirchner, R. N. Schindler and T. Benter, *J. Phys. Chem. B*, 2004, **108**, 16786–16797.

95. F. D. Pope, J. C. Hansen, K. D. Bayes, R. R. Friedl and S. P. Sander, *J. Phys. Chem. A*, 2007, **111**, 4322–4332.

96. M. von Hobe, F. Stroh, H. Beckers, T. Benter and H. Willner, *Phys. Chem. Chem. Phys.*, 2009, **11**, 1571–1580.

97. M. L. Santee, I. A. MacKenzie, G. L. Manney, M. P. Chipperfield, P. F. Bernath, K. A. Walker, C. D. Boone, L. Froidevaux, N. J. Livesey and J. W. Waters, *J. Geophys. Res.*, 2008, **113**.

98. R. M. Stimpfle, D. M. Wilmouth, R. J. Salawitch and J. G. Anderson, *J. Geophys. Res.*, 2004, **109**.

99. K. Frieler, M. Rex, R. J. Salawitch, T. Canty, M. Streibel, R. M. Stimpfle, K. Pfeilsticker, M. Dorf, D. K. Weisenstein and S. Godin-Beekmann, *Geophys. Res. Lett.*, 2006, **33**, doi: 10.1029/2005GL025466.

100. V. Ramaswamy, M. L. Chanin, J. Angell, J. Barnett, D. Gaffen, M. Gelman, P. Keckhut, Y. Koshelkov, K. Labitzke, J. J. R. Lin, A. O'Neill, J. Nash, W. Randel, R. Rood, K. Shine, M. Shiotani and R. Swinbank, *Rev. Geophys.*, 2001, **39**, 71–122.

101. R. J. Salawitch, D. K. Weisenstein, L. J. Kovalenko, C. E. Sioris, P. O. Wennberg, K. Chance, M. K. W. Ko and C. A. McLinden, *Geophys. Res. Lett.*, 2005, **32**.

102. S. Solomon, S. Borrmann, R. R. Garcia, R. Portmann, L. Thomason, L. R. Poole, D. Winker and M. P. McCormick, *J. Geophys. Res.*, 1997, **102**, 21411–21429.

103. A. Gettelman, P. H. Lauritzen, M. Park and J. E. Kay, *J. Geophys. Res.*, 2009, **114**, D13303.

104. J. Aschmann, B.-M. Sinnhuber, M. P. Chipperfield and R. Hossaini, *Atmos. Chem. Phys.*, 2011, **11**, 2671–2687.

105. B. M. Sinnhuber and I. Folkins, *Atmos. Chem. Phys.*, 2006, **6**, 4755–4761.

106. J. Aschmann, B. M. Sinnhuber, E. L. Atlas and S. M. Schauffler, *Atmos. Chem. Phys.*, 2009, **9**, 9237–9247.

107. R. Hossaini, M. P. Chipperfield, B. M. Monge-Sanz, N. A. D. Richards, E. Atlas and D. R. Blake, *Atmos. Chem. Phys.*, 2010, **10**, 719–735.

108. F. Hendrick, M. Van Roozendael, M. P. Chipperfield, M. Dorf, F. Goutail, X. Yang, C. Fayt, C. Hermans, K. Pfeilsticker, J. P. Pommereau, J. A. Pyle, N. Theys and M. De Maziere, *Atmos. Chem. Phys.*, 2007, **7**, 4869–4885.

109. R. Schofield, K. Kreher, B. J. Connor, P. V. Johnston, A. Thomas, D. Shooter, M. P. Chipperfield, C. D. Rodgers and G. H. Mount, *J. Geophys. Res.*, 2004, **109**, D14304.

110. A. Richter, F. Wittrock, A. Ladstatter-Weissenmayer and J. P. Burrows, *Remote Sensing of Trace Constituents in the Lower Stratosphere, Troposphere and The Earth's Surface: Global Observations, Air Pollution and the Atmospheric Correction*, 2002, **29**, 1667–1672.

111. R. J. Salawitch, T. Canty, T. Kurosu, K. Chance, Q. Liang, A. da Silva, S. Pawson, J. E. Nielsen, J. M. Rodriguez, P. K. Bhartia, X. Liu, L. G. Huey, J. Liao, R. E. Stickel, D. J. Tanner, J. E. Dibb, W. R. Simpson, D. Donohoue, A. Weinheimer, F. Flocke, D. Knapp, D. Montzka, J. A.

Neuman, J. B. Nowak, T. B. Ryerson, S. Oltmans, D. R. Blake, E. L. Atlas, D. E. Kinnison, S. Tilmes, L. L. Pan, F. Hendrick, M. Van Roozendael, K. Kreher, P. V. Johnston, R. S. Gao, B. Johnson, T. P. Bui, G. Chen, R. B. Pierce, J. H. Crawford and D. J. Jacob, *Geophys. Res. Lett.*, 2010, **37**, L21805.

112. M. Dorf, J. H. Butler, A. Butz, C. Camy-Peyret, M. P. Chipperfield, L. Kritten, S. A. Montzka, B. Simmes, F. Weidner and K. Pfeilsticker, *Geophys. Res. Lett.*, 2006, **33**.

113. F. Hendrick, P. V. Johnston, M. De Maziere, C. Fayt, C. Hermans, K. Kreher, N. Theys, A. Thomas and M. Van Roozendael, *Geophys. Res. Lett.*, 2008, **35**.

114. S. Borrmann, S. Solomon, L. Avallone, D. Toohey and D. Baumgardner, *Geophys. Res. Lett.*, 1997, **24**, 2011–2014.

115. S. Borrmann, S. Solomon, J. E. Dye and B. P. Luo, *Geophys. Res. Lett.*, 1996, **23**, 2133–2136.

116. B. Bregman, P. H. Wang and J. Lelieveld, *J. Geophys. Res.*, 2002, **107**.

117. E. R. Keim, D. W. Fahey, L. A. DelNegro, E. L. Woodbridge, R. S. Gao, P. O. Wennberg, R. C. Cohen, R. M. Stimpfle, K. K. Kelly, E. J. Hintsa, J. C. Wilson, H. H. Jonsson, J. E. Dye, D. Baumgardner, S. R. Kawa, R. J. Salawitch, M. H. Proffitt, M. Loewenstein, J. R. Podolske and K. R. Chan, *Geophys. Res. Lett.*, 1996, **23**, 3223–3226.

118. B. F. Thornton, D. W. Toohey, L. M. Avallone, A. G. Hallar, H. Harder, M. Martinez, J. B. Simpas, W. H. Brune, M. Koike, Y. Kondo, N. Takegawa, B. E. Anderson and M. A. Avery, *J. Geophys. Res.*, 2005, **110**, D23304.

119. B. F. Thornton, D. W. Toohey, L. M. Avallone, H. Harder, M. Martinez, J. B. Simpas, W. H. Brune and M. A. Avery, *J. Geophys. Res.*, 2003, **108**, 8333.

120. M. von Hobe, J. U. Grooß, G. Günther, P. Konopka, I. Gensch, M. Krämer, N. Spelten, A. Afchine, C. Schiller, A. Ulanovsky, N. Sitnikov, G. Shur, V. Yushkov, F. Ravegnani, F. Cairo, A. Roiger, C. Voigt, H. Schlager, R. Weigel, W. Frey, S. Borrmann, R. Müller and F. Stroh, *Atmos. Chem. Phys.*, 2011, **11**, 241–256.

121. M. Höpfner, J. Orphal, T. von Clarmann, G. Stiller and H. Fischer, *Atmos. Chem. Phys.*, 2009, **9**, 1735–1746.

122. E. Mahieu, P. Duchatelet, P. Demoulin, K. A. Walker, E. Dupuy, L. Froidevaux, C. Randall, V. Catoire, K. Strong, C. D. Boone, P. F. Bernath, J. F. Blavier, T. Blumenstock, M. Coffey, M. De Maziere, D. Griffith, J. Hannigan, F. Hase, N. Jones, K. W. Jucks, A. Kagawa, Y. Kasai, Y. Mebarki, S. Mikuteit, R. Nassar, J. Notholt, C. P. Rinsland, C. Robert, O. Schrems, C. Senten, D. Smale, J. Taylor, C. Tetard, G. C. Toon, T. Warneke, S. W. Wood, R. Zander and C. Servais, *Atmos. Chem. Phys.*, 2008, **8**, 6199–6221.

123. R. Zander, E. Mahieu, P. Demoulin, P. Duchatelet, G. Roland, C. Servais, M. D. Mazière, S. Reimann and C. P. Rinsland, *Science of the Total Environment*, 2008, **391**, 184–195.

124. C. P. Rinsland, E. Mathieu, R. Zander, N. B. Jones, M. P. Chipperfield, A. Goldman, J. Anderson, J. M. Russell, P. Demoulin, J. Notholt, G. C. Toon, J. F. Blavier, B. Sen, R. Sussmann, S. W. Wood, A. Meier, D. W. T. Griffith, L. S. Chiou, F. J. Murcray, T. M. Stephen, F. Hase, S. Mikuteit, A. Schulz and T. Blumenstock, *J. Geophys. Res. [Atmos.]*, 2003, **108**, 4252.

125. T. P. Marcy, D. W. Fahey, R. S. Gao, P. J. Popp, E. C. Richard, T. L. Thompson, K. H. Rosenlof, E. A. Ray, R. J. Salawitch, C. S. Atherton, D. J. Bergmann, B. A. Ridley, A. J. Weinheimer, M. Loewenstein, E. M. Weinstock and M. J. Mahoney, *Science*, 2004, **304**, 261–265.

126. T. P. Marcy, P. J. Popp, R. S. Gao, D. W. Fahey, E. A. Ray, E. C. Richard, T. L. Thompson, E. L. Atlas, M. Loewenstein, S. C. Wofsy, S. Park, E. M. Weinstock, W. H. Swartz and M. J. Mahoney, *Atmos. Environ.*, 2007, **41**, 7253–7261.

127. Y. Mebarki, V. Catoire, N. Huret, G. Berthet, C. Robert and G. Poulet, *Atmos. Chem. Phys.*, 2010, **10**, 397–409.

128. G. Moreau, C. Robert, V. Catoire, M. Chartier, C. Camy-Peyret, N. Huret, M. Pirre, L. Pomathiod and G. Chalumeau, *Appl. Opt.*, 2005, **44**, 5972–5989.

129. L. Froidevaux, Y. B. Jiang, A. Lambert, N. J. Livesey, W. G. Read, J. W. Waters, R. A. Fuller, T. P. Marcy, P. J. Popp, R. S. Gao, D. W. Fahey, K. W. Jucks, R. A. Stachnik, G. C. Toon, L. E. Christensen, C. R. Webster, P. F. Bernath, C. D. Boone, K. A. Walker, H. C. Pumphrey, R. S. Harwood, G. L. Manney, M. J. Schwartz, W. H. Daffer, B. J. Drouin, R. E. Cofield, D. T. Cuddy, R. F. Jarnot, B. W. Knosp, V. S. Perun, W. V. Snyder, P. C. Stek, R. P. Thurstans and P. A. Wagner, *J. Geophys. Res.*, 2008, **113**, D15525.

130. R. M. Stimpfle, R. C. Cohen, G. P. Bonne, P. B. Voss, K. K. Perkins, L. C. Koch, J. G. Anderson, R. J. Salawitch, S. A. Lloyd, R. S. Gao, L. A. D. Negro, E. R. Keim and T. P. Bui, *J. Geophys. Res.*, 1999, **104**, 26705–26714.

131. M. Höpfner, T. von Clarmann, H. Fischer, N. Glatthor, U. Grabowski, S. Kellmann, M. Kiefer, A. Linden, G. M. Tsidu, M. Milz, T. Steck, G. P. Stiller, D. Y. Wang and B. Funke, *J. Geophys. Res.*, 2004, **109**, 11308.

132. W. H. Brune, J. G. Anderson and K. R. Chan, *J. Geophys. Res.*, 1989, **94**, 16649–16663.

133. B. Vogel, R. Müller, A. Engel, J. U. Grooß, D. Toohey, T. Woyke and F. Stroh, *Atmos. Chem. Phys.*, 2005, **5**, 1623–1638.

134. N. Glatthor, T. von Clarmann, H. Fischer, U. Grabowski, M. Hopfner, S. Kellmann, M. Kiefer, A. Linden, M. Milz, T. Steck, G. P. Stiller, G. M. Tsidu, D. Y. Wang and B. Funke, *J. Geophys. Res.*, 2004, **109**, D11307.

135. L. J. Kovalenko, N. L. Livesey, R. J. Salawitch, C. Camy-Peyret, M. P. Chipperfield, R. E. Cofield, M. Dorf, B. J. Drouin, L. Froidevaux, R. A. Fuller, F. Goutail, R. F. Jarnot, K. Jucks, B. W. Knosp, A. Lambert, I. A. MacKenzie, K. Pfeilsticker, J. P. Pommereau, W. G. Read, M. L. Santee,

M. J. Schwartz, W. V. Snyder, R. Stachnik, P. C. Stek, P. A. Wagner and J. W. Waters, *J. Geophys. Res.*, 2007, **112**, D24S41.

136. G. Wetzel, H. Oelhaf, O. Kirner, R. Ruhnke, F. Friedl-Vallon, A. Kleinert, G. Maucher, H. Fischer, M. Birk, G. Wagner and A. Engel, *Atmos. Chem. Phys.*, 2010, **10**, 931–945.
137. S. Kühl, J. Pukite, T. Deutschmann, U. Platt and T. Wagner, *Advances In Space Research*, 2008, **42**, 1747–1764.
138. T. von Clarmann, N. Glatthor, U. Grabowski, M. Höpfner, S. Kellmann, A. Linden, G. M. Tsidu, M. Milz, T. Steck, G. P. Stiller, H. Fischer and B. Funke, *J. Geophys. Res.*, 2006, **111**, D05311.
139. J. G. Anderson, J. J. Margitan and D. H. Stedman, *Science*, 1977, **198**, 501-503.
140. K. A. McKinney, J. M. Pierson and D. W. Toohey, *Geophys. Res. Lett.*, 1997, **24**, 853–856.
141. F. Ferlemann, C. Camy-Peyret, R. Fitzenberger, H. Harder, T. Hawat, H. Osterkamp, M. Schneider, D. Perner, U. Platt, P. Vradelis and K. Pfeilsticker, *Geophys. Res. Lett.*, 1998, **25**, 3847–3850.
142. I. G. Nolt, P. A. R. Ade, F. Alboni, B. Carli, M. Carlotti, U. Cortesi, M. Epifani, M. J. Griffin, P. A. Hamilton, C. Lee, G. Lepri, F. Mencaraglia, A. G. Murray, J. H. Park, K. Park, P. Raspollini, M. Ridolfi and M. D. Vanek, *Geophys. Res. Lett.*, 1997, **24**, 281–284.

CHAPTER 4

Polar Stratospheric Clouds and Sulfate Aerosol Particles: Microphysics, Denitrification and Heterogeneous Chemistry

THOMAS PETER*[a] AND JENS-UWE GROOß[b]

[a] ETH Zürich, Institute for Atmospheric and Climate Science, 8092 Zürich, Switzerland; [b] Forschungszentrum Jülich, IEK-7, 52425 Jülich, Germany

4.1 Historical Overview

There is a long and exciting chronology of scientific discoveries dating from the late 19th century, when polar stratospheric clouds (PSCs) were first observed, *via* the discovery and chemical analysis of the ubiquitous stratospheric aerosol in the middle of the 20th century and the detection of the Antarctic ozone hole in 1985, followed by the understanding that PSCs and cold stratospheric aerosol particles play an important role in chemical ozone destruction. The understanding grew that these particles transform chlorine—mainly of anthropogenic origin—from relatively inert species into ozone depleting species; and further, that these particles can remove nitrogen-containing species by gravitational settling, which could otherwise reduce the effectiveness of the ozone depleting species. These particles are now recognized as converting the anthropogenic and natural chlorine species into active radicals, which may then destroy the ozone layer in both polar regions. Over Antarctica, in spring, massive ozone losses developed since the 1980s in a $20–30 \times 10^6$ km^2 large area, the ozone hole (see Chapter 5).

Stratospheric Ozone Depletion and Climate Change
Edited by Rolf Müller
© Royal Society of Chemistry 2012
Published by the Royal Society of Chemistry, www.rsc.org

In the years 1883–1886 the readers of *Nature* had the opportunity to follow a vivid controversy between observers of an apparently new type of cloud and "non-observers" or "non-believers".[1] What had happened? Clouds of peculiar shapes, with brilliant colors, had been observed all over England, "these iridescent clouds, in brightness and richness of their colouring, reminded one more of mother-of-pearl inlaid in a back tea-tray than any ordinary sunset sky".[2] They seemed to stay motionless at a great height, in spite of stormy weather in the troposphere. Their discovery triggered a number of questions pertaining to their microphysical and optical properties and concerning the underlying meteorology, many of which were answered more than 100 years later, *i.e.* during the past two decades.

Today, we know that these clouds, which soon after their discovery were called "mother-of-pearl clouds" or "nacreous clouds", indeed consist of ice particles that form deep in the stratosphere at altitudes between 15 and 30 km. Normally, stratospheric air only contains about 5 ppmv H_2O, because most of the moisture freezes out as the air transits the tropopical tropopause, where temperatures are extremely low. This air is normally far too dry to render further cloud formation possible, except in the polar winter stratosphere, where ice saturation is approached at temperatures which are typically around 188 ± 4 K. If then, as must have been the case in the years 1883–1886 above England, dynamical perturbations from the troposphere, such as orographically induced mountain lee waves, penetrate the stratosphere and lead to local ascend and cold stationary pools, this can suffice for ice nucleation to take place and actuate the colorful spectacle. The smallness and similar size of the ice particles, resulting from the rapid cooling in the lee waves, explains their pronounced iridescence, in contrast to most tropospheric clouds with larger polydisperse particles. And the quasi-stationarity of the lee waves, linked to the orography below, explains the apparent motionlessness of the clouds.

However, clouds do not usually form via the condensation of vapor to form new particles—so-called gas-to-particle conversion—rather preexisting aerosol particles are required, which act as cloud condensation nuclei. The discovery of the stratospheric aerosol layer constitutes another great story of atmospheric science. While glowing red sunsets and sunrises in the months following a strong volcanic eruption bore testimony to its existence, during volcanically quiescent periods only twilight measurements suggested a "dust layer" in the stratosphere.[3] Direct proof of this layer was, however, missing.

In the late 1950s, Christian Junge launched balloons from Sioux Falls (South Dakota) and Hyderabad (India), while searching for cosmic dust and for debris from nuclear bomb tests. In addition, he was allowed to fly impactor probes on U2 spy planes, at that time unknown to the public, and Junge was only informed of the altitude and approximate location of the measurements.[4] The results of his measurements surpassed all expectations, leading to the notion of "A Worldwide Stratospheric Aerosol Layer".[5] The microscopic particles in this layer between the tropopause and approximately 35 km altitude appeared to be liquid droplets.[6] They were composed of sulfuric acid and water, and formed obviously by the chemical transformation of sulfur-containing gases. Today, this layer of aerosol particles is a key focus of ozone and climate research. When Junge analyzed the chemical composition of the sampled particles by means of electron microscopy, he found predominantly sulfate ions, SO_4^{2-}, a clear indication against an extra-terrestrial origin. Ironically, the success of Junge's research put an unexpected end to the space program that had originally funded him.[4] The aerosol layer he discovered subsequently became known as the "Junge Layer".

Stratospheric particulate matter continued to be seen as meteorological curiosity, until a program of monthly balloon-borne particle counters was begun in 1971 in Laramie, Wyoming. In 1978 the Stratospheric Aerosol Measurement II (SAM II) instrument aboard the Nimbus-7 satellite started to provide daily vertical profiles of aerosol extinction in both the Arctic and Antarctic regions. Figure 4.1 shows these measurements. They corroborated Junge's findings of a quasi-continuous aerosol loading of the stratosphere at mid-latitudes, however, it was found that this loading was strongly modulated by volcanic eruptions.[7–9] Conversely, the polar measurements revealed an annual cycle of cloudiness in both lower winter stratospheres,[10,11] appearing as a regular "heartbeat-like" phenomenon.

The discovery of the Antarctic ozone hole in 1985 during austral spring[12] might have been the most significant event in environmental sciences during the second half of the 20th century. No chemical or physical mechanism was known that could possibly explain the massive ozone loss as observed since about 1980, but Farman and colleagues speculated that anthropogenic chlorine gases

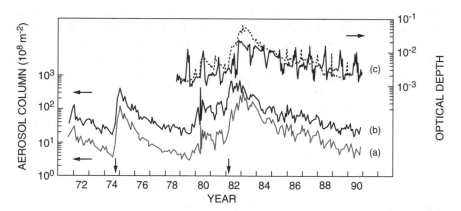

Figure 4.1   Time series of atmospheric aerosol column and optical depth. (a, b) Record of 5 km column integrals (15–20 km) of particle number densities above Laramie, WY (41°N), measured *in situ* with the balloon-borne Optical Particle Counter (OPC): (a) particles with radius $r \geq 0.25\,\mu m$; (b) $r \geq 0.15\,\mu m$.[8] (c) Weekly averaged stratospheric aerosol optical depth (at 1 μm wavelength) integrated from 2 km above the tropopause upward measured by the SAM II satellite instrument.[11] Solid line: Antarctic. Dashed line: Arctic. Vertical arrows: significant volcanic eruptions: 1974—Fuego, Guatemala; 1982—El Chichón, Mexico.

might be involved. The ozone loss region was found to cover most of the Antarctic continent.[13,14] Just one year after Farman's discovery, Solomon and coworkers established the basis for a chemical understanding by postulating PSC particles to be extremely efficient in catalyzing the transformation of photostable chlorine reservoir compounds into photolabile species, which are actively involved in springtime ozone-depletion. Specifically they modeled the potential role of the heterogeneous reaction $HCl + ClONO_2 \xrightarrow{PSC} Cl_2 + HNO_3$ on particle surfaces of Antarctic stratospheric clouds,[15] which they assumed to consist of water ice.[16] The further development of ideas came fast. Not only was it claimed that PSC particles host heterogeneous chemical reactions, but they were suspected to contain nitric acid and to exist at temperatures significantly higher than those required for ice existence.[17,18] These ideas were corroborated by airborne LiDAR (Light Detection And Ranging) observations of Arctic polar stratospheric clouds (PSCs) during January 1984 and January 1986 revealing two distinct PSC growth stages,[19] called Type-I for the more frequent HNO_3-containing clouds and Type-II for the colder, rarer ice clouds. Furthermore, these HNO_3-containing particles were thought to grow to large sizes and sediment out of the stratosphere, thereby denitrifying the air. This irreversible removal of HNO_3 was suggested to reduce the availability of nitrogen species which would deactivate the chlorine, leading to greatly enhanced and persisting ozone depletion.[17,18]

The subsequent years were an Eldorado for laboratory scientists and process-related thermodynamic and kinetic modelers. Many elaborate laboratory experiments were required to verify and quantify the ingenious, but then unproven ideas mentioned above. A famous experiment, performed by Hanson and Mauersberger[20] in 1988, showed that a hydrate of HNO_3, namely nitric acid trihydrate (NAT $\equiv HNO_3 \cdot 3H_2O$), could exist under conditions typical of the lower stratosphere, at temperatures $T \lesssim T_{ice} + 7\,K$, *i.e.* up to seven degrees above the "frost point" (T_{ice}), the temperature to which atmospheric moisture must at least be cooled to reach saturation with respect to ice (and thus enable ice to exist). Subsequently, several other hydrates of nitric and sulfuric acid ($HNO_3 \cdot n\,H_2O$, $H_2SO_4 \cdot m\,H_2O$, n and m being integers or half-integers) were discovered, leading to a multitude of crystalline particles postulated to potentially occur in the stratosphere (see Peter,[21] for an overview). In addition, another surprising discovery was that of liquid PSCs. In 1994, based on thermodynamic modeling of supercooled electrolytic solutions Carslaw *et al.*[22] interpreted certain *in situ* PSC measurements[23] as liquid aqueous HNO_3-H_2SO_4 droplets. This put an end to the myth that under the cold polar stratospheric winter conditions all droplets should freeze as hydrates, and that HNO_3 would never dissolve in aqueous H_2SO_4 solutions. Rather, it became clear that the opposite is true and that the homogeneous nucleation of NAT in HNO_3-H_2SO_4-H_2O solutions is kinetically strongly hampered.[24]

Laboratory experiments investigated the heterogeneous chemistry occurring on ice and acidic hydrates as well as in liquid solution droplets. These experiments confirmed the high reactivities of gases such as $ClONO_2$ or HOCl on HCl-doped surfaces.[25-28] Interestingly, per unit surface area the activation of chlorine in the presence of supercooled liquids was shown to be comparable to that on solid HNO_3 hydrates at the same temperature.[29] Molina and colleagues[30] argued that this might be due to the formation of a disordered surface layer caused by the dissolution of HCl and its subsequent dissociation to $H^+ + Cl^-$, leading to water molecules with a much larger mobility than in the bulk crystal, similar to a liquid. This behavior is often described as the formation of a "quasi-liquid layer".[31] This hypothesis later received support from laboratory work and molecular dynamics calculations.

The fact that chlorine activation is largely independent of the physical state of the surface, together with the kinetic suppression of homogeneous nucleation of NAT in HNO_3-H_2SO_4-H_2O solutions led to the recognition that cold liquid aerosols or liquid PSCs should be the dominant reaction site for chlorine activation[32-36] whereas solid PSCs play a secondary role (at least in the Arctic, where regular warming events evaporate NAT particles, should they have formed). However, while the liquid particles host heterogeneous reactions, they cannot grow to large sizes due to their high number density and mutual competition for gas phase HNO_3. In contrast, denitrification requires the selective nucleation and sufficient longevity of solid hydrate particles, most likely NAT, in low number densities, so that they may grow to diameters of around 10 μm, thus developing sedimentation speeds of around 1 km/d.

Observed trends in PSC occurrence frequencies in the Arctic and Antarctic have been positive during the past decades. Rex et al.[37] demonstrated that measured chemical ozone loss in the Arctic correlates very strongly with integrated volume of stratospheric air below PSC formation temperatures, and that there is a trend towards lower temperatures and more favorable PSC formation conditions.[38] To the degree that ozone loss itself is responsible for this cooling and enhanced PSC formation, the trend is expected to reverse as the stratospheric halogen loading subsides. However, if a part of the cooling trend was a result of climate change (see Chapter 8), then this would have potentially alarming implications in terms of a substantial increase in Arctic denitrification and ozone loss over the next few decades while the stratospheric halogen loading remains high.[39,40]

4.2   Distribution and Composition of PSCs

As shown by the studies discussed above, polar stratosphere clouds (PSCs) consist of different types of particles that are composed of water (H_2O), sulfuric acid (H_2SO_4) and nitric acid (HNO_3). Solid water ice particles can exist only below the frost point T_{ice} (typically below 188 K in the lower stratosphere), because only then the vapor pressure of ice drops below water partial pressures prevalent in the lower stratosphere. However, already below the so-called "NAT-temperature", $T_{NAT} \approx 195$ K, crystalline nitric acid trihydrate (NAT) particles are thermodynamically stable[20] and can exist if they nucleate. Balloon-borne mass spectrometric measurements[41] have confirmed the stoichiometric HNO_3:H_2O ratio inside solid PSCs as 1:3 (to within a few percent), confirming the existence of the long sought solid $HNO_3 \cdot 3H_2O$ crystals well above ice formation temperatures.

Besides these frozen particles, also liquid particles exist in the stratosphere. When temperatures drop below $T_{dew} \approx 192 \, K$ (the "dew point" of HNO_3), the binary H_2SO_4-H_2O droplets with radii $\sim 70 \, nm$ in the ubiquitous Junge layer take up significant amounts of HNO_3 from the gas phase, forming supercooled ternary solution droplets (STS), HNO_3-H_2SO_4-H_2O, with droplet radii growing up to $\sim 0.3 \, \mu m$.[22]

4.2.1 LiDAR-based PSC Classification

Remote sensing techniques have a long history of measuring the altitudes at which PSCs occur and of identifying various types of PSCs. The LiDAR technique with two or more wavelengths is suitable to estimate particle sizes, thus distinguishing HNO_3-containing PSCs (Type-I) from ice PSCs (Type-II). Furthermore, by measuring the depolarization of the backscattered light, they make it possible to distinguish spherical from aspherical particles, *i.e.* to determine their physical state, liquid *vs.* crystalline.[42] PSCs have been classified into three major types: Type-Ia (small backscatter ratio, significant volume depolarization), most likely crystalline NAT particles; Type-Ib (moderate backscatter ratio, no volume depolarization), most likely liquid ternary aerosols; and Type-II (large backscatter ratio, significant volume depolarization), most likely ice particles.[42,43] Table 4.1 shows a classification scheme of PSCs adopted from Biele *et al.*[44] based on the Ny-Ålesund PSC LiDAR climatology. Two additional PSC-Types had been identified: mixed phase clouds, and Type-Ia-enh. Just like Type-Ia, Type-Ia-enh clouds contain NAT particles, but in higher number densities.[45] Based on microphysical modeling, Tsias *et al.*[45] showed that Type-Ia-enh can be explained by NAT nucleation on ice clouds, leading to a cloud in which 10% or more of all stratospheric particles contain NAT. Conversely, in Type-Ia, less than 0.1% of the particles contain NAT.

In Table 4.1 Type-Ib is fully liquid (*i.e.*, no solids can be optically detected). In contrast, all other types denote clouds with optical evidence for both solid and liquid particles (*i.e.* significant aligned *and* cross-polarized backscatter signal). For example, these could be ensembles of NAT crystals and STS droplets. Types Ia, Ia-enh, II and "Mixed" usually only contain a minority of solid particles, externally mixed with small interstitial droplets (consisting mostly of aqueous H_2SO_4 with some HNO_3). This means that "NAT PSCs" are never fully solid, as not all particles have crystallized. Furthermore, those particles that did crystalize are not composed of pure NAT, but contain liquid remnants of H_2SO_4-H_2O with traces of HNO_3 (unless the H_2SO_4 froze as a hydrate $H_2SO_4 \cdot m \, H_2O$, for which observational evidence is lacking). The existence of these mixtures has been postulated by Koop *et al.*,[46] who investigated the thermodynamic stability and the particle formation pathways of PSCs, concluding that liquid particles can coexist with solid particles as long as the H_2SO_4 remains in the aqueous phase. Coexistence of NAT and STS is one

Table 4.1 PSC classification scheme according to Biele et al.[44] relating optical properties (as measured by LiDARs) to spherical liquid particles (binary H_2SO_4-H_2O, ternary HNO_3-H_2SO_4-H_2O), aspherical solid hydrates (i.e., NAT $\equiv HNO_3 \cdot 3H_2O$ or NAD $\equiv HNO_3 \cdot 2H_2O$, etc.), and water ice.

PSC Type	Aligned Backscatter Ratio*	Cross-Polarized Backscatter Ratio*	Class	Likely Physical Cloud Composition	Particle shape	Typical particle radius	Fraction of ptcls in cloud phase	CALIPSO PSC class (Pitts et al., 2011)[72]
None	Small	Small	Background aerosol	Binary H_2SO_4-H_2O droplets	Spherical	~ 0.1 μm	None	None
Ib	Large	Small	STS, liquid cloud	Ternary HNO_3-H_2SO_4-H_2O droplets	Spherical	~ 0.3 μm	1	STS
Ia	Small	Large	Hydrate cloud	Few large solid particles plus binary droplets	Aspherical	1–10 μm	$\ll 1$	Mix 1; $<10^{-3}$ cm^{-3} solid
"Mixed"	Medium	Medium	Mixed phase cloud	Few small solid particles plus ternary droplets	Mixed	~ 0.3 μm	Variable	Mix 2; $>10^{-3}$ cm^{-3} solid
Ia-enh	Medium	Large	Hydrate cloud	Many solid particles plus binary droplets ("Mother Clouds")[59]	Aspherical	~ 0.3 μm	$\lesssim 1$	Mix 2-enh,$>10^{-2}$ cm^{-3} solid
II	Very large	Very large	Ice cloud	Ice particles mixed with liquids/hydrates	Aspherical	$\gtrsim 1.5$ μm	Variable	Ice

* Approximate ratios for 532 nm LiDAR backscatter are as follows, according to Biele et al.[44]
Aligned: small $S_{\parallel}<1.2$; medium $1.2<S_{\parallel}<7$; large $7<S_{\parallel}$. Cross-polarized: small $S_{\perp}<1.36$; medium $1.36<S_{\perp}<1+14(S_{\parallel}-1)$; large $1.36<S_{\perp}$; very large $85<S_{\perp}$.

possibility (which Biele *et al.*[44] called Type "Mixed"). To better emphasize the generally "mixed" character of solid-containing PSCs, Pitts *et al.*[47] introduced a classification scheme, which differs from the Biele *et al.* scheme and is shown in the last column of the table.

4.2.2 Antarctic PSC Diversity

The diversity of phase and composition of stratospheric PSC particles is illustrated in Figure 4.2 using observations by the CALIOP instrument on-board the CALIPSO satellite (Cloud-Aerosol LiDAR and Infrared Pathfinder Satellite Observations). CALIOP actively sounds the atmosphere from space using the LiDAR technique.[48] From these measurements important information about tropospheric and stratospheric particles can be obtained on a global basis.

An example is provided by Figure 4.2, which shows CALIPSO observations obtained in June 2006 over the Antarctic continent. The corresponding flight track is displayed in the top right inset. The top panel displays the total backscattered signal of the laser beam showing the location of PSCs and tropospheric clouds. The center panel shows the perpendicular polarized signal, indicating the presence of aspherical particles, *i.e.* crystalline solids. The bottom panel corresponds to the PSC classification defined by Pitts *et al.*[47], in which four classes of PSCs are deduced from the detected signals at 532 nm wavelength: ice particles, STS particles and mixtures of solids (likely NAT) and STS particles, for low number densities of solid particles ($< 10^{-3}$ cm^{-3}, termed Mix 1), and for high number densities of solid particles ($> 10^{-3}$ cm^{-3}, termed Mix 2). The tropopause altitude (thin white line) and the position of the vortex edge (thick white line calculated according to Nash *et al.*[49]) have been calculated from ECMWF operational analyses. Obviously, there are lots of clouds in the troposphere. On that day, the clouds extend into the stratosphere above the Antarctic continent with only a small seam of reduced cloudiness just above the local tropopause (~ 12 km). The physical phase of the particles can be deduced from the depolarization signal (center panel). PSC types and mixtures along the displayed orbit section are listed in Table 4.1. Ice PSCs in the stratosphere require the lowest temperatures (lower than 188 K, corresponding to the frost point for typical humidities in the stratosphere). Although temperatures are low enough for potential ice existence in a wide region, homogeneous nucleation of ice requires a supercooling of ~ 3 K below the frost point (~ 185 K, or 60–70% supersaturation above ice vapor pressure), which prevents the ice from nucleating in a large part of the cold area. Thus, in these regions, in the absence of heterogeneous nuclei, the aerosol does not crystallize,[24] but liquid STS clouds are observed. There are also extensive regions filled with mixtures of solution droplets (binary or ternary) with solids at low or high particle number densities (Mix 1 or 2, respectively), which are present at temperatures below the NAT temperature of about 195 K. The evaluation of the CALIPSO image suggests the presence of mixtures of liquid and solid

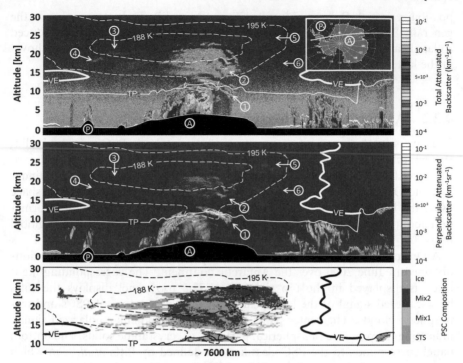

Figure 4.2   PSC observations by the spaceborne aerosol LiDAR on the CALIPSO
satellite over the Antarctic continent on 22 June 2006 (20:05-20:25 UTC).
Top panel: 532 nm total attenuated backscatter. Insert: flight track of
CALIPSO traversing the Antarctic main land (A) and the Antarctic
Peninsula (P). Center panel: 532 nm perpendicular attenuated backscatter
indicating the presence of aspherical particles (ice or HNO_3 hydrates).
Bottom panel: PSC classification according to Pitts *et al.* (2009). Top and
center panels: dark blue = molecular backscatter; light blue/green-
= aerosols or thin cloud fields; yellow/red = dense aerosols or cloud fields.
Encircled numbers: (1) tropospheric cirrus reaching up to tropopause; (2)
ice PSCs; (3) large fields of STS (HNO_3-H_2SO_4-H_2O droplets); (4) STS lee
wave clouds above the mountain chain of the Antarctic Peninsula (P); (5)
Mix2: extensive fields of tenuous PSC mixtures containing solid (likely nitric
acid trihydrate, NAT) crystals in significant number density ($< 10^{-3}$ cm^{-3})
externally mixed with droplets (STS or binary H_2SO_4-H_2O, ~ 10 cm^{-3}); (6)
Mix1: large fields of low number density crystals (likely NAT, $< 10^{-3}$)
externally mixed with droplets (~ 10 cm^{-3}). Thin white line marked "TP":
tropopause (WMO definition), from ECMWF operational data. Thin
dashed lines: 195 K and 188 K temperature contours indicative for T_{NAT}
and T_{ice}, respectively. Thick white line marked "VE": polar vortex edge
(Nash criterion). Upper and center panels: non-validated data from http://
www-calipso.larc.nasa.gov/products/lidar/browse_images. Lower panel:
courtesy of Michael Pitts (NASA Langley).

particles with very low NAT number density even in regions with temperatures
above 195 K, in particular at altitudes below 15 km. Most likely this is caused
by sedimentation of large HNO_3-containing particles, which, upon

evaporation, leads to enhanced HNO_3 in the gas phase and increases T_{NAT}, the temperature at which NAT can exist. This interpretation is corroborated by an enhanced HNO_3 layer observed by the MLS instrument on the AURA satellite (not shown).

4.2.3 Arctic Measurements of STS and NAT

As indicated above, liquid binary sulfate aerosol particles can take up the HNO_3 from the gas phase if temperatures become low enough. Particle measurements in the Arctic by Dye et al.[23] clearly indicated the growth of particles with decreasing temperatures. Figure 4.3 shows these particle size data together with simulations of the composition of the liquid stratospheric aerosols (solid curves) calculated by means of thermodynamic electrolyte modeling by Carslaw et al.[22] Also shown are the expected particle volumes that would result, if NAT nucleated at T_{NAT} without a kinetic barrier, and subsequently grew assuming equilibrium between the HNO_3 partial pressure in the gas phase and NAT vapor pressure (dashed curve). Furthermore, the growth of the aerosol is shown for the (fictitious) case that HNO_3 was absent or would not be taken up at all, and only H_2O was taken up (dotted curve). From these comparisons it is evident that the observed temperature-dependent growth of the particles is best explained when assuming liquid supercooled aerosol particles, which gradually take up HNO_3 below the "dew point" of HNO_3 ($T_{dew} \approx 192\,K$). Laboratory

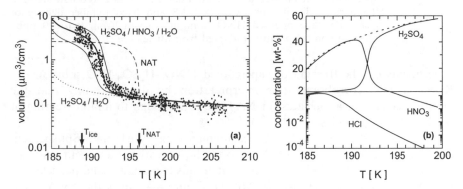

Figure 4.3 Volumes and concentrations of supercooled ternary solution (STS $\equiv$ HNO_3-H_2SO_4-H_2O) droplets as function of temperature. (a) Symbols: Arctic measurements by Dye et al.[23] on 24 January 1989. Lines: model calculations of the volumes of various types of stratospheric particles in thermodynamic equilibrium with the gas phase: dotted line = supercooled binary solution (SBS $\equiv$ H_2SO_4-H_2O); dashed line = NAT particles;[20] thick solid line = STS.[22] Calculated volumes assume 5 ppmv H_2O and 10 ppbv total HNO_3 at 55 hPa, i.e. about 19 km altitude. Thin solid lines correspond to 5 ppbv and 15 ppbv HNO_3. (b) Weight fractions in STS droplets corresponding to thick solid line in (a). Solid curves: H_2SO_4, HNO_3, HCl (remaining fraction to balance 100 wt-% is H_2O) according to Carslaw et al.[22] Dotted line corresponds to H_2SO_4 in binary H_2SO_4-H_2O droplets. From Peter.[21]

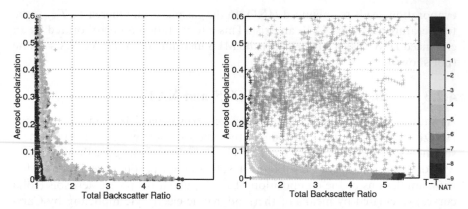

Figure 4.4   Scatter plots of aerosol depolarization versus total backscatter ratio of PSCs above Ny-Ålesund (Spitsbergen). Left panel: LiDAR observations during winters 1995–1997. Color coding is $T - T_{NAT}$ (in degrees Kelvin). Right panel: gray symbols are theoretical T-matrix simulations for pure NAT PSCs corresponding to 5 ppbv condensed HNO_3 (at 50 hPa); particles are monodisperse but occur with varying radius (and corresponding number density) and aspect ratios between 0.5 (prolate spheroids) and 1.5 (oblate spherioids). Color symbols: T-matrix calculations of mixed PSCs assuming a fixed NAT distribution but varying (temperature-dependent) STS abundance. Number density, radius and condensed HNO_3 of NAT particles are 0.005^{-3}, $1.56\,\mu m$, and 0.335 ppbv, respectively, while their aspect ratios are allowed to vary from 0.5–1.5 (prolate to oblate shapes). STS particle distribution (width $\sigma = 1.8$ and fixed number density of $10\,cm^{-3}$) with condensed HNO_3 amount varying from 0 to 10 ppbv depending on temperature (adapted from Biele *et al.*)[44]

experiments on the freezing of supercooled HNO_3-H_2SO_4-H_2O solutions by Koop *et al.*[24] corroborate this interpretation by demonstrating that the homogeneous nucleation rates for NAT are low enough to inhibit freezing of the liquid STS particles.

Ground-based aerosol LiDAR measurements in Ny-Ålesund, Spitsbergen, confirm the conception that PSCs are often liquid without apparent evidence for solid (aspherical) admixtures, and that clouds containing solid particles are mostly very faint (*i.e.*, Type-Ia or Mix 1). Figure 4.4 shows the Ny-Ålesund measurements as scatter plots of aerosol depolarization versus total backscatter ratio. The left panel shows 99% of all observations in the winters 1995–1997. The remaining 1% correspond to Type-Ia-enh PSCs that scatter over the entire plane (in Figure 4.4 suppressed for clarity). Most data points are found either near the depolarization axis (Type-Ia) or near the backscatter ratio axis (Type-Ib), while background aerosols are located near the origin. The right panel shows theoretical T-matrix simulations of backscatter and depolarizations of pure NAT clouds (gray symbols) or mixed STS/NAT clouds with only small fractions of the HNO_3 in the NAT (colored symbols). Gray symbols are for NAT particle number density of $10\,cm^{-3}$ and 5 ppbv condensed HNO_3 (at 50 hPa). Such particles, owing to their asphericity, would cause large depolarization, and depending on their size also large backscatter. However, the modeled distribution of gray points bears no obvious resemblance to the measurements at Ny-Ålesund. In contrast, colored symbols in the right panel display T-matrix calculations of mixed PSCs assuming $0.005 \, cm^{-3}$ NAT particles containing only 3% of the total HNO_3, while the other 97% are partitioned in varying amounts between the gas phase and STS particles (depending on temperature, see color bar). This assumption enables the model to reproduce the observations satisfactorily. It indicates that the NAT in the condensed phase does not readily reach equilibrium with the gas phase. Rather, the HNO_3 in the gas phase stays highly supersaturated with respect to NAT. The supersaturation is maintained by the strongly impeded uptake of HNO_3 onto the very small number of NAT particles.

There are only very few *in situ* composition measurements of PSCs, which could directly confirm the existence of NAT and STS. Using balloon-borne mass spectrometry Voigt *et al.*[41] found the stoichiometry $H_2O:HNO_3$ of solid PSC particles above T_{ice} to be 3:1 ($\pm 15\%$) as expected for NAT, whereas liquid clouds showed molar ratios larger than 4.5:1, consistent with calculations for ternary solution droplets (cp. Figure 4.3b). These field observations are in agreement with the laboratory-based understanding that most of the HNO_3-containing PSC particles consist either of NAT particles, which is the thermodynamically stable phase (but rarely equilibrated because of low NAT number density), or of supercooled ternary solution droplets (STS), or of NAT/STS mixtures.

Information on the composition of the particles can also be derived spectroscopically from remote sensing instruments, *e.g.* from the spectral features in the IR observed by MIPAS onboard the ENVISAT satellite,[50] see the next section.

Furthermore, nitric acid hydrates other than NAT (hereafter termed "NAX") may be important as intermediates in the conversion process from metastable STS to thermodynamically stable NAT. This concerns, in particular, the dihydrate NAD ($HNO_3 \cdot 2H_2O$). Experiments with binary HNO_3-H_2O droplets in AIDA, the world's largest low-temperature aerosol chamber, identified the first stage of the solidified particles unequivocally as strongly aspherical nitric acid dihydrate crystals (NAD), while there was no indication of direct nucleation of NAT or conversion of NAD into NAT within a period of one hour following the initial nucleation of NAD.[51]

The existence of NAD in the atmosphere is supported by Kim *et al.*,[52] who identified dual mixtures of liquid + NAT, liquid + ice, ice + NAT, and NAT + NAD in solar occultation data measured by the ILAS-II instrument on the ADEOS-II platform during the period of June to August 2003.

4.3 Nucleation Mechanisms of NAT-PSCs

Figure 4.5 illustrates two potential pathways for PSC formation. The main difference between both is the presence or absence of efficient heterogeneous

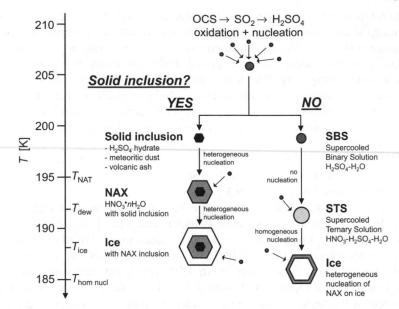

Figure 4.5   Sketch of potential PSC formation pathways. Left branch: nitric acid
hydrate, NAX (*e.g.* NAT or NAD), nucleates heterogeneously at or
slightly below $T_{NAT} \approx 195\,K$ on pre-existing solid inclusion (*e.g.*, a dust or
ash particle or an H_2SO_4 hydrate particle); subsequently ice nucleates
heterogeneously on NAX at or slightly below $T_{ice} \approx 188\,K$. Right branch:
SBS droplets form STS droplets at $T_{dew} \approx 192\,K$; subsequently ice
nucleates homogeneously in STS below $T \approx T_{ice} - 3\,K$; finally NAT may
nucleate heterogeneously on ice. Values for the frost point (T_{ice}), the
equilibrium NAT temperature (T_{NAT}), the STS formation temperature
(T_{dew}) and the temperature for homogeneous nucleation of ice ($T_{hom\ nucl}$)
refer to typical conditions in the polar lower stratosphere.

nuclei. The left branch allows heterogeneous nucleation of nitric acid hydrate
NAX (*e.g.* NAT or NAD) on pre-existing solid inclusions (*e.g.*, dust, ash or
H_2SO_4 hydrates) at or slightly below $T_{NAT} \approx 195$ K. This branch corresponds
to the "3-stage model" of PSC development:[19,21,53] first, formation of an H_2SO_4
hydrate, second, formation of NAT on this H_2SO_4 hydrate, third, formation of
ice on the NAT particles. In contrast, the right branch assumes the absence of
such nuclei, so that the binary H_2SO_4-H_2O droplets remain liquid and grow
into an STS cloud upon cooling due to uptake of HNO_3 and H_2O. In the
absence of solid inclusions homogeneous nucleation of NAX is indeed negligible, as Koop *et al.*[24] demonstrated by means of bulk phase laboratory
experiments. These experiments are convincing evidence against homogeneous
nucleation, as they operated with bulk samples of several milliliters of STS in
test tubes at temperatures using compositions described in Figure 4.3 (right
panel), hence eliminating not only the possibility of homogeneous nucleation
but also of heterogeneous nucleation on the many (unidentified) heterogeneous nuclei that unavoidably must have been present in such bulk samples.

In support of Koop's work, MacKenzie *et al.*,[54] using classical nucleation theory also arrived at the conclusion that homogeneous nucleation of NAT does not occur under stratospheric conditions, while heterogeneous nucleation could not be ruled out. Later, Biermann *et al.*[55] based on bulk samples including solid meteoritic nuclei found that meteoritic nuclei could at best have freezing times around 20 months, and hence would not be very efficient for PSC freezing during a single winter.

These laboratory and modeling investigations must be seen against the background that detailed *in situ* measurements and multi-year remote sensing, such as the ER-2 data in Figure 4.3 and the Ny-Ålesund LiDAR data in Figure 4.4, also disfavor the presence of efficient heterogeneous nuclei. The Ny-Ålesund climatology based on three winters showed that 50% of all measured PSCs were STS (Type-Ib) without evidence for solids, 25% were thin NAX (Type-Ia), and 25% were STS with some solid particles (Type Mixed).

However, the resulting suppression of nitric acid hydrate (NAX) nucleation (Figure 4.5, right branch) has peculiar consequences: solid particles occur only after the homogeneous nucleation of ice, *i.e.* at $T \lesssim T_{ice} - 3\,K \approx 185\,K$. Only after ice nucleated may NAT or other hydrates nucleate *via* heterogeneous nucleation on ice, as demonstrated by Koop *et al.*[24] Although NAT is the thermodynamically most stable stratospheric cloud phase, it nucleates on ice only $\sim 10\,K$ below the NAT equilibrium temperature ($T \lesssim T_{NAT} - 10\,K \approx 185\,K$). This in turn has been made responsible for the frequently observed Type-Ib clouds,[21] and the laboratory bulk experiments by Koop and colleagues[24] fully back this interpretation up.

4.3.1 Ice-assisted NAT Nucleation

Laboratory experiments on NAT nucleation from STS in thermodynamic equilibrium with polar winter stratospheric conditions showed that the homogeneous nucleation rates of stratospheric aerosols are exceedingly low, ruling out homogeneous freezing as a pathway for PSC formation.[24,56] The same experimental work suggests that the heterogeneous nucleation rate of NAX on ice in the immersion mode is also small, whereas it is large in deposition mode (*i.e.* when the ice is in contact with the supersaturated gas phase). Furthermore, airborne LiDAR observations of NAX wakes downstream of ice clouds[57] clearly support the ice-assisted pathway for NAX formation (right-hand side of Figure 4.5). An important discovery made during the SOLVE/THESEO campaign in the Arctic stratospheric winter 1999/2000 was the *in situ* measurement of NAX particles (likely NAT, possibly NAD), which occurred only in very low number densities of a few times $10^{-4}\,cm^{-3}$, but because of these low densities and the resulting lack of competition for gas phase HNO_3 could grow to extremely large radii, $r \approx 10\,\mu m$, leading to gravitational settling velocities $v_{sed} \approx 60\,m/h$.[58] Earlier evaluations of LiDAR observations, including the Ny-Ålesund climatology, had provided evidence for solid particles at very low number densities, see Type-Ia in Table 4.1 and Figure 4.4; however, only the measurements of Fahey *et al.*[58]

proved that these particles had sizes sufficiently large to enable rapid denitrification of the gas phase with repercussions for the ozone layer in polar regions. Fueglistaler *et al.*[59] suggested a mechanism by which very large NAT particles might sediment out of dense clouds of very small NAT particles, and termed this the "Mother Cloud/NAT Rock mechanism", because the Type-Ia-enh clouds serve as "mother clouds", producing very large NAT particles (the "NAT rocks") by sedimentation and dilution. The following PSC formation sequence was then conceivable:

(1) ice PSCs (Type-II), *e.g.* formed in small-scale cold pools in lee waves above polar mountain ranges,
(2) high number density NAT clouds (Type-Ia-enh) nucleate on the ice and travel far distances as "mother clouds",
(3) low number density NAT particles (Type-Ia) sediment out of the Type-Ia-enh into supersaturated regions and grow to the "NAT rocks".

Via this sequence, the occurrence of large NAT particles could be explained by a mechanism taking place fully along the right branch in Figure 4.5. While the apparent patchiness of the wave ice clouds was first taken as an argument against its effectiveness,[60] later more detailed vortex-wide simulations by the same authors showed that this mechanism could potentially indeed explain up to 80% of the observed denitrification,[61] see next section.

The large NAT particles may cause significant vertical transport of HNO_3 by virtue of rapid sedimentation and can explain the observed denitrification, *i.e.* the permanent removal of reactive nitrogen, NO_y, in the stratosphere ($NO_y = NO + NO_2 + HNO_3 +$ other N-containing species). The denitrification occurs annually in the cold Antarctic and less frequently in the warmer Arctic. It is important as it directly influences the polar ozone loss, especially as the main chlorine deactivation reaction, $ClO + NO_2 + M \rightarrow ClONO_2 + M$, is significantly slowed or inhibited, resulting in long-lasting chlorine-catalyzed ozone loss even after PSCs have disappeared.

Waibel *et al.*[39] used balloon-borne measurements of denitrification during the cold Arctic winter 1994–1995 to constrain a microphysical model, corroborating the sequence "ice PSCs → dense NAT clouds → sedimenting large NAT particles". Conversely, Waibel *et al.* showed that neither sedimentation of NO_y on ice particles only, nor nucleation of NAT particles independent of ice formation was able to model the observed NO_y. Waibel *et al.*[39] predicted that the Arctic winter stratosphere might behave more Antarctic-like and possibly develop an ozone hole due to enhanced denitrification, if the cooling of the Arctic stratosphere continues as observed in the 1990s.

Antarctic PSC information has been mostly derived from satellite observations (such as Figure 4.2). Also in Antarctica mountain waves have been made responsible for PSC formation,[62] and ice-assisted NAT nucleation has been invoked to explain NAT downstream of mountain waves.[63] From IR spectral information it is possible to detect different particle types and composition.[64,50] Höpfner *et al.*[63] used this property of MIPAS (a Michelson interferometer)

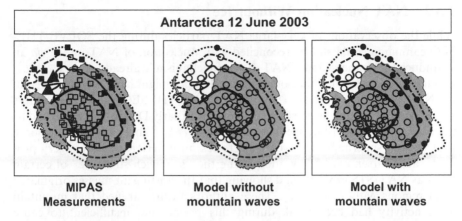

Figure 4.6 Left panel: PSC composition measurements at ~21 km altitude over Antarctica on 12 June 2003 by the Michelson interferometer MIPAS onboard the ENVISAT satellite. Solid triangles: ice clouds in the lee of the Antarctic Peninsula. Solid squares: NAT particles with radii <3 μm that appear downstream of the ice clouds. Open squares: most likely STS droplets (or much less likely large NAT particles with radii >3 μm or very thin ice clouds, which cannot be distinguished from the MIPAS spectral information). Contour lines: ECMWF analyses of regions with potential ice occurrence (T_{ice}, solid curve), STS existence (T_{dew}, dashed curve), and potential NAT occurrence (T_{NAT}, dotted curve). Center panel: microphysical results at the time and location of the MIPAS observations using a PSC box model driven directly by ECMWF temperatures along air parcel trajectories (air moves clockwise approximately along the temperature contours). Right panel: same as center panel but with mountain wave corrections from a mountain wave forecast model. Adapted from Höpfner *et al.*[63]

onboard the satellite ENVISAT to investigate Antarctic PSCs. They showed that the sudden onset of formation of NAT particles in June 2003 could be localized above the Antarctic peninsula, leading to the formation of a belt-like structure shown in Figure 4.6. They traced the observed onset of NAT PSCs back to large-amplitude stratospheric mountain waves over the Antarctic Peninsula and the Ellsworth Mountains introducing temperature fluctuations suitable for homogeneous ice formation and subsequent heterogeneous nucleation of NAT on ice. In contrast, MIPAS observations of PSCs before this event provided indication for the presence of only supercooled droplets of ternary HNO_3-H_2SO_4-H_2O solution (STS).

Eckermann *et al.*[65] confirmed this mechanism by showing the presence of mountain-waves above the Antarctic peninsula with amplitudes of 10–15 K at 40 hPa. They further tested this explanation by analyzing perturbations in stratospheric radiances from the Atmospheric Infrared Sounder (AIRS), which showed mountain waves directly preceding the event, but no resolved wave activity before or after. Again, these observations support the notion that a NAT wake was produced in a small region of mountain wave activity acting as source region of a circumpolar NAT outbreak.

4.3.2   NAT Nucleation Without Ice

While the observations of very large NAT particles during the SOLVE/THE-
SEO campaign could still be reconciled with nucleation of NAT on ice *via* an
ice-induced "Mother Cloud/NAT Rock mechanism", already Fahey *et al.*[58]
pointed to the possibility that other nuclei than ice might have been involved in
the NAT nucleation process. Two other SOLVE/THESEO papers cast doubts
on the conception that large-scale NAT clouds in early December 1999 could be
traced back to pre-existing ice particles: Pagan *et al.*,[66] by using aircraft-borne
LiDAR and superimposing back trajectories on AVHRR cloud imagery products, and Larsen *et al.*,[67] by interpreting the "sandwich structure" of certain
PSCs as NAT/STS mixtures and showing that the hosting air masses originated
from regions with $T > T_{NAT}$. These authors further argued that mountain
wave activity had been weak during this period and insufficient to cause
intermediate ice formation.

Some remote sensing studies using LiDAR measurements argued that also
over Antarctica Type-Ia PSCs—and thus NAT—might indeed nucleate dif-
ferently than on ice. Santacesaria *et al.*[68] pointed to regular observations at
Dumont D'Urville of Type-Ia clouds before Type-II (ice) occurrence, however
could not exclude that ice might have formed upwind, which at the time of their
publication was difficult to verify or falsify using backward trajectories due to
the limited "quality of the meteorological analysis in the Antarctic region".
Also from today's perspective the caveats against ice-induced NAT expressed
by Santacesaria *et al.*[68] remain weak, given the location of Dumont D'Urville
located on the shore of Adélie Land, 66°S, on the brink of the NAT-belt
identified by Höpfner *et al.* (2006b) some 7000 km downstream of the Antarctic
Peninsula and the Ellsworth Mountains with their frequently strong orographic
lee waves.[69] Similar concerns apply to an eight-year PSC climatology measured
by the McMurdo LiDAR presented by Adriani *et al.*[70] They used six-day
backward trajectories based on NCEP data ending in NAT clouds above
McMurdo, finding that about 60% did indeed encounter $T < T_{ice}$, and still 15%
of the all cases showed $T < T_{ice} - 3\,K$. Adriani *et al.*[70] therefore rated their
trajectory results as inconclusive.

Finally, an *in situ* observation that allowed a chemical characterization of
the faint hazes of large solid particles with unclear origin was achieved during
the EUPLEX campaign based in Kiruna, Sweden. During an aircraft flight on
6 February 2003, three weeks after a sudden major warming, small NAX
particles were detected in situ around 19 km altitude with $n_{NAX} \approx 2 \times 10^{-4}\,cm^{-3}$
and $r \approx 3\,\mu m$,[71] containing roughly 0.1 ppb HNO_3 or only 1% of the total
available HNO_3. While the number density of these particles reminds of the
observations by Fahey *et al.*,[58] their size was much smaller. Moreover, the
vortex in this phase of the winter was already for three weeks much warmer
than the generally cold SOLVE/THESEO winter 1999/2000. Backward trajectory calculations for the EUPLEX observations indicate that these air masses
originated from a warm region with $T > T_{NAT} + 10\,K$ only 36 hours before the
measurement, and subsequently had been less than 3 K below T_{NAT} for less

than 20 hours. Based on highly resolved ECMWF temperatures analyses unresolved mountain wave activity sufficient for ice nucleation could be clearly excluded, showing that the air masses remained at least 4 K above T_{ice}. This demands an ice-free NAX nucleation mechanism at low supersaturation with respect to NAT. Voigt *et al.*[71] derived an air volume averaged NAT nucleation rate of about $J_{NAX} = n_{NAX}/24\,h \approx 10^{-5}\,cm^{-3}$ air h^{-1} and argued that the heterogeneous NAT nucleation rates on meteoritic material measured by Biermann *et al.*,[55] while too low to be responsible for dense NAT clouds, might actually be sufficient to produce the low NAT number density of $\sim 10^{-4}\,cm^{-3}$ within hours. Since the nucleation rates published by Biermann *et al.*[55] were upper limits and the real rates might be considerably smaller, the suitability of meteoritic dust remains an open issue which calls for further investigation.

A temporary apex in discoveries of ice-free NAT nucleation has been reached with recent spaceborne LiDAR measurements from CALIPSO, providing a vortex-wide perspective of the 2009–2010 Arctic PSC season[72]. The 2009–2010 Arctic winter developed widespread patchy fields of low number density liquid/NAT mixtures already during the early season, 15–30 December 2009, clearly some ten days before any ice clouds were detected, see Figure 4.7 (panels marked "Ice" and "Mix 1"). This strongly suggests that these early season NAT clouds were formed through a non-ice nucleation mechanism, as is also supported by temperatures staying about the maximum temperature departures below T_{NAT} as determined by the Goddard Earth Observing System Model, (GEOS-5, see uppermost panel in Figure 4.7).

In contrast to the localized *in situ* measurements and occasional remote sensing of unexplained NAX, the CALIPSO measurements provide unprecedented evidence that non-ice nucleation mechanisms have led to a patchy, but widespread and persistent occurrence of optically thin NAX clouds. These clouds were detected between 15 and 30 December 2009 in the cold pool centered over northern Greenland, with patches stretching from the Canadian Arctic Archipelago across Greenland into the Northern Atlantic towards Spitsbergen, covering a region of a few million square kilometers. After 30 December 2009, the scene changed: CALIPSO identified mountain wave-induced ice clouds over Greenland and Novaya Zemlya, which—as expected—nucleated widespread NAX particles throughout the vortex, including Mix 2-enh (or Type-Ia-enh). Later, the winter remained unusually cold and was one of only a few winters from the past 52 years with synoptic-scale regions of temperatures below the frost point.[72] Synoptic scale ice clouds, possibly nucleating heterogeneously on the pre-existing NAT particles and/or other heterogeneous nuclei, were the consequence.[72]

4.3.3 Summary on NAT Nucleation

The important question whether NAT nucleation and subsequent denitrification are governed predominantly by ice-assisted or ice-free processes (see Figure 4.5) remains unsettled. Based mainly on the interpretation of laboratory

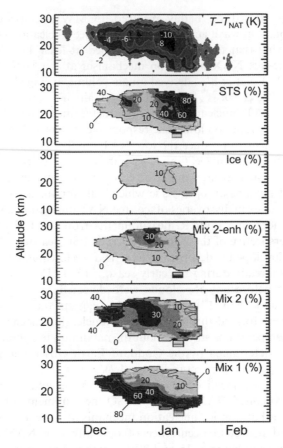

Figure 4.7 Relative proportion of PSCs observed by the CALIOP LiDAR onboard the CALIPSO satellite in the following composition classes: STS, ice, Mix 2-enh, Mix 2 and Mix 1 as a function of altitude and time (contour lines and shading in steps of 10%). The sequence (from top to bottom) corresponds to the chronological occurrence of PSC classes expected from the ice-induced NAX nucleation (right pathway in Figure 4.5), *i.e.* first STS, then homogeneous nucleation of ice, and finally heterogeneous nucleation of hydrates on the ice. For reference, the maximum temperature departure from T_{NAT} derived from GEOS-5 analyses is shown in uppermost panel. Adapted from Pitts *et al.*[72]

and field observations the following NAT nucleation mechanisms have been suggested, in chronological order:

(1) *Heterogeneous nucleation on SAT:* binary H_2SO_4-H_2O solution droplets freeze out sulfuric acid tetrahydrate (SAT), then NAT nucleates on SAT directly from the vapor;[73]

(2) *Homogeneous nucleation of STS:* the binary droplets initially remain liquid and take up large amounts of HNO_3 when $T \lesssim 193\,K$, and subsequently NAT nucleates in these ternary solutions;[28]

(3) *Heterogeneous nucleation on ice:* even the ternary droplets remain liquid, unless the temperature drops sufficiently below T_{ice}, so that water ice precipitates from the solutions and serves as nucleus for NAT;[24]

(4) *Homogeneous nucleation of glassy aerosols:* at $T \lesssim 194\,K$ the ternary liquid transforms into a glassy state, which crystallizes upon warming;[74]

(5) *Homogeneous nucleation of non-equilibrium aerosol:* rapid temperature fluctuations may lead to quasi-binary HNO_3-H_2O solutions with high HNO_3 concentrations, in which homogeneous nucleation of NAT becomes possible;[75]

(6) *Heterogeneous nucleation on solid inclusions:* solid particulates such as ash, soot, rocket exhaust particles or meteoritic debris may act as NAT nuclei, as postulated by Iraci *et al.*;[76]

(7) *Heterogeneous nucleation after preactivating SAT:* formation of NAT on preactivated (as opposed to pristine) SAT surfaces;[77]

(8) *Homogeneous nucleation interface-induced:* occurrence of NAD or NAT nucleation "pseudoheterogeneously" at the air-aqueous nitric acid solution interface of the droplet.[78]

Since these mechanisms have been suggested, the following new findings or considerations have been contributed to their verification or falsification:

ad (1) *Heterogeneous nucleation on SAT:* this mechanism is hampered by the absence of convincing observational evidence for the existence of SAT or other H_2SO_4 hydrates in the atmosphere. Although Rosen *et al.*[79] measured a lack of water uptake by background sulfate particles during cooling phases in the Antarctic winter and interpreted this as due to frozen solids rather than liquids, Dye *et al.*[23] found clear evidence for liquid H_2SO_4-H_2O droplets (Figure 4.3a). From laboratory experiments it is known for long that H_2SO_4 hydrates tend to supercool massively, even in bulk samples in the presence of an unspecified variety of heterogeneous nuclei,[80] and that barium sulfate is the only known nucleus. So why should these hydrates form in the stratosphere under apparently much cleaner and less suitable conditions?

ad (2) *Homogeneous nucleation of STS:* this possibility was basically eliminated by bulk experiments of Koop *et al.*,[24] demonstrating that HNO_3-H_2SO_4-H_2O with stratospheric composition could be supercooled to at least T_{ice}—$1\,K$.

ad (3) *Heterogeneous nucleation on ice:* this mechanism has been corroborated by many field observations of NAT wakes downstream of ice clouds.[57] It is the only mechanism for which a quantitative formulation for the NAT nucleation rate is available.[81] It is found from bulk phase experiments[24,46] that deposition nucleation of NAT on bare ice surfaces is the most likely nucleation mechanism.

ad (4) *Homogeneous nucleation of glassy aerosols:* This has been challenged by Koop *et al.*[46] showing that the rate of crystal growth is large enough that stratospheric aerosol droplets would freeze as soon as nucleation has occurred.

ad (5) *Homogeneous nucleation of non-equilibrium aerosol:* although still perfectly feasible and corroborated by laboratory experiments for pure binary HNO_3-H_2O bulk solutions[56] and aerosols,[51] this mechanism lacks support by field observations, as it remains unclear how often the required fast heating rates actually occur.

ad (6) *Heterogeneous nucleation on solid inclusions:* this mechanism has received support by measurements using condensation particle counters showing that inside the Arctic vortex more than half of the stratospheric aerosol particles contain non-volatile residues, considerably more than outside. Most likely this is due to a strongly increased fraction of meteoric material in the particles, which from the mesosphere is transported downward preferably into the polar vortex.[82]

ad (7) *Heterogeneous nucleation after preactivating SAT:* this serves to support mechanism (1); however, given the little backup for (1) from field and lab work, mechanism (7) remains just as uncertain.

ad (8) *Homogeneous nucleation interface-induced:* the rates of this mechanism, should it apply at all, were found to be too low by Knopf,[83] who argued that interface-induced nucleation would yield at most $6 \times 10^{-6}\,cm^{-3}$, too small by 2 orders of magnitude to produce the low NAT number densities in Type-Ia or Mix 1 PSCs (see also commentaries by Tabazadeh[84] and Knopf).[85]

In summary, only mechanism (3) is supported by both laboratory measurements and field observations. Conversely, mechanisms (2), (4), and (8) can be excluded. Mechanisms (1), (5), and (7) remain possible, but lack observational evidence. Finally, field measurements require mechanism (6) or will remain unexplained, although laboratory support for this mechanism is missing (or at best weak, giving the upper limits for heterogeneous nucleation established by Biermann *et al.*)[55] There are neither field studies nor laboratory studies specifically on the deposition mode of solid particles nucleation, which could be a potential pathway for mechanism (6).

4.4   Simulation of NAT-Rock Formation and Denitrification

Toon *et al.*[17] and Crutzen and Arnold[18] proposed the vertical redistribution of reactive nitrogen (NO_y) by large sedimenting HNO_3 containing particles (usually thought to be NAT) to explain the enhanced ozone loss in the Antarctic polar vortex. This mechanism has been coined "denitrification".[86] It was first observed in situ by Fahey *et al.*[87] using NO_y measurements onboard the ER-2 high-altitude research aircraft in the Antarctic and Arctic. In retrospect it is clear that the pronounced negative spikes (Figure 4.1, curve (c)) in the aerosol optical depth measured each year by SAM II[11] during the austral spring can be explained as massive denitrification of the Antarctic stratosphere. The active chlorine (Cl, ClO, Cl_2O_2) is less quickly deactivated in denitrified air masses, because they contain less NO_x ($= NO + NO_2$), and therefore the deactivation reaction $ClO + NO_2 + M \rightarrow ClONO_2 + M$ is slowed or suppressed. As a consequence, the period of chlorine-catalyzed ozone loss will be extended and more ozone may subsequently be destroyed, even after temperatures have increased above the chlorine activation threshold. Upon evaporation at lower altitudes, the sedimenting HNO_3-containing particles release HNO_3 back to the gas phase, causing an increase in NO_y. This mechanism has been termed "renitrification" and has been observed by *in situ* measurements and by remote sensing in the Arctic.[39,88]

4.4.1 Simple Fixed Grid Simulations

While denitrification in the Antarctic vortex is a strong, annually reoccurring phenomenon (Figure 4.1), Waibel *et al.*[39] suggested that denitrification in the northern hemisphere is at the threshold of being important for ozone depletion. This means that, if the Arctic winter stratosphere continued to cool in the future, denitrification might kick in and significantly enhance Arctic ozone loss, similar to the Antarctic. For determining the extent of de- and renitrification, the number density of nucleated NAX particles (n_{NAX}) is essential: too few NAX particles are inefficient, as they transport only little HNO_3 despite rapid sedimentation; too many NAX particles are inefficient as well, as they compete for HNO_3 uptake and cannot grow to large sizes, *i.e.* their sedimentation is slow and they are subject to evaporation in the next warm phase before much NO_y could be dislocated. Already Waibel *et al.*[39] suggested that number densities of NAT as low as $n_{NAT} \approx 5 \times 10^{-3}$ cm^{-3} were required for effective denitrification. The *in situ* measurements by Fahey *et al.*[58] of very large HNO_3-containing particles— most likely NAT—were the clue to understanding Arctic denitrification, showing $r \approx 10$ μm and even lower number densities, $n_{NAT} \approx 2.3 \times 10^{-4}$ cm^{-3}.

This called for more accurate numerical simulations of the vertical HNO_3 fluxes by dynamically calculating the sizes and locations of the NAT particles. They will grow as long as they fall through altitudes in which the HNO_3 partial pressure in the gas phase is larger than the HNO_3 vapor pressure over NAT. The latter is a function of the prevailing temperature and humidity.[20,21] Conversely, they will shrink and finally evaporate, when the HNO_3 partial pressure in the gas phase is lower than the HNO_3 vapor pressure over NAT. These processes have been calculated in a Lagrangian way by Davies *et al.*[89] and Grooß *et al.*,[88] who both calculated trajectories of single representative NAT particles that undergo growth/evaporation and gravitational settling, and thus derived the vertical transport of NO_y. The particles along their trajectories interact with the gas phase concerning HNO_3 uptake and release and concerning heterogeneous chemical reactions, such as $ClONO_2 + Cl^- \rightarrow Cl_2 + NO_3^-$, and the entire gas phase chemistry is part of a full-fledged stratospheric chemistry model.

4.4.2   Lagrangian Modeling with Constant Volume or Ice-induced NAT Nucleation

The calculated vertical flux of HNO_3 through sedimentation may be expected to sensitively depend on the details of the nucleation rate of these NAT particles. Only one observation has been published so far from which a nucleation rate can be directly derived,[71] because the temperature history of the measured NAT particles could be accurately determined. The air masses were only for about one day below T_{NAT}. The estimated NAT nucleation rate from this study is $J_{NAX} = n_{NAX} / 24\,h \approx 10^{-5}\,cm^{-3}$ air h^{-1}, and was assumed to apply wherever temperatures are lower than the NAT equilibrium temperature ($T < T_{NAT}$). This rate is higher by about 5 orders of magnitude compared to the upper limit of homogeneous nucleation of NAT in STS measured by Koop *et al.*,[24] clearly pointing to the heterogeneous nature of the process at work.

As outlined above, different hypotheses exist for the mechanisms behind these low nucleation rates. Even though the nucleation process is not identified, the estimated nucleation rates have been used to simulate the vertical redistribution of HNO_3. The simplest assumption is to have a constant NAT nucleation rate wherever $T < T_{NAT}$. The studies by Davies *et al.*[89] and Grooß *et al.*[88] are based on this assumption.

Figure 4.8 shows profiles of NO_y redistribution measured (crosses) in January 2003 and simulated by CLaMS (Chemical Lagrangian Model of the Stratosphere; lines and gray circles). The vertical axis is roughly linear in geometric altitude with 350 K potential temperature corresponding to $\sim 14\,km$, and 600 K to $\sim 24\,km$. The model uses a constant nucleation rate of $7.8 \times 10^{-6}\,cm^{-3}$ air h^{-1}. The thick gray line corresponds to the modeled vortex average and the dotted lines indicate the standard deviation ($\pm 1\ \sigma$, revealing the large variability caused by the sedimentation process, despite the generally widespread nucleation wherever $T < T_{NAT}$. Crosses show measurements onboard the Geophysica high-altitude aircraft. Results from the CLaMS simulations for the location of these observations are shown as gray circles. This figure shows that the vertical redistribution of NO_y can be generally well described by the model, although the model can miss specific measurement points by more than 50%.

In a similar study, Mann *et al.*[61] performed simulations with the DLAPSE/SLIMCAT model and investigated the sensitivity with respect to the nucleation mechanism. If the mother cloud mechanism for nucleating NAT particles[59] is used instead of a tuned constant nucleation rate at all locations below T_{NAT}, still about 80% of the denitrification of the reference run are reproduced (compare Figure 4.9). From the available limited number of NO_y measurements, the natural variability and the overall uncertainties, a clear conclusion about the NAT nucleation mechanism can presently not be drawn from these

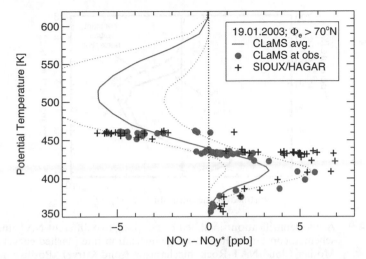

Figure 4.8 Denitrification derived from measurements (black crosses) and model simulations (lines and gray circles). Black crosses: denitrification derived from HAGAR and SIOUX *in situ* measurements in the Arctic onboard the Geophysica aircraft on 19 January 2003. NO_y^* denotes the NO_y expected if no denitrification had occurred (derived from the Geophysica N_2O measurements). Gray circles: corresponding values at the measurement location simulated by CLaMS (Chemical Lagrangian Model of the Stratosphere, shown as the simulated change in NO_y due to vertical redistribution of HNO_3 by sedimenting NAT particles). Solid grey line: modeled denitrification on 19 January 2003 averaged across the polar vortex core (equivalent latitude > 70°N). Dotted lines: standard deviation ($\pm 1\ \sigma$). Adapted from Grooß *et al.*[88]

simulations. Most likely a combination of an ice-free mechanism leading to very low number densities, and an ice-assisted "mother-cloud" mechanism with high NAT number densities downstream of wave ice clouds are at work.

In addition to these Lagrangian simulations there are also Eulerian studies that simulate the vertical redistribution of NO_y. Daerden *et al.*[90] use a detailed microphysical model coupled with a global 3D Chemical Transport Model for the simulation of the Antarctic winter 2003. Particles are divided into 36 size bins and the sedimentation of the particles are calculated in an Eulerian way. They used NAT nucleation rates adopted from Tabazadeh *et al.*,[78] however after reducing the rates by a factor of 100, which highlights again the large uncertainties in NAT nucleation.

4.5 Heterogeneous Reaction Rates on PSCs and Cold Sulfate Aerosols

Heterogeneous reactions on liquid and solid particles are of importance for stratospheric ozone chemistry, since they are responsible for chlorine

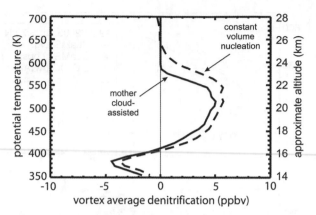

Figure 4.9   Arctic denitrification simulated by adopting two different NAT nucleation schemes: constant volume average nucleation rate (dashed curve) and the Mother-Cloud/NAT-Rock mechanism (solid curve). Positive numbers indicate denitrification, negative numbers renitrification, *i.e.* reductions and increases in NO_y, respectively. Simulations are for 20 January 2000, performed by the Lagrangian particle sedimentation (DLAPSE) model coupled with the 3D chemical transport model SLIMCAT. Dashed curve: simulations with a constant NAT nucleation rate (per volume of air) of $J_{NAT} = 8.1 \times 10^{-10}$ particles $^{-3}s^{-1}$, which has been chosen as best fit of the NAT number densities measured during the SOLVE/THESEO campaign (in the period January to March 2000). Solid curve: NAT nucleation on lee wave ice clouds leading to wave-induced Type-1a-enh NAT clouds, which then serve as "mother clouds" for the release of low concentrations of sedimenting NAT particles. Adapted from Mann *et al.*[61]

activation, *i.e.* the conversion of the chlorine reservoir species HCl and $ClONO_2$ to active chlorine (Cl, ClO, Cl_2O_2), which drives the ozone destroying catalytic cycles (Chapter 3). Figure 4.10 summarizes the principle of the interaction between PSCs and cold aerosols particles (shown as grey cloud) and the ozone loss cycles (left). For simplicity only the ClO dimer cycle is depicted for chlorine-catalyzed ozone loss (compare Chapter 3, Section 3.4 for a complete list of ozone loss cycles). In the absence of heterogeneous chlorine activation, in the polar summer stratosphere, the major ozone loss cycle is the NO_x-cycle. However, in the polar winter stratosphere, the NO_x-cycle is of very little importance, as the heterogeneous reaction $N_2O_5 + H_2O$ has converted the major fraction of NO_x into HNO_3 during polar night, and the activation reactions in Figure 4.10 add to this effect. Table 4.2 lists the heterogeneous reactions that are most important for stratospheric ozone chemistry.

The heterogeneous reactions are either surface reactions on the surface of the individual particles or reactions in the bulk phase of a liquid particle. For the determination of the heterogeneous reaction rate, the phase, the composition and the size of the particles must be known. Information about these properties is mostly limited due to uncertainties in particle nucleation or phase transitions, but also a lack of the exact knowledge of stratospheric temperatures. The

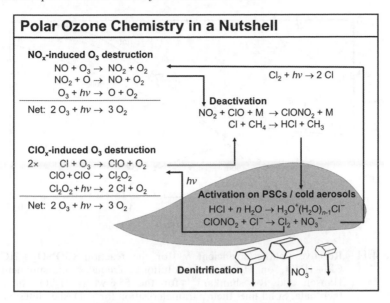

Figure 4.10 Sketch of selected chemical ozone loss cycles interacting with each other *via* homogeneous deactivation reactions and heterogeneous activation reactions on surfaces or in volumes of cold aerosols or PSCs (light gray shading). As examples, NO_x- and ClO_x-catalyzed cycles are shown on the left. Due to deactivation reactions (*e.g.*, $NO_2 + ClO$ and $Cl + CH_4$), the simultaneous presence of NO_x and ClO_x catalysts can lead to slower ozone destruction than if only one catalyst was present. Rapid activation, *e.g. via* $ClONO_2 + Cl^-$, on cold aerosols or PSCs leads to restoration of active chlorine (ClO_x), whereas nitrogen remains deactivated as nitrate (NO_3^-). Formation of large NAT particles leads to gravitational removal of NO_y, which inhibits chlorine deactivation and amplifies chemical ozone destruction.

Table 4.2 Important heterogeneous reactions in the polar stratosphere. These reactions are known to occur rapidly both on liquid (cold binary and ternary) aerosol particles and on solid (NAT and ice) particles.

R1	$ClONO_2 + HCl \rightarrow Cl_2 + HNO_3$
R2	$ClONO_2 + H_2O \rightarrow HOCl + HNO_3$
R3	$HOCl + HCl \rightarrow Cl_2 + H_2O$
R4	$N_2O_5 + H_2O \rightarrow 2\,HNO_3$
R5	$BrONO_2 + H_2O \rightarrow HOBr + HNO_3$

heterogeneous reaction rates are determined by the rate of collision of gas molecules with the particle and the uptake coefficient γ that describes the probability of a reaction taking place after that collision (*i.e.*, $\gamma = 0$ for no reaction upon collision, $\gamma = 1$ when each collision results in a reaction). The uptake coefficient γ is determined from laboratory measurements in

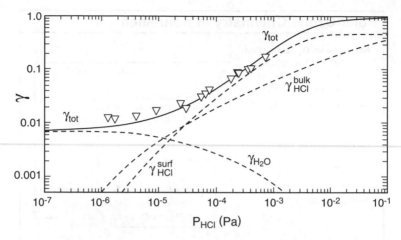

Figure 4.11   Reactive uptake coefficient γ, for the reaction $ClONO_2 + HCl \rightarrow$ $Cl_2 + HNO_3$ on H_2SO_4-H_2O solutions. Triangles: measurements by Hanson and Ravishankara[93] for the 55.6-wt-% H_2SO_4 subset of their data. Solid line: their parameterization for γ. Dashed lines: γ_{H2O}—fraction of $ClONO_2$ that reacts with water and produces HOCl; γ_{HCl}^{bulk}—fraction of $ClONO_2$ reacting with HCl in the bulk of the solution; γ_{HCl}^{surf}—fraction reacting with HCl on the surface. From Peter.[21]

combination with theory. For liquid phase reactions, the uptake of a given gas into the bulk phase is determined by the solubility (Henry's law), liquid phase diffusivity and reactivity of the reacting species. Hanson *et al.*[91] derived the reactive uptake coefficients for the reaction $ClONO_2 + HCl \rightarrow Cl_2 + HNO_3$ on liquid binary aerosol using a theoretical framework.[92] The interplay between bulk phase and surface reactions for the reaction of $ClONO_2 + HCl$ is displayed in Figure 4.11.[21,93] The figure is for sulfuric acid/water particles with 55.6 wt% sulfuric acid corresponding to binary liquid aerosol in the atmosphere at 50 hPa and 197 K showing that the reactive uptake coefficient γ as a function of HCl partial pressure can be represented by the sum of different surface and bulk phase reactions.[93] For HCl partial pressures below about 10^{-5} Pa, the uptake coefficient is dominated by the reaction of $ClONO_2$ with H_2O.

A more recent parameterization by Shi *et al.*[94] for the reaction of $ClONO_2$ with H_2O and HCl and HOCl with HCl in sulfuric acid solutions takes into account the competition between the reactions of $ClONO_2$ with HCl or H_2O on liquid aerosol. It combines various laboratory measurements for the heterogeneous reaction rates based on a detailed modeling of liquid phase solubility, diffusion and reaction kinetics. Changes of this updated parameterization compared with the previous version by Hanson and Ravishankara[93] (Figure 4.11) are most significant at relatively high temperatures above 205 K, whereas for the temperature range 195–202 K, for the reaction of $ClONO_2$ with HCl, the new parameterization agrees rather closely with the formulation of Hanson and Ravishankara.[93] The parameterization by Shi *et al.*[94] is currently recommended for the use in models.[95]

For reactions on solid particles, laboratory studies are complicated by the preparation of the substrate.[21,25,96,97] From this circumstance uncertainties arise. In particular, it was pointed out that the derived reactive uptake coefficient for the key reaction for chlorine activation, $HCl + ClONO_2$, on NAT particles differs by orders of magnitude between the measurements by Hanson and Ravishankara[26,97] on the one hand and Abbatt and Molina[27] and Zhang *et al.*[98] on the other hand.[99,100] Until now, this discrepancy has not yet been resolved. Examples can be found where this difference has significant impact on simulations of polar ozone.[100]

However, despite the large uncertainties in particle nucleation, particle phase and size distribution, as well as the uncertainties in heterogeneous reaction rates on the various particle types, the impact of all these uncertainties are often limited concerning the resulting chlorine activation and ozone loss. This is, for example, the case for circumstances when chlorine activation is saturated in the sense that the rate of chlorine activation is determined by the supply of reaction partners for HCl (the most abundant chlorine reservoir). Such conditions occur, for example, at the early stages of chlorine activation in polar winter, when temperatures first drop below the threshold for rapid heterogeneous reactions. On the other hand, in late winter and early spring, when due to increasing solar elevation, the rate of production of $ClONO_2$ increases and thus competes with the rate of heterogeneous chlorine activation, an accurate knowledge of the heterogeneous reaction rates is essential for a quantitative description of polar ozone chemistry.[101]

The exact simulation of the chlorine chemistry is difficult for temperature conditions near the onset of chlorine activation. Comparisons with observations of chlorine compounds by the satellite instruments MLS and ACE-FTS with simulations performed using the model SLIMCAT show that the degree of chlorine activation is simulated to be too large and with too early an onset.[102] However, this over-estimated chlorine activation might be due to the fact that in this model NAT PSCs are assumed to be formed directly at the threshold temperature T_{NAT}. As described in the context of the Ny-Ålesund PSC climatology (Section 4.3), this will lead to a gross overestimation of surface area. This illustrates the difficulties especially near the threshold temperature. Furthermore, due to the strong temperature dependence of chlorine activation rates, a small deviation in temperature near the threshold temperature may result in very different chlorine activation.

4.6 Which Type of Particles—Solid NAT *vs.* Liquid STS—Control Chlorine Activation?

An important prerequisite for chlorine-catalyzed ozone loss is the heterogeneous chlorine activation on particles. Historically, in the late 1980s and early 1990s, as an explanation for the Antarctic ozone hole the picture had emerged that chlorine activation occurs on the surfaces of solid PSCs consisting of either NAT or ice.[103] Soon it became clear that also heterogeneous reactions on the surfaces of the liquid aerosol particles (that consist of binary sulfate or ternary H_2SO_4-H_2O-HNO_3 solution) play an important role.[29,32,103] As discussed in Section 4.2, the particle volume increase with decreasing temperatures observed in the Arctic stratosphere in 1989 could often be better explained by HNO_3 uptake into the liquid aerosol[21] (compare to Figure 4.3). Under such conditions, liquid particles can support rapid heterogeneous chlorine activation, before NAT particles form.

Due to the strong temperature dependence of heterogeneous reaction rates, the chlorine activation reactions (R1–R3 in Table 4.2) have the character of a threshold process occurring below a certain threshold temperature but hardly above, even if no phase transition is involved. In many contexts the threshold temperature T_{NAT} has been used as a proxy for the onset of chlorine activation, *e.g.* in studies that link ozone depletion to the areas where temperatures are below T_{NAT}.[37] As a first approximation, the use of T_{NAT} yields a reasonable proxy for chlorine activation, since the threshold temperature T_{NAT} is close to the temperature range, where chlorine activation on liquid aerosols starts to become important.

Figure 4.12 shows the chlorine activation rate as inverse time on liquid and on NAT particles for the two most important heterogeneous chlorine activation reactions, R1 and R2 in Table 4.2, as a function of temperature. The reaction $ClONO_2 + HCl$ often dominates the chlorine activation, except for temperatures above T_{dew}, when the solubility of HCl in the aerosol becomes small (Figures 4.3b and 11). The inverse of the rates shown in Figure 4.11 are the typical times needed for activation through the individual reactions. Clearly, the rate of chlorine activation increases strongly with decreasing temperatures both for liquid and NAT particles. For the reaction of $ClONO_2 + HCl$ on NAT, the results of the two different parameterization discussed above (Hanson and Ravishankara *vs.* Abbatt and Molina) are plotted; the large difference between the two parameterizations is obvious and remains unresolved.[99] Furthermore, the chlorine activation rate on NAT is shown in Figure 4.12 for two assumed NAT number densities, $1 \, cm^{-3}$ and $10^{-4} \, cm^{-3}$, reflecting the Type-Ia-enh and tenuous Type-Ia conditions, respectively. This illustrates that only in dense "mother clouds" downstream of ice clouds could NAT compete with STS in terms of chlorine activation; however, under such conditions chlorine activation will be complete anyway, because of the ice clouds. Despite the differences between the parameterizations for NAT reactivity, it is important to note that NAT is unlikely to form at T_{NAT} and that supercooling by about 3 K below T_{NAT} is necessary for NAT formation.[23,47,104] Therefore, heterogeneous reaction rates on liquid binary aerosol particles reach substantial values before NAT particles form,[35,36] see Figure 4.12. Thus, cold binary liquid particles are expected to be solely responsible for the onset of chlorine activation in the early winter.[35,36] Despite the remaining discrepancies in the different parameterizations, it can be concluded from the temperature dependence shown in Figure 4.12 that chlorine activation is driven not by solid PSC formation, but rather already by the onset of activation on the binary cold sulfate aerosol particles. Therefore, the unresolved discrepancies between the

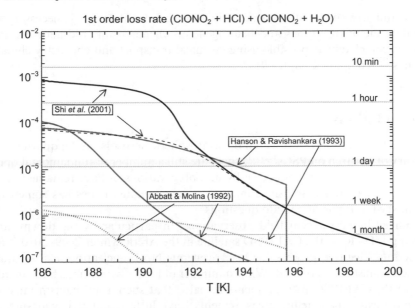

Figure 4.12 Heterogeneous chlorine activation rate as function of temperature on STS and NAT PSCs for the sum of the two most important chlorine activation reactions, $HCl + ClONO_2$ and $ClONO_2 + H_2O$, for typical conditions (55 hPa, 5 ppmv H_2O, 10 ppbv HNO_3, liquid aerosol number density $10\,cm^{-3}$). Thick black line: the parameterizations for reactions on liquid ternary aerosol from Shi *et al.*[94] Thin dashed line: chlorine activation if HNO_3 were not taken up by the liquid (binary sulfate aerosol, H_2O uptake only). Solid gray lines: reactions on NAT with number densities $1\,cm^{-3}$ for two parameterizations (Hanson and Ravishankara,[97] Abbatt and Molina).[27] Dashed gray lines: same, but for $10^{-4}\,cm^{-3}$ NAT particles. Values are first order loss rates divided by the $ClONO_2$ concentration. Corresponding times are marked by horizontal lines. Courtesy of Tobias Wegner (manuscript for *Atmos. Chem. Phys.*, in preparation, 2011).

parameterizations for this reaction on NAT are of minor importance. This argument of cold binary liquid aerosol being responsible for chlorine activation was put forward by Drdla[35] and Drdla and Müller,[36] who also provided a fit formula for a threshold temperature T_{ACl} below which the chlorine activation rate is faster than $0.1\,day^{-1}$, which should be used in place of T_{NAT}. For current atmospheric conditions the thresholds for NAT (T_{NAT}) and cold liquid aerosol (T_{ACl}) are both around $195\,K$.[36] However, this may change in a future atmosphere.

On the basis of the concept of T_{ACl}, projections about changing ozone depletion were made for currently debated geoengineering ideas, which might be applied to cool the planet,[105] and for H_2O increase through hydrogen emissions by a possible future hydrogen-based economy.[106] This shows that there are applications in which a correct threshold temperature for chlorine activation is required to obtain accurate results for the expected ozone loss.

For a full assessment of chlorine activation and ozone loss it is necessary to simulate the microphysics and heterogeneous chemistry of cold aerosols and PSC as completely as possible using chemical transport and chemistry climate models[107] (compare Chapter 9).

4.7   Outlook

Despite half a century of research on stratospheric aerosols[5] and a quarter of a century of research on PSCs[15,17,18] there are still a number of fundamental open questions, which are either driven by new observations or have remained open during the past ten years, since interest in ozone-related issues has somewhat diminished in favor of climate questions.

Spurred by very recent field observations, in particular by the downward-looking LiDAR on the CALIPSO satellite in the Arctic winter 2009–2010,[72] an unidentified ice-free process for heterogeneous NAT nucleation, *e.g.* on dust particles, must be postulated. While a number of local observations had already before the CALIPSO measurements revealed that such a mechanism must in principle exist, the satellite measurements leave little doubt that widespread, almost vortex-filling ice-free nucleation can occur. The NAT nucleus is unknown. While heterogeneous NAT nucleation on water ice is accepted with respect to both, laboratory experiments and field observations,[21] the CALIPSO measurements are puzzling and reverse our previous understanding, because there is currently no laboratory evidence that could explain this finding. A nucleation mechanism on meteoritic dust has received support by measurements using condensation particle counters, which show that inside the Arctic vortex more than half of the stratospheric aerosol particles contain non-volatile residues, considerably more than outside.[82] However, previous laboratory studies—though possibly not entirely conclusive—would rather suggest that meteoritic dust in immersion mode is not effective enough to cause this nucleation.[55] Meteoritic dust has not been examined in deposition mode. This uncertainty seriously compromises our ability to predict future denitrification and polar ozone loss. Therefore, laboratory studies on meteoritic or volcanic dust should be undertaken.

Furthermore, there are open issues concerning chlorine activation on solid and liquid PSC and cold stratospheric aerosol particles, which are known for more than a decade.[99] Activation reactions on nitric acid hydrate particles measured by two laboratories[26,27,97,98] remain uncertain by almost 2 orders of magnitude. However, while this uncertainty remains physico-chemically unsatisfactory, it appears not to curtail our ability to understand and predict chlorine activation and ozone loss in the polar stratosphere. This is because liquid particles are usually the dominant chlorine processors, reducing the importance of accurate descriptions of heterogeneous chemistry on solid particles.[32–36] Heterogeneous and multi-phase chemistry in the liquid phase is much less uncertain. Although it would be desirable to clarify the activation processes on NAT particles, there is no immediate need to undertake these complex laboratory measurements. Moreover, it appears that under many circumstances the uptake of HNO_3 to form STS clouds is not even a requirement for rapid chlorine activation,[36] but already the heterogeneous chemistry on sufficiently cold H_2SO_4-H_2O background aerosol ensures a sufficiently high solubility of HCl and production of active chlorine via $Cl^- + ClONO_2 \rightarrow Cl_2 + NO_3^-$.

Finally, it should be noted that in contrast to the Lagrangian modeling studies mentioned in Section 4.4, many of the global scale models still rely on a simple switch-on/switch-off activation below/above T_{NAT},[107] which is not state of the art. Moreover, the sedimentation routines implemented in many of the large scale models might be not sufficient, in particular if the condensation of HNO_3 onto NAT particles is assumed to be fast and thermodynamic equilibrium between the NAT particles and the HNO_3 in the gas phase is applied. In reality, the depletion of the gas phase HNO_3 by NAT particles in low number densities is very slow, so that the adopted parameterizations may lead to too large NAT particles, which then sediment and denitrify too rapidly. This does not allow to properly capture changes in denitrification in a changing future atmosphere. It would be useful to improve this state through a concerted activity.

Acknowledgements

We thank Tobias Wegner for fruitful discussions and for providing Figure 4.12. We further acknowledge Mike Pitts and Lamont Poole, who provided CALIPSO data including the derived PSC composition in Figure 4.2. We also thank Christopher Hoyle, Beiping Luo, Ines Engel, Ulrich Krieger, Simone Tilmes and Rolf Müller for fruitful discussions and for proofreading this chapter. Part of the scientific content of this chapter was supported by the project RECONCILE that is funded under the European Commission Seventh Framework Programme (FP7) under the Grant number RECONCILE-226365-FP7-ENV-2008-1.

References

1. H. Dieterichs, *Pure Appl. Geophys.*, 1950, **16**, 128–132.
2. C. Piazzi-Smythe, *Nature*, 1884, **51**, 148.
3. E. K. Bigg, *Nature*, 1956, **177**, 77–79.
4. R. Jaenicke, International Aerosol Research Assembly, "Aerosol Pioneers: Christian Junge, 2 July 1912–18 June 1996", ed. G. J. Sem, 2009, http://www.iara.org/newsfolder/pioneers/6AerosolPioneerEditedAugJunge.pdf (downloaded on 29 June 2011).
5. C. E. Junge, M. J. E. and C. C. W., *Science*, 1961, **133**, 1478–1479.
6. C. E. Junge, C. W. Changnon and J. E. Manson, *J. Meteorol*, 1961, **18**, 81–108.
7. D. J. Hofmann, J. M. Rosen, T. J. Pepin and R. G. Pinnick, *J. Atmos. Sci.*, 1975, **32**, 1446–1456.

8. T. Deshler, M. E. Hervig, D. J. Hofman and J. B. Liley, *J. Geophys. Res.*, 2003, **108**, 4167, DOI: 10.1029/2002JD002514.

9. *SPARC Assessment of Stratospheric Aerosol Properties*, ed. L. Thomason and T. Peter, WCRP-124, WMO/TD-No.1295, 2006.

10. M. P. McCormick, P. Hamill, T. J. Pepin, W. P. Chu, T. J. Swissler and L. R. McMaster, *Bull. Am. Meteorol. Soc.*, 1979, **60**, 1038–1046.

11. M. P. McCormick, P. H. Wang and M. C. Pitts, *Adv. Space Res.*, 1993, **13**, 7–29.

12. J. C. Farman, B. G. Gardiner and J. D. Shanklin, *Nature*, 1985, **315**, 207–210.

13. R. S. Stolarski, A. J. Krueger, M. R. Schoeberl, R. D. McPeters, P. A. Newman and J. C. Alpert, *Nature*, 1986, **322**, 808–811.

14. M. R. Schoeberl, A. J. Krueger and P. A. Newman, *Geophys. Res. Lett.*, 1986, **13**, 1217–1220.

15. S. Solomon, R. R. Garcia, F. S. Rowland and D. J. Wuebbles, *Nature*, 1986, **321**, 755–758.

16. S. Solomon, personal communication, 1997.

17. O. B. Toon, P. Hamill, R. P. Turco and J. Pinto, *Geophys. Res. Lett.*, 1986, **13**, 1284–1287.

18. P. J. Crutzen and F. Arnold, *Nature*, 1986, **342**, 651–655.

19. L. R. Poole and M. P. McCormick, *J. Geophys. Res.*, 1988, **93**, 8423–8430.

20. D. R. Hanson and K. Mauersberger, *Geophys. Res. Lett.*, 1988, **15**, 855–858, DOI: 10.1029/88GL00209.

21. T. Peter, *Ann. Rev. Phys. Chem.*, 1997, **48**, 785–822.

22. K. S. Carslaw, B. P. Luo, S. L. Clegg, T. Peter, P. Brimblecombe and P. J. Crutzen, *Geophys. Res. Lett.*, 1994, **21**, 2479–2482, DOI: 10.1029/94GL02799.

23. J. E. Dye, D. Baumgardner, B. W. Gandrud, S. R. Kawa, K. K. Kelly, M. Loewenstein, G. V. Ferry, K. R. Chan and B. L. Gary, *J. Geophys. Res.*, 1992, **97**, 8015–8034.

24. T. Koop, U. M. Biermann, W. Raber, B. P. Luo, P. J. Crutzen and T. Peter, *Geophys. Res. Lett.*, 1995, **22**, 917–920, DOI: 10.1029/95GL00814.

25. D. R. Hanson and A. R. Ravishankara, *J. Geophys. Res.*, 1991, **96**, 17307–17314.

26. D. R. Hanson and A. R. Ravishankara, *J. Phys. Chem.*, 1992, **96**, 2682–2691.

27. J. P. D. Abbatt and M. J. Molina, *J. Phys. Chem.*, 1992, **96**, 7674–7679.

28. M. J. Molina, R. Zhang, P. J. Wooldridge, J. R. McMahon, J. E. Kim, H. Y. Chang and K. D. Beyer, *Science*, 1993, **261**, 1418–1423.

29. A. R. Ravishankara and D. R. Hanson, *J. Geophys. Res.*, 1996, **101**, 3885–3890.

30. M. J. Molina, L. T. Molina and C. E. Kolb, *Ann. Rev. Phys. Chem.*, 1996, **47**, 327–367, DOI: 10.1146/annurev.physchem.47.1.327.

31. T. Huthwelker, M. Ammann and T. Peter, *Chemical Reviews*, 2006, **106**, 1375–1444.

32. R. W. Portmann, S. Solomon, R. R. Garcia, L. W. Thomason, L. R. Poole and M. P. McCormick, *J. Geophys. Res.*, 1996, **101**, 22991–23006.

33. F. Lefevre, F. Figarol, K. S. Carslaw and T. Peter, *Geophys. Res. Lett.*, 1998, **25**, 2425–2428.
34. M. Dameris, T. Peter, U. Schmidt and R. Zellner, *Chemie in unserer Zeit*, 2007, **41**, 152–168, DOI: 10.1002jciuz.200700418.
35. K. Drdla, *Eos Trans. AGU*, 2005, **86**, Fall Meet. Suppl., Abstract A31D-03.
36. K. Drdla and R. Müller, *Atmos. Chem. Phys. Discuss.*, 2010, **10**, 28687–28720.
37. M. Rex, R. J. Salawitch, P. von der Gathen, N. R. P. Harris, M. P. Chipperfield and B. Naujokat, *Geophys. Res. Lett.*, 2004, **31**, L04116, DOI: 10.1029/2003GL018844.
38. M. Rex, R. J. Salawitch, H. Deckelmann, P. von der Gathen, N. R. P. Harris, M. P. Chipperfield, B. Naujokat, E. Reimer, M. Allaart, S. B. Andersen, R. Bevilacqua, G. O. Braathen, H. Claude, J. Davies, H. De Backer, H. Dier, V. Dorokov, H. Fast, M. Gerding, S. Godin-Beekmann, K. Hoppel, B. Johnson, E. Kyrö, Z. Litynska, D. Moore, H. Nakane, M. C. Parrondo, A. D. Risley Jr., P. Skrivankova, R. Stübi, P. Viatte, V. Yushkov and C. Zerefos, *Geophys. Res. Lett.*, 2006, **33**, L23808, DOI: 10.1029/2006GL026731.
39. A. E. Waibel, T. Peter, K. S. Carslaw, H. Oelhaf, G. Wetzel, P. J. Crutzen, U. Pöschl, A. Tsias, E. Reimer and H. Fischer, *Science*, 1999, **283**, 2064–2069.
40. P. Hitchcock, T. G. Shepherd and C. McLandress, *Atmos. Chem. Phys.*, 2009, **9**, 483–495.
41. C. Voigt, J. Schreiner, A. Kohlmann, P. Zink, K. Mauersberger, N. Larsen, T. Deshler, C. Kröger, J. Rosen, A. Adriani, F. Cairo, G. D. Donfrancesco, M. Viterbini, J. Ovarlez, H. Ovarlez, C. David and A. Dörnbrack, *Science*, 2000, **290**, 1756–1758.
42. O. B. Toon, E. V. Browell, S. Kinne and J. Jordan, *Geophys. Res. Lett.*, 1990, **17**, 393–396.
43. E. V. Browell, C. F. Butler, S. Ismail, P. A. Robinette, A. F. Carter, N. S. Higdon, O. B. Toon, M. Schoeberl and A. F. Tuck, *Geophys. Res. Lett.*, 1990, **17**, 385–388, DOI: 1029/90GL00333.
44. J. Biele, A. Tsias, B. P. Luo, K. S. Carslaw, R. Neuber, G. Beyerle and T. Peter, *J. Geophys. Res.*, 2001, **106**, 22991–23007.
45. A. Tsias, M. Wirth, K. S. Carslaw, J. Biele, H. Mehrtens, J. Reichardt, C. Wedekind, V. Weiss, W. Renger, R. Neuber, U. von Zahn, B. Stein, V. Santacesaria, L. Stefanutti, F. Fierli, J. Bacmeister and T. Peter, *J. Geophys. Res.*, 1999, **104**, 23961–23969.
46. T. Koop, B. P. Luo, U. M. Biermann, C. P. J. and T. Peter, *J. Phys. Chem. A*, 1997, **101**, 1117–1133.
47. M. C. Pitts, L. R. Poole and L. W. Thomason, *Atmos. Chem. Phys.*, 2009, **9**, 7577–7589.
48. D. M. Winker, W. H. Hunt and M. McGill, *Geophys. Res. Lett.*, 2007, **34**, L19803, DOI: 10.1029/2007GL030135.
49. E. R. Nash, P. A. Newman, J. E. Rosenfield and M. R. Schoeberl, *J. Geophys. Res.*, 1996, **101**, 9471–9478.

50. M. Höpfner, B. P. Luo, P. Massoli, F. Cairo, R. Spang, M. Snels, G. Di Donfrancesco, G. Stiller, T. von Clarmann, H. Fischer and U. Biermann, *Atmos. Chem. Phys.*, 2006, **6**, 1201–1219.

51. O. Stetzer, O. Moehler, R. Wagner, S. Benz, H. Saathoff, H. Bunz and O. Indris, *Atmos. Chem. Phys.*, 2006, **6**, 3023–3033.

52. Y. Kim, W. Choi, K. M. Lee, J. H. Park, S. T. Massie, Y. Sasano, H. Nakajima and T. Yokota, *J. Geophys. Res.*, 2006, **111**, D13S90, DOI: 10.1029/2005JD006445.

53. D. Lowe and A. R. MacKenzie, *J. Atmos. Solar Terr. Phys.*, 2008, **70**, 13–40, DOI: 10.1016/j.jastp.2007.09.011.

54. A. R. MacKenzie, M. Kulmala, A. Laaksonen and T. Vesala, *J. Geophys. Res.*, 1995, **100**, 11, 275–11, 288.

55. U. M. Biermann, T. Presper, T. Koop, J. Mossinger, P. J. Crutzen and T. Peter, *Geophys. Res. Lett.*, 1996, **23**, 1693–1696.

56. T. Koop, K. S. Carslaw and T. Peter, *Geophys. Res. Lett.*, 1997, **24**, 2199–2202, DOI: 10.1029797GL02148.

57. K. S. Carslaw, M. Wirth, A. Tsias, B. P. Luo, A. Dörnbrack, M. Leutbecher, H. Volkert, W. Renger, J. T. Bacmeister, E. Reimer and T. Peter, *Nature*, 1998, **391**, 675–678.

58. D. W. Fahey, R. S. Gao, K. S. Carslaw, J. Kettleborough, P. J. Popp, M. J. Northway, J. C. Holecek, S. C. Ciciora, R. J. McLaughlin, T. L. Thompson, R. H. Winkler, D. G. Baumgardner, B. Gandrud, P. O. Wennberg, S. Dhaniyala, K. McKinley, T. Peter, R. J. Salawitch, T. P. Bui, J. W. Elkins, C. R.Webster, E. L. Atlas, H. Jost, J. C. Wilson, R. L. Herman, A. Kleinböhl and M. von König, *Science*, 2001, **291**, 1026–1031.

59. S. Fueglistaler, B. P. Luo, H. Wernli, C. Voigt, M. Müller, R. Neuber, C. A. Hostetler, L. Poole, H. Flentje, D. W. Fahey, M. J. Northway and T. Peter, *Geophys. Res. Lett.*, 2002, **29**, 1610, DOI: 10.1029/2001GL014548.

60. K. S. Carslaw, J. A. Kettleborough, M. J. Northway, S. Davies, R. Gao, D. W. Fahey, D. G. Baumgardner, M. P. Chipperfield and A. Kleinböhl, *J. Geophys. Res.*, 2002, **107**, 8300, DOI: 10.1029/2001JD000467.

61. G. W. Mann, K. S. Carslaw, M. P. Chipperfield, S. Davies and S. D. Eckermann, *J. Geophys. Res.*, 2005, **110**, D08202, DOI: 10.1029/2004JD005271.

62. D. Cariolle, S. Muller, F. Cayla and M. McCormick, *J. Geophys. Res.*, 1989, **94**, 11233–11240.

63. M. Höpfner, N. Larsen, R. Spang, B. P. Luo, J. Ma, S. H. Svendsen, S. D. Eckermann, B. Knudsen, P. Massoli, F. Cairo, G. Stiller, T. Von Clarmann and H. Fischer, *Atmos. Chem. Phys.*, 2006, **6**, 1221–1230.

64. R. Spang and J. Remedios, *Geophys. Res. Lett.*, 2003, **30**, 1875, DOI: 10.1029/2003GL017231.

65. S. D. Eckermann, L. Hoffmann, M. Höpfner, D. L. Wu and M. J. Alexander, *Geophys. Res. Lett.*, 2009, **36**, L02807, DOI: 10.1029/2008GL036629.

66. K. L. Pagan, A. Tabazadeh, K. Drdla, M. E. Hervig, S. Eckermann, E. Browell, M. Legg and P. Foschi, *J. Geophys. Res.*, 2004, **109**, D04312, DOI: 10.1029/2003JD003846.

67. N. Larsen, B. M. Knudsen, S. H. Svendsen, T. Deshler, J. M. Rosen, R. Kivi, C. Weisser, J. Schreiner, K. Mauersberger, F. Cairo, J. Ovarlez, H. Oelhaf and R. Spang, *Atmos. Chem. Phys.*, 2004, **4**, 2001–2013.
68. V. Santacesaria, A. R. MacKenzie and L. Stefanutti, *Tellus B*, 2001, **53**, 306–321.
69. S. Alexander, A. Klekociuk and D. Murphy, *J. Geophys. Res.*, 2011, in press, DOI: 10.1029/2010JD015164.
70. A. Adriani, P. Massoli, G. Di Donfrancesco, F. Cairo, M. Moriconi and M. Snels, *J. Geophys. Res.*, 2004, **109**, D24211, DOI: 10.1029/2004JD004800.
71. C. Voigt, H. Schlager, B. Luo, A. Dörnbrack, A. Roiger, P. Stock, J. Curtius, H. Vössing, S. Borrmann, S. Davies, P. Konopka, C. Schiller, G. Shur and T. Peter, *Atmos. Chem. Phys.*, 2005, **5**, 1371–1380.
72. M. C. Pitts, L. R. Poole, A. Dörnbrack and L. W. Thomason, *Atmos. Chem. Phys.*, 2011, **11**, 2161–2177.
73. R. P. Turco, O. B. Toon and P. Hamill, *J. Geophys. Res.*, 1989, **94**, 16493–16510, DOI: 10.1029/89JD00998.
74. A. Tabazadeh, O. B. Toon and P. Hamill, *Geophys. Res. Lett.*, 1995, **22**, 1725–1728.
75. S. Meilinger, K. Koop, B. P. Luo, T. Huthwelker, K. S. Carslaw, U. Krieger, P. J. Crutzen and T. Peter, *Geophys. Res. Lett.*, 1995, **22**, 3031–3034, DOI: 10.1029/95GL03056.
76. L. T. Iraci, A. M. Middlebrook and M. A. Tolbert, *J. Geophys. Res.*, 1995, **100**, 20969– 20977.
77. R. Y. Zhang, M. T. Leu and M. J. Molina, *Geophys. Res. Lett.*, 1996, **23**, 1669–1672.
78. A. Tabazadeh, Y. S. Djikaev, P. Hamill and H. Reiss, *J. Phys. Chem. A*, 2002, **106**, 10238–10246.
79. J. M. Rosen, N. T. Kjome and S. J. Oltmans, *J. Geophys. Res.*, 1993, **98**, 12741–12751.
80. O. Hülsmann and W. Blitz, *Z. anorg. u. allg. Chem.*, 1934, **218**, 369–378.
81. B. P. Luo, C. Voigt, S. Fueglistaler and T. Peter, *J. Geophys. Res.*, 2003, **108**, 4443, DOI: 10.1029/2002JD003104.
82. J. Curtius, R. Weigel, H. J. Vössing, H. Wernli, A. Werner, C. M. Volk, P. Konopka, M. Krebsbach, C. Schiller, A. Roiger, H. Schlager, V. Dreiling and S. Borrmann, *Atmos. Chem. Phys.*, 2005, **5**, 3053–3069.
83. D. A. Knopf, *J. Phys. Chem. A*, 2006, **110**, 5745–5750.
84. A. Tabazadeh, *J. Phys. Chem. A*, 2007, **111**, 1374–1375.
85. D. A. Knopf, *J. Phys. Chem. A*, 2007, **111**, 1376–1377.
86. R. J. Salawitch, G. P. Gobbi, S. C. Wofsy and M. B. McElroy, *Nature*, 1989, **339**, 525–527.
87. D. W. Fahey, K. K. Kelly, S. R. Kawa, A. F. Tuck, M. Loewenstein, K. R. Chan and L. E. Heid, *Nature*, 1990, **344**, 321–324.
88. J.-U. Grooß, G. Günther, R. Müller, P. Konopka, S. Bausch, H. Schlager, C. Voigt, C. M. Volk and G. C. Toon, *Atmos. Chem. Phys.*, 2005, **5**, 1437–1448.

89. S. Davies, M. P. Chipperfield, K. S. Carslaw, B.-M. Sinnhuber, J. G. Anderson, R. M. Stimpfle, D. M. Wilmouth, D. W. Fahey, P. J. Popp, E. C. Richard, P. von der Gathen, H. Jost and C. R. Webster, *J. Geophys. Res.*, 2003, **108**, 8322, DOI: 10.1029/2001JD000445.

90. F. Daerden, N. Larsen, S. Chabrillat, Q. Errera, S. Bonjean, D. Fonteyn, K. Hoppel and M. Fromm, *Atmos. Chem. Phys.*, 2007, **7**, 1755–1772.

91. D. R. Hanson, A. R. Ravishankara and S. Solomon, *J. Geophys. Res.*, 1994, **99**, 3615–3629.

92. P. V. Danckwerts, *Gas-liquid reactions*, McGraw-Hill Book Co., New York, 1970.

93. D. R. Hanson and A. R. Ravishankara, *J. Phys. Chem.*, 1994, **98**, 5728–5735.

94. Q. Shi, J. T. Jayne, C. E. Kolb, D. R. Worsnop and P. Davidovits, *J. Geophys. Res.*, 2001, **106**, 24259–24274, DOI: 10.1029/2000JD000181.

95. S. P. Sander, R. R. Friedl, J. R. Barker, D. M. Golden, M. J. Kurylo, P. H. Wine, J. P. D. Abbatt, J. B. Burkholder, C. E. Kolb, G. K. Moortgat, R. E. Huie and V. L. Orkin, *Chemical Kinetics and Photochemical Data for Use in Atmospheric Studies*, JPL Publication 10–6, 2011.

96. L. F. Keyser, M. T. Leu and S. B. Moore, *J. Phys. Chem.*, 1993, **97**, 2800–2801.

97. D. R. Hanson and A. R. Ravishankara, *J. Geophys. Res.*, 1993, **98**, 22931–22936.

98. R. Zhang, J. T. Jayne and M. J. Molina, *J. Phys. Chem.*, 1994, **98**, 867–874.

99. K. S. Carslaw and T. Peter, *Geophys. Res. Lett.*, 1997, **24**, 1743–1746.

100. K. S. Carslaw, T. Peter and R. Müller, *Geophys. Res. Lett.*, 1997, **24**, 1747–1750.

101. J.-U. Grooß, K. Brautzsch, R. Pommrich, S. Solomon and R. Müller, *Atmos. Chem. Phys. Discuss.*, 2011, **11**, 22173–22198.

102. M. L. Santee, I. A. MacKenzie, G. L. Manney, M. P. Chipperfield, P. F. Bernath, K. A. Walker, C. D. Boone, L. Froidevaux, N. J. Livesey and J. W. Waters, *J. Geophys. Res.*, 2008, **113**, D12307, DOI: 10.1029/2007JD009057.

103. S. Solomon, *Rev. Geophys.*, 1999, **37**, 275–316, DOI: 10.1029/1999RG900008.

104. H. Schlager, F. Arnold, D. J. Hofmann and T. Deshler, *Geophys. Res. Lett.*, 1990, **17**, 1275–1278, DOI: 10.1029/90GL01621.

105. S. Tilmes, R. Müller and R. J. Salawitch, *Science*, 2008, **320**, 1201–1204, DOI: 10.1126/science.1153966.

106. T. Feck, J.-U. Grooß and M. Riese, *Geophys. Res. Lett.*, 2008, **35**, L01803, DOI: 10.1029/2007GL031334.

107. *SPARC report on the evaluation of chemistry-climate models*, ed. V. Eyring, T. G. Sheperd and D. W. Waugh, World Meteorol. Organ., Geneva, 2010.

CHAPTER 5

Ozone Loss in the Polar Stratosphere

NEIL R. P. HARRIS*[a] AND MARKUS REX[b]

[a] European Ozone Research Coordinating Unit, University of Cambridge Department of Chemistry, Lensfield Road, Cambridge, CB2 1EW, United Kingdom; [b] Alfred Wegener Institute for Polar and Marine Research, P. O. Box 60 01 49, 14401 Potsdam, Germany

5.1 Introduction

One of the beguiling features of studying ozone in the polar stratosphere is the interplay of a wide range of factors which affect the natural levels of ozone as well as the chlorine and bromine catalysed ozone loss. Stratospheric dynamics and transport, microphysics, photochemistry, and radiative processes all play critical roles in determining both the natural ozone amounts and the anthropogenic ozone loss. In this chapter we will concentrate on quantitative studies of polar ozone loss, using both models and observations, so as to understand the main factors which influence ozone loss from year to year and from decade to decade. Before doing so, we summarize the most important features of the wintertime stratosphere and show how they influence polar ozone and polar ozone loss in the two hemispheres.

The stratospheric (Brewer–Dobson) circulation is characterized by upward motion of air in the tropics, flow toward the winter pole at all altitudes and downward motion of air at high latitudes in the winter hemisphere. A westerly circulation (the polar vortex) forms over the wintertime polar region (Chapter 1). In the Antarctic it is centred over the pole and its edge is at about 65°S. The core of the Antarctic vortex is cold, reaching as low as 185 K, with transport

Stratospheric Ozone Depletion and Climate Change
Edited by Rolf Müller
Published by the Royal Society of Chemistry, www.rsc.org

into and out of the vortex being significantly inhibited.[1] While also clearly identifiable, the Arctic vortex is not as strong as its Antarctic counterpart. The northern hemisphere winter stratosphere is much more disturbed than the southern hemisphere stratosphere as a result of a greater number of mountain ranges and a more active tropospheric meteorology, giving rise to greater and more variable planetary wave activity. It is also significantly warmer. As a result, there are much larger interannual variations in vortex stability in the Arctic than in the Antarctic causing it to be shorter-lived than its Antarctic counterpart. These differences lead to higher natural average amounts of total ozone over the Arctic than over the Antarctic, as well as to higher interannual variability[2] (Figure 5.1). The values for the Antarctic are shown for 1992–2010, a period when the ozone hole was present. This feature can be seen in the much lower total ozone values in the September to December period. This contrasts to a clear springtime (January to April) maximum in the northern hemisphere. The other notable difference is the much higher variability in the northern hemisphere on inter- and intra-annual timescales. Any chemical ozone loss needs to be understood in the light of the natural variation of ozone over the two polar regions (Figure 5.1).

The basic mechanism leading to polar ozone loss is the same over both poles and is generally well understood (see earlier chapters). Briefly, as the polar vortex is established, the temperatures drop and polar stratospheric clouds (PSCs) can form (Chapter 4). The PSCs provide surfaces on which HCl and $ClONO_2$ can react to form Cl_2 and HOCl, which are rapidly photolysed to form the chemically active species ClO_x ($= Cl + ClO + Cl_2O_2$), *i.e.* the reservoir chlorine species are converted into the active forms capable of destroying ozone. The ClO_x then destroys ozone in the presence of sunlight, either through the dimer cycle or through the ClO/BrO cycle (Chapter 3). Once the vortex warms and the PSCs evaporate, the ClO_x gradually reverts back to the reservoir species, HCl and $ClONO_2$. In cold winters, the PSC particles have long enough to grow and fall to lower altitudes. This reduces the ambient HNO_3 (denitrification), which is the source, through photolysis, of the NO_2 which removes ClOx through reaction with ClO to form $ClONO_2$.

In order to understand the chemical ozone loss over the two poles, the stratospheric dynamics and the photochemical processes must be considered together, and almost every step in the ozone loss process is favored in the south. The colder, longer-lived Antarctic vortex is more conducive to the occurrence of denitrification and hence slower deactivation of ClO_x in spring. In addition, the vortex lasts longer into spring so that more ozone can be lost through the prolonged exposure to sunlight. The observed evolution of the main chemical species in the Arctic and Antarctic can be seen in Figure 5.2 for the 2004/2005 and 2003 winters, respectively. The top panels show the fraction of the polar vortices at 460 K (~ 18 km altitude) (a) containing PSCs (turquoise) and (b) exposed to sunlight (pink). In the Antarctic, PSCs cover over half the vortex for 120 days, while in the Arctic they cover about half the vortex for 50–60 days. In both hemispheres the timing of the decline in HCl (and, by inference, the increase in ClO_x) coincides with the onset of PSCs. The decrease in HCl occurs

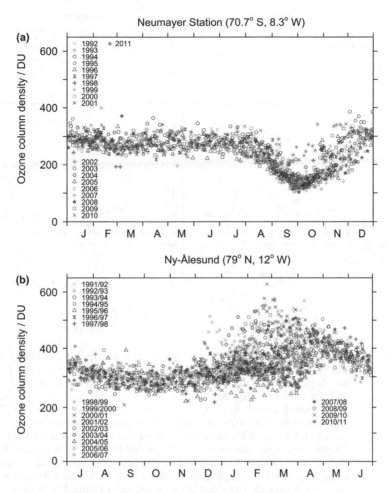

Figure 5.1 Total ozone over (a) the Antarctic (Neumayer, 70.6°S, 8.2°W) from 1992 to 2011; and (b) the Arctic (Koldewey, 78.9°N, 11.9°E) from 1991/92 to 2010/11.

more slowly in the Arctic and (when measurements are available) there is less ClO_x. Once formed, ClO_x remains high for much longer in the Antarctic than the Arctic. The resulting ozone loss is correspondingly greater (>95% *vs.* 50–60%). In looking at this comparison, it is important to remember that 2004/2005 was one of the coldest Arctic winters in the stratosphere with the largest ozone loss at this altitude. In other winters the contrast is even greater.

The nearly complete removal of ozone in the Antarctic vortex means that detailed understanding or even knowledge of the ozone loss rates there has not been a high priority. The loss was occurring and it was more important to find what caused it. As a result there are relatively few measurement-based studies of ozone loss rates in the Antarctic. However in the Arctic, where most steps in the

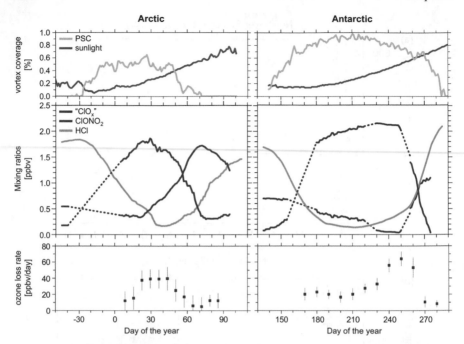

Figure 5.2 Evolution of chlorine species at 460 K (a) inside the Arctic vortex in the 2004/2005 winter, and (b) inside the Antarctic in the 2003 winter. The top panels show the area of PSCs as a percentage of the vortex area (light blue) and the vortex average sunlit time per day in percent (pink). The middle panel (based on ref. 3) is the vortex average HCl from Aura MLS, the vortex average $ClONO_2$ from ACE FTS, and an estimate of the vortex average ClO_x found by subtracting the sum of the HCl and $ClONO_2$ from 2.8 ppb, a representative value of Cl_y for 460 K for those winters. $ClONO_2$ and ClO_x have been smoothed with a ten-day running mean in order to compensate for the sampling biases from ACE FTS, which has the typical sampling issues of any solar occultation instrument. The bottom panel shows the ozone loss rates found from the Match campaign for those years.[4,5]

ozone loss process are near a threshold in most winters, more sophisticated approaches of determining ozone loss from measurements are required in order to assess the size of the problem each winter. Since a combination of measurements and models is needed to understand the situation in the Arctic completely, the situation was confused for many years by the inability of the photochemical models to reproduce the ozone loss rates deduced from measurements.

The differences between Antarctic and Arctic ozone loss are thus marked and revealing. In this chapter we look first at the processes which determine the extent of ozone loss in the Antarctic (Section 5.2) and then the Arctic (Section 5.3). To do so we summarize the available observational evidence and explain it in the light of theoretical understanding and modelling studies. In Section 5.4 we discuss the similarities and contrasts between the two hemispheres.

5.2 Antarctic Ozone Loss

5.2.1 Main Features of the Antarctic Ozone Hole

The ozone hole first became apparent in the early 1980s as a sharp downward trend in the Antarctic springtime column ozone values over the station at Halley[6] (Figure 5.3(a)). Between 1960 and the mid-1990s the October monthly mean dropped by over 50%. The minimum values occurred in 1993–1995 and since then the amount has not changed much, with the obvious exception of 2002 when there was an anomalously early breakdown of the Antarctic vortex. The minimum value observed by satellite in the vortex and the size of the ozone hole exhibit a similar evolution of time (Figure 5.3(b) and (c)). Since the mid-1990s the area has maximized at a little over 20 million km^2 (the size of Antarctica; $\sim 10\%$ of the southern hemisphere; or the size of north America), while the minimum total ozone value in the vortex each year has been roughly 100 DU. In all time series, 2002 stands out as the anomaly. There has been considerable discussion as to whether any sign of recovery is apparent in the available data,[7] and the current consensus is that no definite sign of recovery has yet been detected. Two factors make it hard to be definitive: (a) the amount of chlorine and bromine (Effective Equivalent Stratospheric Chlorine—EESC) available to deplete ozone is not changing much; and (b) the sensitivity of column ozone loss to changes in EESC is small when the ozone depletion is large.

Farman *et al.*[6] also showed that the ozone loss occurs between late August and the end of September. Figure 5.4(a) shows the average seasonal evolution of ozone starting in late August for the two 17 year periods 1957–1973 (unperturbed) and 1993–2009 (peak ozone hole). The ozone declines dramatically in September, coincident with the increasing insolation, before recovering and reaching the lower end of the unperturbed range around the turn of the year. This increase coincides with erosion and break-up of the Antarctic vortex during which the ozone-depleted air in the vortex is mixed with mid-latitude air with higher ozone amounts. The size of the ozone hole means that its effect can be seen even after it has been mixed into the surrounding atmosphere, with total ozone amounts in January to April still lower than in the unperturbed atmosphere. The resulting, though less dramatic cooling in the lower stratosphere is apparent in Figure 5.4(b).

Ozone is destroyed pretty much completely at altitudes between 15 and 20 km (Figure 5.5), so much so that there are large parts of the Antarctic stratosphere in which it cannot be measured by conventional instruments. This altitude range is the coldest part of the wintertime stratosphere and so is where PSCs can form. The loss of ozone occurs over ~ 6 week period. It has occurred every year for the last 20 or so years—even in 2002 when the vortex broke up anomalously early. Some ozonesonde measurements were made from Antarctica in the 1960s and early 1970s.[2] These show no such loss in late August and September, confirming that this loss has only occurred since the ozone hole appeared.

5.2.2   Chemical Ozone Loss in the Antarctic Vortex

The evidence for the involvement of chlorine compounds released from CFC and other ozone depleting substances in the Antarctic ozone loss is unambiguous (Chapters 3 and 4). The coincidence of the ozone loss and high ClO

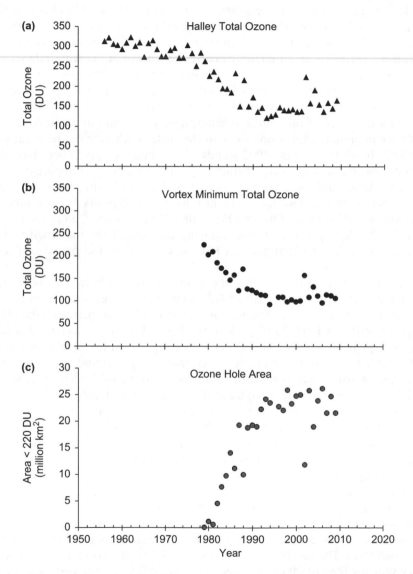

Figure 5.3   Long-term evolution of the Antarctic ozone hole: (a) the October monthly mean total ozone at Halley; (b) the minimum total ozone over Antarctica; and (c) the area with total ozone less than 220 DU south of 45°S. (Data courtesy J. D. Shanklin, British Antarctic Survey, and P. Newman, NASA.)

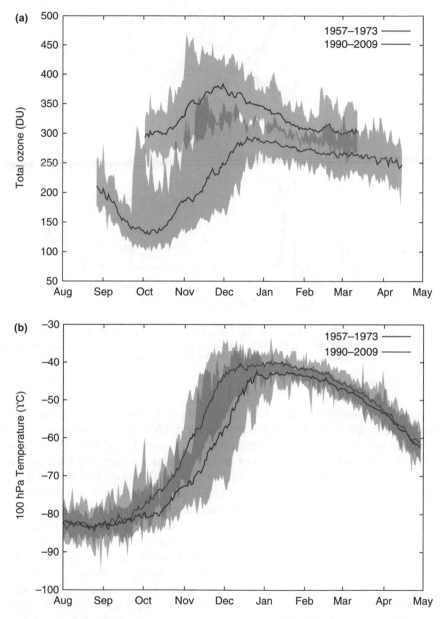

Figure 5.4 Measurements of (a) total ozone and (b) temperature at Halley (75.6°S, 26.6°W) for 1957–1973 and 1990–2009. The central line in each dataset is the mean for the period and the shaded area shows the complete range of observations for that day. (Data courtesy J. D. Shanklin, British Antarctic Survey; plot courtesy L. Abraham, Cambridge University.)

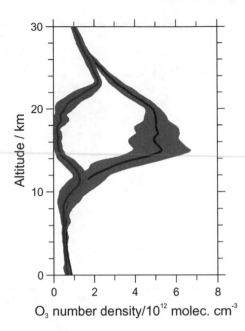

Figure 5.5   Mid-winter (blue, June/July mean) and late winter (red, October mean)
ozone profiles in the 2006 Antarctic winter measured by ozonesondes from
the station Neumayer (70.6°S, 8.2°W). The shading indicates the variability of the profiles (1 σ).

periods, the anti-correlation at the vortex edge, as well as the geographic extent
revealed by satellites is extremely compelling evidence. But can we also say that
the ozone loss rates match the observed chemical evolution? The answer in the
Antarctic, curiously, is only "to some degree". There has certainly been nothing
like the degree of attention there has been in the Arctic (see Section 5.3.2)
because understanding the causes and processes involved in Arctic ozone loss
was more important since the loss is complete each year.

Ozone concentrations at any particular location are determined by chemical
and dynamical factors. In order to estimate the chemical ozone loss from
observations, it is necessary to quantify the changes due to dynamics. The
"Match" technique[8] is one such method for estimating ozone loss rates using
meteorological information. The various techniques for estimating chemical
ozone loss are discussed in more detail in section 5.3 as they were mainly
developed for use in the Arctic.

In the Antarctic the Match approach has been used with satellite measurements of the vertical distribution of ozone[9] and with ozonesondes.[5,10] The
bottom panel of Figure 5.2 shows the estimated ozone loss rate at 460 K for the
winter/spring 2003. Values between 20 ppbv/day were found from mid-June to
early August with most of the ozone loss occurring in the edge of the vortex
which was the only region exposed to sunlight. The ozone loss then peaked at
∼70 ppbv/day in late September, by which time nearly all the ozone had been

removed. The ozone loss rate thus drops close to zero. The interannual variability in dynamics does lead to interannual variability in the ozone loss, but this is not large as the Antarctic vortex is relatively stable during the period in which ozone loss takes place.[9,11]

The calculated ozone loss rates have been compared with model calculations in order to test the understanding of the photochemistry.[5,10] Unfortunately, the estimates of the loss rates proved sensitive to the meteorological re-analyses used, which are less accurate in the Antarctic stratosphere than elsewhere as fewer measurements are made by meteorological radiosondes. However, this situation is improving with the production of the new generation of meteorological re-analyses,[12] which are producing temperature and wind fields capable of being used for calculations of chemical activation and vortex motions. Even so, these ozone loss studies did provide evidence for using higher photolysis cross-sections for Cl_2O_2, as supported by more recent laboratory studies (Chapter 3), and higher BrO levels than expected from just the longer-lived Br source gases.

One cause of uncertainty in the ozone loss rates calculated by photochemical models arises from the uncertainties in the laboratory measurements of the individual photochemical processes. Figure 5.6 shows calculations of ozone by a box model (red line and grey shaded area) along realistic trajectories in the Antarctic vortex. The model was driven by Monte Carlo simulations varying the photochemical parameters within their assessed uncertainties,[13] *i.e.* with a high recommended value for the photolysis cross-section of Cl_2O_2. The grey

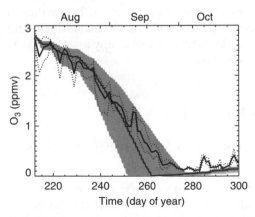

Figure 5.6 Comparison of MLS O_3 data averaged for 2005–2007 (solid blue lines) with model simulations. Maxima and minima of three years of MLS data for each day are shown by dotted lines. Equivalent latitude/potential temperature averaged MLS data are mapped to the equivalent latitude and potential temperature of the model trajectory for each day. Also shown are Monte Carlo ozone loss calculations made with a photochemical box model. Gray shaded areas encompass the median 95% of O_3 mixing ratio values from the model scenario distribution on each day of the model runs. The red line is the calculation using the nominal rates from ref. 13. Figure adapted from ref. 14.

area shows the uncertainty in ozone arising from the photochemical parameters. The calculated ozone values compare reasonably well, though certainly
not perfectly, with MLS ozone measurements (blue line). Even so, the slowest
possible ozone loss rate is about half the fastest. A full discussion of the
uncertainties in the halogen chemistry can be found in Chapter 3.

It is hard to assess how well the coupled chemistry-climate models (CCM)
represent past and current conditions in the Antarctic vortex due to the large
scatter between the models (Chapter 9). This scatter is caused by a range of
factors connected to errors or simplifications in the dynamic and microphysical
schemes as well in the photochemistry. Individual models have typically not
compared too well when compared with observations, often underestimating the
ozone loss.[15] However, as a group the models appear to be improving their
estimates of past ozone loss, and there are certainly improvements in their
descriptions of individual facets such as lower stratospheric temperatures and
the amount of available chlorine (Cly).[16] The heterogeneous and gas-phase
chemistry schemes have been tested against the chemical measurements using
box models and chemical transport models which contain either the same or
more sophisticated descriptions than the CCMs. For example, the $ClONO_2$, HCl
and ClO measurements from AURA-MLS and ACE-FTS in 2004 and 2005 were
compared with the calculated fields from the SLIMCAT CTM[3]. In this case, the
model was found to overestimate the magnitude, spatial extent and duration of
chemical activation, partly due to the simplified, equilibrium treatment of PSCs.

When all the ozone in large parts of the column is destroyed, these differences
do not matter too much. However they are important when considering regions
where there is incomplete loss (*e.g.* above and below the main region of loss).
These areas of incomplete loss will become increasingly important as chlorine
levels decline and so play a greater role in any assessment of the likely recovery
of the Antarctic ozone hole.

5.3   Arctic Ozone Loss

5.3.1   Natural Variability in Stratospheric Ozone over the Arctic

The natural variability of the ozone over the Arctic is much greater than over
the Antarctic. While a vortex does form over the Arctic, it is much less stable
(and more mobile) as a result of much greater dynamical forcing from the
troposphere. As a result, the Arctic vortex is warmer, shorter-lived and less isolated. All of these properties lead to reduced ozone loss in the Arctic than the
Antarctic: fewer PSCs form and are present for less time; denitrification is less
common and less extensive; and the activated air is less exposed to sunlight. The net
effect can be seen in Figure 5.7, which shows ozonesonde profiles from three days in
late January 1992. The dominant reason for the variability is dynamical activity,
with the edge of the Arctic vortex moving over the Koldewey station. Photochemical ozone loss plays only a small role in the differences shown in Figure 5.7.

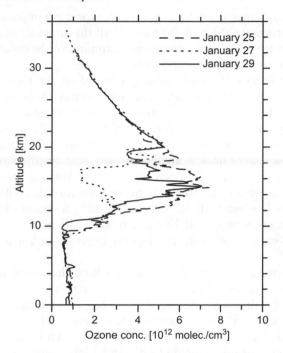

Figure 5.7 The natural variability of ozone over the Arctic. The vertical distribution of ozone at Koldewey station (78.9°N, 11.9°E) is shown on 25th, 27th and 29th January 1992. The typical altitude region for the formation of PSCs is shown by the solid vertical line. Adapted from ref. 17.

The movement of air masses is clearly the most important natural influence on the ozone fields at high northern latitudes. Measurement-based estimates of ozone loss must be able to distinguish between these motions and photochemical loss. In the lower stratosphere, the fastest air motions are horizontal and often reach 100 ms^{-1}. On timescales of a few days, these horizontal motions occur on isentropic surfaces and are well captured in standard meteorological analyses. The slower motion is vertical: local, adiabatic (reversible) vertical velocities are of the order of 1 ms^{-1}, while the slow radiative cooling of the air in the vortex (1 K/day), the diabatic descent, is equivalent to a downward velocity of the order of 0.1 cm s^{-1}. As a result a great deal of work has gone in to developing techniques to estimate ozone loss rates and seasonally integrated ozone losses in the Arctic vortex. The main approaches are now discussed.

5.3.2 Chemical Ozone Loss in the Arctic Vortex

Broadly speaking, the various approaches can be split into two categories: (1) studies where the effects of transport are calculated explicitly using transport calculations driven by winds, temperatures, *etc.*, based on meteorological analyses; and (2) studies where the effects of transport are implicitly allowed for by using measurements of long-lived tracers. All the approaches require spatial and temporal averaging in order to provide estimates of the ozone loss which is occurring on timescales of days and weeks.

Match is a pseudo-Lagrangian technique based on the identification of air parcels whose ozone amount is measured twice within a ten-day period. It has mainly been used with networks of ozonesonde stations which are coordinated to ensure a large number of "matches" (e.g. ref. 8 and17), although the technique has also been used with satellite measurements.[9,18] The trajectories are calculated using meteorological analyses which realistically describe the fast, quasi-horizontal motions in the polar vortex. The slower descent is calculated using radiative heating rates from 3D atmospheric models. With a sufficiently large number (a few hundred) of "matches" during a winter, the evolution of the vortex average ozone loss can be reconstructed with high vertical (10 K) and temporal (two-week) resolution, the highest time resolution of the available techniques.

A related method has been developed in which the ozone fields observed by satellites are reconstructed recurrently through the winter using a 3D Lagrangian model driven by meteorological analyses. The differences between the observed field and those reconstructed from earlier measurements is ascribed to ozone loss.[19,20] This is a forerunner of data assimilation studies (ref. 21) which can also be compared with ozone loss estimates found from the photochemical calculations in the same model.

A similar conceptual approach ("vortex average") based on ozonesonde measurements involves calculation of the bulk vertical advection of the average ozone profile inside the vortex from diabatic cooling rates. The difference between the advected ozone profile and the observed ozone profile is attributed to chemical ozone loss. This technique has been used with a wide variety of ozone measurements,[4,22,23] and it is probably the simplest method and thus is applicable in most winters.

3D chemical transport models driven by meteorological analyses contain chemically active ozone (i.e. depletable) and passive ozone which is not affected by chemistry from the start of the simulation. This passive ozone can be compared to observations to infer chemical ozone loss during a winter. High latitude measurements from the ground-based SAOZ instruments that are inside the Arctic vortex have been combined with calculations from the REBPROBUS and SLIMCAT CTMs, to give a multi-year record of estimates of column ozone losses.[24,25] Ozone profile measurements from balloons and the POAM satellite have also been used.[26]

The value of all the techniques described so far depend on the quality of the meteorological analyses. While these are considered to be good in the Arctic region, there have been more concerns about the quality in the Antarctic stratosphere primarily as a result of fewer radiosondes in the region.[12,27] There is thus great value in the ozone/tracer correlation method which does not use meteorological analyses. In this technique, changes in the compact correlations between ozone and long-lived tracers over the winter are ascribed to chemical loss. This approach requires that a reference ozone/tracer relation representative of the early winter polar vortex can be established and assumes that transport and mixing processes do not significantly alter the validity of the early winter reference. It was originally applied to aircraft measurements of O_3 and N_2O[28,29] and later to satellite measurements to provide a multi-year record.[30,31]

The different nature of the techniques and of the quantities they produce led to large apparent differences in the ozone loss estimates. However, a detailed comparison revealed remarkably good agreement between the various estimates as long as the same temporal and spatial averaging was applied.[32] This is shown for the accumulated chemical loss in the total column ozone in Figure 5.8. Similarly, good agreement between techniques is found for the loss at particular altitudes[32] and for more recent winters.[4,33]

So, what losses have actually occurred in the Arctic vortex? The most intensively studied winter was in 1999/2000 when the SOLVE and THESEO-2000 field campaigns took place involving research aircraft, balloons, intensified ozonesondes and ground-based instruments with modeling and satellite

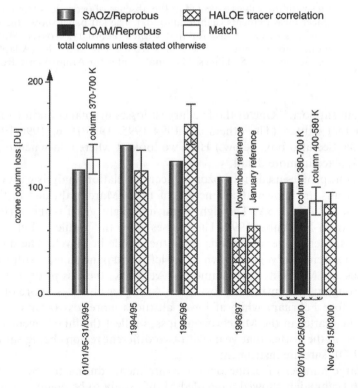

Figure 5.8 Comparison of ozone loss techniques. Column ozone losses are shown from (i) the SAOZ/REPROBUS; (ii) POAM/REPROBUS; (iii) Match; and (iv) HALOE tracer correlation approaches. For more details see Figure 3 in ref. 32.

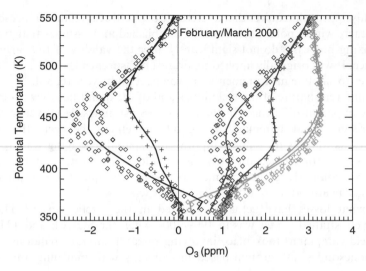

Figure 5.9   Late winter vertical profile of ozone mixing ratios (February, crosses; March, diamonds) measured by HALOE in the 1999/2000 winter, together with the values expected in the absence of ozone loss from long-lived tracer measurements (light diamonds). The left panel shows the resulting estimates of ozone loss for that winter (February, crosses; March, diamonds) for potential temperatures from 350 K to 550 K. (Adapted from ref. 30; courtesy S. Tilmes, National Center for Atmospheric Research).

instrument support.[34] One of the five largest losses in total column ozone took place in that winter. (The others are 1994/1995, 1995/1996, 1996/1997, and 2004/2005. See also box, below.) Here we look at where and when the ozone loss took place in more detail.

The ozone profiles used in the ozone/tracer correlation analysis with HALOE measurements are shown as a function of final (March) altitude in the right hand panel of Figure 5.9. The right-hand line (light gray) is the early winter reference from November 1999. The crosses show the profile in February and the more scattered open diamonds show the profile in March. The differences from the reference curve are shown in the left hand panel, again with February as crosses and March as open diamonds. At 450 K, the loss prior to February 2000 is roughly the same as the loss after. At higher altitudes, more of the loss occurs before February while at lower altitudes more loss occurs after February. The scatter in the March ozone losses reflect the inhomogeneity in the ozone loss in the vortex that year, with two different regions being sampled by the HALOE satellite instrument.

The large number of ozonesonde measurements during the SOLVE-THE-SEO 2000 campaign allowed a detailed Match study to be made.[35] The results are summarized in Figure 5.10. Chemical ozone loss depends on the exposure to sunlight, and Match analyses have always found ozone loss to correlate well with the number of sunlit hours with zero ozone loss in the dark.[17] Panel (a)

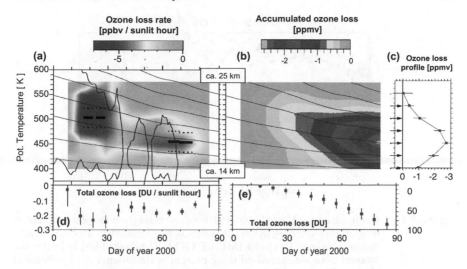

Figure 5.10   (a) Instantaneous, vortex averaged ozone loss rates (ppbv/sunlit hour) estimated from the Match ozonesonde network for 1999/2000. The dashed lines indicate the potential temperatures of air masses as they cool through the winter using cooling rates from the SLIMCAT CTM. (b) The ozone loss (ppmv) accumulated through the winter. (c) The temporal evolution of the ozone loss rate for the air which ended the winter at 475 K. (d) The accumulated column ozone loss for the partial column between 400 and 575 K. (e) The end of winter accumulated ozone loss in ppmv as a function of altitude. Adapted from ref. 17.

shows the ozone loss rate in ppbv/sunlit hour. Two main periods of ozone loss are seen. The first, in January, occurred at a higher altitude than the one in March, consistent with the altitudes of the PSCs which activated the chlorine in these two periods. Panel (b) shows the accumulated ozone loss, with the end of winter value show in panel (c). The values are similar to those shown in Figure 5.9 with a peak at 450 K and the same vertical distribution. Panel (d) shows the ozone loss rate (again in ppbv/sunlit hour) for air which has descended to 475 K by the end of March. While ozone is destroyed faster toward the end of January, the days are longer in March and so more ozone is lost in the second period of chemical destruction (panel (e)).

3D chemical transport models can now reproduce the observed ozone evolution in Arctic winters reasonably well, as can be seen in Figure 5.11, which shows the modeled ozone for the 1999/2000 winter at about 18 km altitude. Similar results are found at other altitudes. This situation is encouraging given the remaining uncertainties in chemistry, microphysics and dynamics. In addition it was found that a higher model resolution gives larger ozone losses in better agreement with the observations.[36] Additional, significant model improvements reported in this study were the radiation scheme used (which is used to diagnose the vertical motion in the stratosphere) and a NAT-based denitrification scheme. These results highlight the importance of making sure

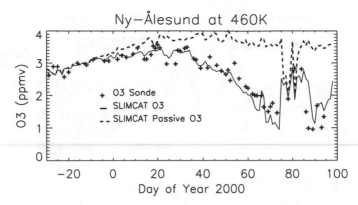

Figure 5.11   Comparison of O_3 sonde observations (+ marks) on the 460 K potential
temperature level (~ 18 km altitude) at Ny-Ålesund in 1999/2000 with
ozone calculated by the SLIMCAT CTM. The dashed line indicates the
amount of ozone calculated to be present in the absence of any chemical
change. (Figure adapted from ref. 36 courtesy W. Feng, University of
Leeds.)

that all the relevant processes are described correctly in the complex 3D models
whether CTMs or CCMs.

5.3.3   Interannual Variability in Ozone Loss in the Arctic Vortex

The results presented in Section 5.3.2 are primarily about the chemical ozone
loss in the 1999/2000 Arctic winter. Some idea of the interannual variability in
ozone loss can be seen in Figure 5.8. However, the winters shown in this figure
are all winters in which there were relatively large chemical losses of ozone. In
this section, the interannual variability of the chemical ozone loss is discussed.
    The large natural variability in Arctic ozone means that it has to be con-
sidered at the same time as the variability in the chemical ozone loss. The main
cause of natural interannual variability in Arctic ozone is the variation in the
descent rates in the vortex. These depend on the vortex temperatures. In
warmer years, there is greater radiative cooling, greater descent and higher
ozone amounts at given altitudes and in the column. (The mixing ratio of ozone
increases with altitude in the lower stratosphere.) The converse is true in colder
winters. Thus, in a chemically unperturbed stratosphere, low temperatures
would lead to low ozone, and high temperatures would lead to high ozone. In
the recent and current stratosphere with high chlorine levels, chemical ozone
loss is greatest in cold winters and smallest in warm winters. The chemical
impact on ozone thus tends to reinforce the dynamical impact, enhancing the
natural variability.
    The relative sizes of the chemical and dynamical influences on Arctic ozone
have been investigated[37] and are shown in Figure 5.12 (lower two panels). In
every winter the dynamical supply is either larger than or equal to the chemical

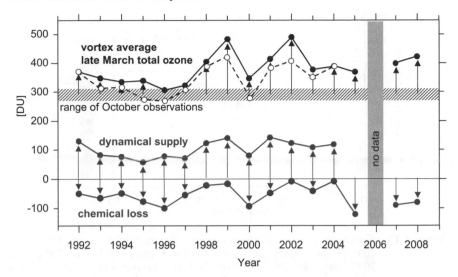

Figure 5.12 Interannual variability in the chemical and dynamical influence on ozone in the Arctic vortex. The top panel shows the observed ozone, the middle panel shows the estimated dynamical supply of ozone; and the bottom panel shows chemical ozone loss. (Updated from ref. 37.)

loss, which leads to the observed net gain in column ozone in the vortex over the course of the winter (top panel). Further, the chemical and dynamical terms contribute equally to the interannual variability in the Arctic ozone column. The correlation between the two terms is evident. Both are found to correlate with the variability in the planetary wave activity, which is one of the major influences on the strength of the stratospheric circulation (leading to dynamical variations in ozone) and on lower stratospheric temperatures (leading to chemical variations in ozone) in the Arctic.

Arctic column ozone loss has been shown empirically to have a compact, linear relation with the volume of PSCs inferred from meteorological analyses integrated over the course of the winter[4,38,39] (Figure 5.13). Since the inferred PSC volume depends on stratospheric temperatures, this relation implies a sensitivity of column ozone loss to any change in stratospheric temperatures with an additional 80 DU ozone loss for a 5–6 K cooling. While no long-term cooling of the Arctic lower stratosphere has been observed, there is evidence that the cold winters are getting colder[4] so that the ozone losses are increasing in those cold winters (about one in five). This apparent trend is unexplained and not replicated in current CCMs (SPARC CCMVal, 2010). However, it has significant implications for the next few decades while stratospheric chlorine levels remain high enough to cause ozone loss.

The compactness and linearity of the relationship in Figure 5.13 is surprising given the complexity of the processes involved. However, it has been found to hold over a wide range of ozone losses and PSC volumes, implying that the

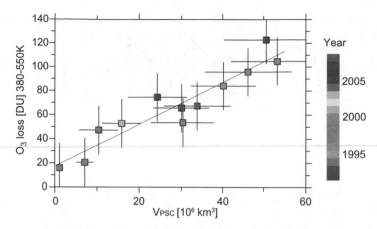

Figure 5.13 Seasonally integrated ozone loss as a function of V_{PSC} for 1992/93 to 2008/09. No values are shown for the winters 2000/2001, 2001/2002, 2003/2004, 2005/2006 and 2008/2009 due to major warmings and/or lack of ozonesonde data. Ozone losses are derived from the Arctic ozonesonde network using the vortex average approach. The error bars on V_{PSC} are based on the sensitivity to a temperature perturbation of $\pm 1\,K$ in the ECMWF data. V_{PSC} is derived from ECMWF ERA-Interim reanalyses using the temperature of formation of nitric acid trihydrate. (From ref. 39.)

Box 5.1: An Arctic Ozone Hole?

Chemical destruction of ozone occurs over both polar regions in winter/spring. While in the southern hemisphere, near complete removal of ozone in the lower stratosphere has resulted in an ozone hole virtually every year since the early eighties, the degree of loss in the Arctic is very variable and is typically much smaller. However in early 2011 chemical destruction of ozone over the Arctic was comparable to that in Antarctic ozone holes for the first time (Figure Box). Unprecedented, nearly complete complete ($\sim 80\%$) loss occurred in a broad vertical range in the lower stratosphere. The degree of loss, its vertical distribution and the evolution of chemical species responsible for it echoed Antarctic ozone hole conditions and were outside the range previously observed in the Arctic. As in the Antarctic a well-defined sharp minimum in total ozone formed, replacing the normal seasonal maximum of the thickness of the ozone layer over the Arctic in spring.[40]

The unprecedented degree of ozone loss was caused by an unusually long period of low temperatures in the Arctic lower stratosphere during winter. These conditions led to record levels of ozone destroying forms of chlorine, which were maintained longer into spring than ever in the Arctic before.[40] The unusually cold meteorological conditions in the stratosphere during winter 2011 are part of a long-term tendency toward lower temperatures during the cold Arctic winters.[4,38] The first occurrence of an ozone hole above the Arctic long after CFC and halon production have been banned highlights the importance of potential links between changes in climate and Arctic ozone loss, which are the subjects of ongoing research. It raises the question whether increasing degrees of Arctic ozone loss can occur in some winters during the next decades, despite slowly decreasing levels of ozone depleting substances in the atmosphere. The warmer, more disturbed Arctic vortex will not become cold and stable like its Antarctic equivalent every year but, if it happens occasionally, then there will be years when the Arctic ozone shows strong similarities to the Antarctic ozone hole.

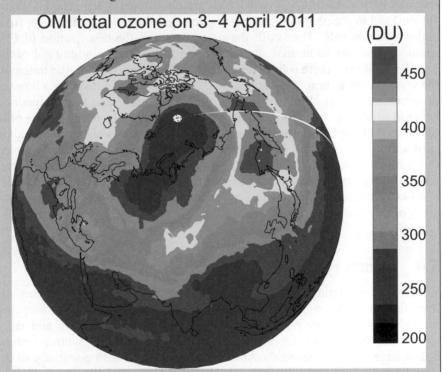

Figure Box (also front cover). *Total ozone observed by the OMI satellite instrument on April 3/4 2011. The area where ozone has been chemically depleted can be clearly seen between western Russia and the north Pole. For more details, see ref. 40.*

effects of many of the processes (*e.g.* initial chemical fields, descent rates, vertical extent, denitrification, solar exposure, vortex inhomogeneities) offset each other to some degree. More detailed investigation of this relation shows that the compactness can be explained by the chemical timescales involved.[39,41] Extensive activation in the Arctic occurs with a timescale of days to weeks associated with the re-supply of $ClONO_2$, which is limited by the photolysis of HNO_3. In the continued, though not necessarily continuous, presence of PSCs, $ClONO_2$ subsequently reacts heterogeneously with HCl. In other words, the extent of activation depends on the length of the period PSCs are present. V_{PSC} is thus a sensible way to quantify the activation process.

In any particular winter the ozone loss is found to depend most strongly on the degree of initial activation.[39,41] The speed of the ozone loss cycle depends principally on the photolysis rate of Cl_2O_2, while the deactivation depends principally on the photolysis rate of HNO_3, which produces NO_2. (The production of NO_2 through the reaction $OH + HNO_3$ is of minor importance in the winter vortex.) NO_2 can subsequently react with ClO forming the reservoir species $ClONO_2$. Both Cl_2O_2 and HNO_3 are photolysed in the near UV (320–400 nm) and so the photolysis rates both increase as the sunlight increases from mid-winter onwards. Thus both the ozone loss and the deactivation of ClO speed up, and the additional sunlight does not give rise to additional ozone loss.[41] In addition, there is little dependence of the ozone loss on the timing or the vertical distribution of the activation.[39] This occurs as a result of a number of offsetting factors. For example, in any given air mass, the competition between ozone loss and chlorine deactivation, both of which rely on the photolysis in the near UV and so accelerate as the sun becomes higher in the sky in early spring, almost cancel. In the vertical, Cly increases with altitude; however, this is offset by the decreasing number density with altitude. In the few winters where denitrification occurs the enhanced ozone loss at the denitrified altitudes is offset by the reduced ozone loss at the lower altitudes where re-nitrification takes place.

5.4   Summary

Winter- and springtime ozone over the Arctic is much more variable than it is over the Antarctic, whether day-to-day or year-to-year variations are considered. This difference results from the greater dynamic stability and more consistently low temperatures of the Antarctic vortex. The Antarctic ozone hole is large, nearly covering the Antarctic continent and with essentially all the ozone removed at altitudes between 15 and 20 km. It develops in August/September each year, with total column ozone values now approximately half what they were before its onset in the late 1970s and early 1980s. Since the early 1990s, the Antarctic ozone hole has become an annually recurring feature of the southern hemisphere, with the only year-to-year difference being about its dynamical stability, size and duration.

By way of contrast, the Arctic wintertime stratosphere has a less stable vortex with more variable temperatures. This difference results from the enhanced planetary wave driving in the northern hemisphere. As a result, ozone amounts vary greatly from day-to-day and from year-to-year. This variability requires more sophisticated analyses of chemical ozone loss to estimate the ozone changes that would anyway have occurred as a result of the atmospheric dynamics. A number of techniques now exist (some using meteorological analyses, some using long-lived tracer measurements) which give reasonably good agreement in their ozone loss estimates. Reliable estimates of ozone loss (to, say, within 10–20%) are now available for most recent winters.

The processes and mechanisms leading to the ozone loss in the Antarctic and Arctic stratosphere are well understood, if not always well quantified. In the Antarctic, the ozone loss is complete over a significant altitude range every winter (15–21 km), and follows a prolonged period of very low temperatures during which PSCs form allowing conversion of inactive forms of chlorine (HCl and $ClONO_2$) to photolabile and active forms (HOCl, Cl_2, Cl_2O_2, ClO, Cl). As sunlight returns, the active forms deplete ozone over a few weeks. Since the ozone loss is pretty much complete, only a few studies of ozone loss rates have been performed and these show rates consistent with current understanding. However, uncertainties in the rates for some critical reactions limit our ability to be able to model the ozone loss accurately. In the Arctic, the larger variability leads to some years with large losses (70% at some altitudes; 30% in the column) and others with hardly any loss at all. This sensitivity to the stratospheric meteorology means that the Arctic is more sensitive to changes in climate, and it is estimated that a 5 K decrease in temperature would lead to an additional 80 DU chemical loss of ozone. An unexplained trend has been observed in the volume of PSCs in the coldest winters, but not in the warmer ones.

Acknowledgements

Neil Harris thanks the UK Natural Environment Research Council for an Advanced Research Fellowship.

References

1. J. C. Farman, *Phil. Trans. Roy. Soc. Series B*, 1977, **279**, 963, 261.
2. S. Solomon, R. W. Portmann and D. W. J. Thompson, *Proc. Nat. Acad. Sci.*, 2007, **104**, 2, 445.
3. M. L. Santee, I. A. MacKenzie, G. L. Manney, M. P. Chipperfield, P. F. Bernath, K. A. Walker, C. D. Boone, L. Froidevaux, N. J. Livesey and J. W. Waters, *J. Geophys. Res.*, 2008, **113**, D12307.
4. M. Rex, R. J. Salawitch, H. Deckelmann, P. von der Gathen, N. R. P. Harris, M. P. Chipperfield, B. Naujokat, E. Reimer, M. Allaart, S. B. Andersen, R. Bevilacqua, G. O. Braathen, H. Claude, J. Davies, H. De Backer, H. Dier, V. Dorokov, H. Fast, M. Gerding, S. Godin-Beekmann, K. Hoppel, B. Johnson, E. Kyrö, Z. Litynska, D. Moore, H. Nakane, M. C. Parrondo, A. D. Risley, P. Skrivankova, R. Stübi, P. Viatte, V. Yushkov and C. Zerefos, *Geophys. Res. Lett.*, 2006, **33**, L23808.

5. K. Frieler, M. Rex, R. J. Salawitch, T. Canty, M. Streibel, R. M. Stimpfle, K. Pfeilsticker, M. Dorf, D. K. Weisenstein and S. Godin-Beekmann, *Geophys. Res. Lett.*, 2006, **33**, L10812.

6. J. C. Farman, B. G. Gardiner and J. D. Shanklin, *Nature*, 1985, **315**, 207.

7. WMO (World Meteorological Organization), *Scientific Assessment of Ozone Depletion: 2006*, Global Ozone Research and Monitoring Project—Report No. 50, 572 pp., Geneva, Switzerland, 2007.

8. P. von der Gathen, M. Rex, N. R. P. Harris, D. Lucic, B. Knudsen, G. O. Braathen, H. de Backer, R. Fabian, H. Fast, M. Gil, E. Kyro, I. S. Mikkelsen, M. Rummukainen, J. Staehelin and C. Varotsos, *Nature*, 1995, **375**, 131.

9. K. Hoppel, R. Bevilacqua, T. Canty, R. Salawitch and M. Santee, *J. Geophys. Res.*, 2005, **110**, D19304.

10. O. P. Tripathi, S. Godin-Beekmann, F. Lefèvre, A. Pazmino, A. Hauchecorne, M. Chipperfield, W. Feng, G. Millard, M. Rex, M. Streibel and P. von der Gathen, *J. Geophys. Res.*, 2007, **112**, D12307.

11. D. J. Hofmann, B. J. Johnson and S. J. Oltmans, *Int. J. Rem. Sens.*, 2009, **30**, 3995.

12. A. J. Simmons, S. M. Uppala, D. Dee and S. Kobayashi: ERA Interim: New ECMWF reanalysis products from 1989 onwards, *ECMWF News Lett.*, 2006, **110**, 25–35.

13. S. P. Sander, R. R. Friedl, D. M. Golden, M. J. Kurylo, G. K. Moortgat, H. Keller-Rudek, P. H. Wine, A. R. Ravishankara, C. E. Kolb, M. J. Molina, B. J. Finlayson-Pitts, R. E. Huie and V. L. Orkin, *Chemical Kinetics and Photochemical Data for Use in Atmospheric Studies, Evaluation Number 15*, JPL Publication 06-02, Jet Propulsion Laboratory, Pasadena, CA, USA, 2006.

14. S. R. Kawa, R. S. Stolarski, P. A. Newman, A. R. Douglass, M. Rex, D. J. Hofmann, M. L. Santee and K. Frieler, *Atmos. Chem. Phys.*, 2009, **9**, 8651.

15. C. Lemmen, M. Dameris, R. Müller and M. Riese, *Geophys. Res. Lett.*, 2006, **33**, L15820.

16. SPARC CCMVal, *SPARC CCMVal Report on the Evaluation of Chemistry-Climate Models*, V. Eyring, T. G. Shepherd, D. W. Waugh (Eds.), SPARC Report No. 5, WCRP-132, WMO/TD-No. 1526, http://www.atmosp.physics.utoronto.ca/SPARC, 2010.

17. M. Rex, P. von der Gathen, N.R.P. Harris, D. Lucic, B. M. Knudsen, G. O. Braathen, H. De Backer, R. Fabian, H. Fast, M. Gil, E. Kyro, I. S. Mikkelsen, M. Rummukainen, J. Staehelin and C. Varotsos, *J. Geophys. Res.*, 1998, **103**, 5843.

18. Y. Sasano, Y. Terao, H. L. Tanaka, T. Yasunari, H. Kanzawa, H. Nakajima, T. Yokota, H. Nakane, S. Hayashida and N. Saitoh, *Geophys. Res. Lett.*, 2000, **27**, 213.

19. G. L. Manney, L. Froidevaux, M. L. Santee, N. J. Livesey, J. L. Sabutis and J. W. Waters, *J. Geophys. Res.*, 2003, **108 (D4)**, 4149.

20. G. L. Manney, M. L. Santee, L. Froidevaux, K. Hoppel, N. J. Livesey and J. W. Waters, *Geophys. Res. Lett.*, 2006, **33**, L04802.

21. D. R. Jackson and Y. J. Orsolini, *Q. J. R. Meteorol. Soc.*, 2008, **134**, 1833.

22. B. M. Knudsen, N. Larsen, I. S. Mikkelsen, J-J. Morcrette, G. O. Braathen, E. Kyrö, H. Fast, H. Gernandt, H. Kanzawa, H. Nakane, V. Dorokhov, V. Yushkov, G. Hansen, M. Gil and R. J. Shearman, *Geophys. Res. Lett.*, 1998, **25**, 627.

23. D. Lucic, N. R. P. Harris, J. A. Pyle and R. L. Jones, *J. Atmos. Chem.*, 1999, **34**, 365.

24. F. Goutail, J.-P. Pommereau, C. Phillips, C. Deniel, A. Sarkissian, F. Lefèvre, E. Kyro, M. Rummukainen, P. Ericksen, S. Andersen, B.-A. Kåstad Høiskar, G. Braathen, V. Dorokhov and V. Khattatov, *J. Atmos. Chem.*, 1999, **32**, 1.

25. F. Goutail, J.-P. Pommereau, F. Lefèvre, M. Van Roozendael, S. B. Andersen, B.-A. Kåstad Høiskar, V. Dorokhov, E. Kyrö, M. P. Chipperfield and W. Feng, *Atmos. Chem. Phys.*, 2005, **5**, 665.

26. C. Deniel, J. P. Pommereau, R. M. Bevilacqua and F. Lefèvre, *J. Geophys. Res.*, 1998, **103**, 19231.

27. S. M. Uppala, P. W. Kallberg, A. J. Simmons, U. Andrae, V. da Costa Bechtold, M. Fiorino, J. K. Gibson, J. Haseler, A. Hernandez, G. A. Kelly, X. Li, K. Onogi, S. Saarinen, N. Sokka, R. P. Allan, E. Andersson, K. Arpe, M. A. Balmaseda, A. C. M. Beljaars, L. van de Berg, J. Bidlot, N. Bormann, S. Caires, F. Chevallier, A. Dethof, M. Dragosavac, M. Fisher, M. Fuentes, S. Hagemann, E. Holm, B. J. Hoskins, L. Isaksen, P. A. E. M. Janssen, R. Jenne, A. P. McNally, J.-F. Mahfouf, J.-J. Morcrette, N. A. Rayner, R. W. Saunders, P. Simon, A. Sterl, K. E. Trenberth, A. Untch, D. Vasiljevic, P. Viterbo and J. Woollen, *Q. J. Roy. Meteor. Soc.*, 2005, **131**, 2961.

28. M. H. Proffitt, J. J. Margitan, K. K. Kelly, M. Lowenstein, J. R. Podolske and K. Chan, *Nature*, 1990, **347**, 31.

29. M. H. Proffitt, K. Aikin, J. J. Margitan, M. Loewenstein, J. R. Podolske, A. Weaver, K. R. Chan, H. Fast and J. W. Elkins, *Science*, 1993, **261**, 1150.

30. S. Tilmes, R. Müller, J.-U. Grooß and J. M. Russell III, *Atmos. Chem. Phys.*, 2004, **4**(8), 2181.

31. R. Müller, S. Tilmes, P. Konopka, J.-U. Grooß and H.-J. Jost, *Atmos. Chem. Phys.*, 2005, **5**, 3139.

32. N. R. P Harris, M. Rex, F. Goutail, B. M. Knudsen, G. L. Manney, R. Müller and P. von der Gathen, Comparison of Empirically Derived Ozone Loss in the Arctic Vortex, *J. Geophys. Res.*, 2002, 107(D20), doi:10.1029/ 2001JD000482.

33. P. A. Newman and M. Rex, *Chapter 8 in Scientific Assessment of Ozone Depletion: 2006*, Global Ozone Research and Monitoring Project—Report No. 50, 572 pp., World Meteorological Organization, Geneva, Switzerland, 2007.

34. P. A. Newman, N. R. P. Harris, A. Adriani, G. T. Amanatidis, J. G. Anderson, G. O. Braathen, W. H. Brune, K. S. Carslaw, M. S. Craig,

P. L. DeCola, M. Guirlet, R. S. Hipskind, M. J. Kurylo, H. Küllmann, N. Larsen, G. J. Mégie, J-P. Pommereau, L. R. Poole, M. R. Schoeberl, F. Stroh, O. B. Toon, C. R. Trepte and M. Van Roozendael, *J. Geophys. Res.*, 2002, **107**, 10.1029/2001JD001303.

35. M. Rex, R. J. Salawitch, N. R. P. Harris, P. von der Gathen, G. O. Braathen, A. Schulz, H. Deckelmann, M. Chipperfield, B.-M. Sinnhuber, E. Reimer, R. Alfier, R. Bevilacqua, K. Hoppel, M. Fromm, J. Lumpe, H. Kuellmann, A. Kleinbohl, H. Bremer, M. von Konig, K. Kunzi, D. Toohey, H. Vömel, E. Richard, K. Aikin, H. Jost, J. B. Greenblatt, M. Loewenstein, J. R. Podolske, C. R. Webster, G. J. Flesch, D. C. Scott, R. L. Herman, J. W. Elkins, E. A. Ray, F. L. Moore, D. F. Hurst, P. Romashkin, G. C. Toon, B. Sen, J. J. Margitan, P. Wennberg, R. Neuber, M. Allart, B. R. Bojkov, H. Claude, J. Davies, W. Davies, H. De Backer, H. Dier, V. Dorokhov, H. Fast, Y. Kondo, E. Kyro, Z. Litynska, I. S. Mikkelsen, M. J. Molyneux, E. Moran, T. Nagai, H. Nakane, C. Parrondo, F. Ravegnani, P. Skrivankova, P. Viatte and V. Yushkov, Chemical loss of Arctic ozone in winter 1999/2000, *J. Geophys. Res.*, 2002, **107**, D20, 8276.

36. W. Feng, M. P. Chipperfield, S. Davies, B. Sen, G. Toon, J. F. Blavier, C. R. Webster, C. M. Volk, A. Ulanovsky, F. Ravegnani, P. von der Gathen, H. Jost, E. C. Richard and H. Claude, *Atmos. Chem. Phys.*, 2005, **5**, 139.

37. S. Tegtmeier, M. Rex, I .Wohltmann and K. Krüger, *Geophys. Res. Lett.*, 2008, **35**, L17801.

38. M. Rex, R. J. Salawitch, P. von der Gathen, N. R. P. Harris, M. P. Chipperfield and B. Naujokat, *Geophys. Res. Lett.*, 2004, **31**, LO4116.

39. N. R. P. Harris, R. Lehmann, M. Rex and P. von der Gathen, *Atmos. Chem. Phys. Disc.*, 2010, **10**, 8499.

40. G. L. Manney, M. L. Santee, M. Rex, N. J. Livesey, M. C. Pitts, P. Veefkind, E. R. Nash, I. Wohltmann, R. Lehmann, L. Froidevaux, L. R. Poole, M. R. Schoeberl, D. P. Haffner, J. Davies, V. Dorokhov, H. Gernandt, B. Johnson, R. Kivi, E. Kyrö, N. Larsen, P. F. Levelt, A. Makshtas, C T. McElroy, H. Nakajima, M. Conceptión Parrondo, D. W. Tarasick, P. von der Gathen, K. A. Walker and N. S. Zinoviev, *Nature*, 2011, in press.

41. N. R. P. Harris, R. Lehmann, M. Rex and P. von der Gathen, *Int. J. Rem. Sens.*, 2009, **30**, 15, 4065.

CHAPTER 6
Mid-latitude Ozone Depletion

M. P. CHIPPERFIELD

Institute for Climate and Atmospheric Science, School of Earth and Environment, University of Leeds, Leeds, U.K.

6.1 Introduction

The distribution of ozone in the stratosphere is maintained by a balance between photochemical production, transport and photochemical loss. As described in Chapter 1, the source of ozone in the stratosphere is the dissociation of O_2 molecules by short-wavelength ultraviolet radiation. The production of ozone is therefore strongest in the middle stratosphere in tropical regions. However, the largest stratospheric ozone columns are observed at higher latitudes, which, in itself, illustrates the important role of atmospheric dynamics in transporting ozone from the tropical source region polewards (see Chapter 1). Ozone is destroyed by a variety of catalytic chemical loss cycles which involve radicals from the O_x, NO_x, HO_x, Cl_x and Br_x chemical families (see Chapters 1 and 3). These chemical species can be present in the stratosphere naturally (*e.g.* water vapor is the source of HO_x) or their abundance can be greatly enhanced through human activity (*e.g.* human activity has increased the chlorine loading by about a factor 6 compared to natural levels). The efficiency of these loss cycles varies with altitude, latitude and season.

The distribution of ozone in the stratosphere may therefore change if either the rate of production, the speed of stratospheric circulation or the rate of chemical loss varies. As ozone production depends only on sunlight and O_2 then the only process which can (slightly) affect this is natural variations in solar radiation (*e.g.* the 11-year cycle). The most significant changes to the ozone layer will occur through changes in atmospheric transport or

Stratospheric Ozone Depletion and Climate Change
Edited by Rolf Müller
© Royal Society of Chemistry 2012
Published by the Royal Society of Chemistry, www.rsc.org

photochemical loss (*i.e.* destruction by thermal or photolytic chemical reactions). We use the term *depletion* to describe a situation where the local concentration or total column ozone has decreased over time. This may be due to either chemical or dynamical processes, or both. The term *loss* is used to describe the chemical destruction of ozone within an air mass.

The middle latitudes are considered to be the regions between 35° and 60° latitude in each hemisphere. The mid-latitudes are coupled to the large tropical regions (35°N–35°S) and are also affected by changes which occur in the respective polar regions. Changes in mid-latitude ozone can therefore occur due to (i) changes in the rate of transport of ozone from the tropical source to higher latitudes, (ii) changes in the rate of mixing of air between polar and mid-latitude regions, or (iii) changes in photochemical ozone loss. Relevant changes in photochemical loss may occur within the mid-latitudes, or may occur in regions which mix with mid-latitude air (*e.g.* dilution of low-ozone air from the polar region). In practice, observed changes in mid-latitude ozone are driven by a combination of many contributing factors.

The global abundance of stratospheric ozone is affected by many processes, both natural and related to human-induced atmospheric changes. The upper panel in Figure 6.1 shows the mean variation of extra-polar (60°S–60°N) column ozone from 1964 to 2006. There are variations on a range of timescales. Using a statistical model, this global mean variation has been decomposed into different contributions (or explanatory variables), which are also shown in Figure 6.1. The statistical fit shows a large annual cycle, a contribution from the 11-year solar cycle, variations due to the quasi-biennial oscillation (QBO) of the equatorial stratospheric winds, enhanced depletion after strong volcanic eruptions (which inject emissions directly into the stratosphere) and a downward trend until the late 1990s due to increasing stratospheric halogen loading (shown as Effective Equivalent Stratospheric Chlorine (EESC), see Chapter 2). The lower panel of Figure 6.1 shows the deviations in global mean column ozone after the effects of the solar cycle, volcanic eruptions and QBO have been removed. There is a clear correlation between this ozone change and the increasing abundance of EESC. However, it should be noted that obtaining a fit with a statistical model does not imply we understand the correct process causing the change, or even that the attribution of the model is correct. We also need to quantitatively understand the observed ozone changes using physical models which contain parameterisations of the relevant processes and which can be integrated forward in time to simulate the past variations and predict the future evolution of stratospheric ozone.

Section 6.2 of this chapter describes observations of recent changes in mid-latitude ozone both in the total column and in the profile. Section 6.3 discusses our current understanding of these observed changes in terms of chemical processes, dynamics and transport and other factors. For more detailed information on advances in our understanding of mid-latitude ozone the reader is referred to the WMO/UNEP Ozone Assessments.[1,2,3] A summary, with an outlook on how climate change may interact with mid-latitude ozone depletion, is given in Section 6.4. The future evolution of mid-latitude ozone is discussed in Chapter 9.

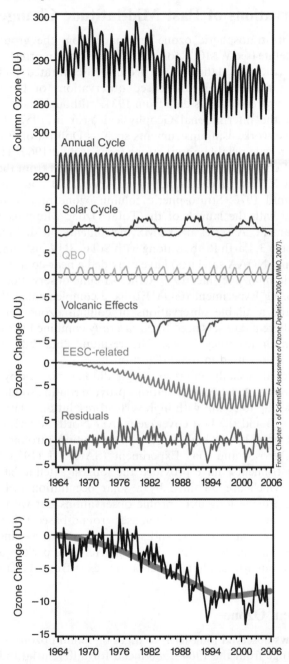

Figure 6.1 (Top) Column ozone variations for 60°S–60°N estimated from ground-based data and individual components that comprise ozone variations as derived from a statistical model. (Bottom) Deseasonalized, area-weighted total ozone deviations estimated from ground-based data adjusted for solar, volcanic, and QBO effects, for 60°S–60°N. The thick yellow line represents the EESC curve scaled to fit the data from 1964–2005. Prepared by V. Fioletov for WMO (2007).[1]

6.2   Observations of Past Mid-latitude Changes

Observations of stratospheric ozone are made from the ground, balloons, aircraft and satellite (see $WMO^{1,3}$ for a full discussion). In order to detect long-term changes, extensive data records from well-calibrated and understood instruments are required. Ground-based observations of the total column ozone suitable for trend studies date from 1931,[4] although the data calibration improved after the International Geophysical Year in 1957. Therefore, the ground-based network, using instruments such as Dobson or Brewer spectro-photometers, provide reliable observations from around 1960. The first global observations of column ozone from space were obtained from the Backscatter Ultraviolet (BUV) instrument launched in April 1970 and which provided reliable data until 1972. Stratospheric column ozone monitoring started in earnest in 1979 with the launch of the Total Ozone Mapping Spectrometer (TOMS) instrument. A series of TOMS instruments on different spacecraft (Nimbus 7, Meteo 3, Earth Probe), along with Solar BUV instruments (Nimbus 7 and operational NOAA satellites) have provided a near-continuous record of daily global column ozone since 1979. Other satellite observations (*e.g.* Global Ozone Monitoring Experiment (GOME)) have complemented this record. In order to use the available observations of column ozone for detection of changes, the different datasets need to be carefully combined taking account of any calibration or accuracy issues. The resultant 'merged' column ozone datasets are generally used in trend detection.[5]

Ozone profiles are usually observed using balloon sondes or by limb-viewing satellite instruments. A network of stations provide routine sonde observations of ozone profiles up to 33 km with high vertical resolution. Data are available from around 1970 and the best coverage is over North America, Europe and Japan. Sondes provide the highest vertical resolution (around 100 m). The Stratospheric Aerosol and Gas Experiment (SAGE) I (1979–1981) and II (1984–2005) instruments[6] have provided the longest satellite-based record of ozone profiles which are obtained by a solar occultation technique. Other techniques can provide long-term profile observations but with poor vertical resolution (*e.g.* Umkehr technique which uses ground-based Brewer or Dobson spectrophotometer data at sunrise and sunset,[7] or microwave instruments) or high vertical resolution but so far only for shorter time periods and for a few locations (*e.g.* ground-based lidar observations exist from the late 1980s).[8]

6.2.1   Column Ozone

Figure 6.2 shows observed long-term changes of column ozone, expressed as a percentage change from the 1964–1980 mean for different latitude regions. In each case the data has been deseasonalised (*i.e.* average seasonal cycle sub-tracted) and the plots show both the smoothed monthly changes and the annual averages. For global ozone (90°S–90°N) depletion of ozone appears to begin around 1980. Ozone decreases through 1993, after which time it increases, though there is the indication of increased depletion around 2006.

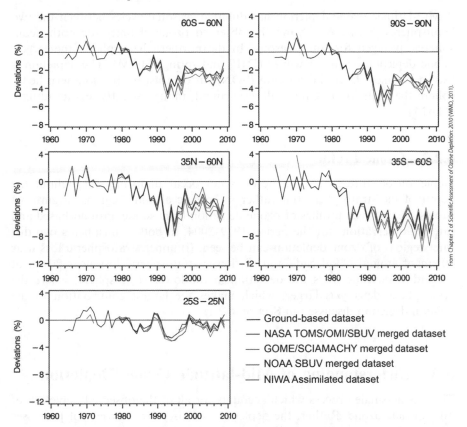

Figure 6.2 Annual mean area-weighted total ozone deviations from the 1964–1980 means for the latitude bands 90°S–90°N, 60°S–60°N, 25°S–25°N, 35°N–60°N, and 35°S–60°S, estimated from different global datasets: ground-based (black), NASA TOMS/OMI/SBUV(/2) merged satellite data set (red), National Institute of Water and Atmospheric Research (NIWA) assimilated data set (magenta), NOAA SBUV(/2) (blue), and GOME/SCIAMACHY merged total ozone data (green). Each dataset was deseasonalized with respect to the period 1979–1987. The average of the monthly mean anomalies for 1964–1980 estimated from ground-based data was then subtracted from each anomaly time series. Deviations are expressed as percentages of the ground-based time average for the period 1964–1980. Updated from Fioletov *et al.*[9] and taken from WMO (2011).[3]

A component of this global ozone depletion is the springtime loss, which occurs in the polar regions (see Chapter 5), and the plot for the extra-polar regions (60°S–60°N) shows correspondingly less depletion—a peak of around 5% in the mid-1990s. The middle latitudes contain the bulk of this extra polar column ozone depletion. In the northern mid-latitudes ozone depletion peaked at around 8% in the early 1990s, while at southern mid-latitudes, ozone depletion continued increasing through to the early 2000s.

Mid-latitude ozone depletion has different seasonal cycles between the two hemispheres. Figure 6.3 shows the observed ozone changes for both hemispheres averaged for four seasons. In the northern hemisphere, the largest ozone depletion occurs in winter (DJF) and spring (MAM). The depletion is less in summer (JJA) and autumn (SON). In contrast, in the southern hemisphere the depletion is similar all year round, though slightly less in autumn (MAM).

6.2.2 Ozone Profile

Depletion of stratospheric ozone does not occur uniformly throughout the depth of the stratosphere but rather in specific altitude regions. Figure 6.4 shows the observed profiles of ozone depletion from sonde, ground-based and satellite observations for the period 1979–2004. In both hemispheres two distinct regions of ozone depletion can be seen: (i) upper stratospheric loss near 40 km of around 6%/decade and (ii) lower stratospheric loss near 20 km of around 4%/decade. It is the depletion in the lower stratosphere, where the atmospheric density is larger, which makes the largest contribution to the column depletion discussed in Section 6.2.1.

6.3 Understanding of Mid-latitude Ozone Depletion

There is no single process which accounts for all of the observed depletion of mid-latitude ozone. Rather, the depletion is caused by a variety of processes whose relative importance varies as a function of time, hemisphere and/or altitude. In the upper stratosphere, where ozone has a photochemical lifetime of days to weeks, then photochemical processes will always dominate. However, in the lower stratosphere, where the photochemical lifetime of ozone can be months to years, both chemical and dynamical processes are important (see Figure 1.2 in Chapter 1).

6.3.1 Chemical Processes

Upper Stratosphere

In the photochemically controlled upper stratosphere region above about 25 km, ozone trends may be driven by trends in the gases that provide the sources for reactive radicals (such as halocarbons, which produce ClOx, N_2O, which produces NO_x, and H_2O, which produces HO_x, *etc.*). However, the dominant driver for ozone trends in this region has been the observed trends in reactive chlorine (ClO_x), which is due in turn to the known trends in halocarbons. Although most chlorine released from halocarbons would exist as the stable reservoir HCl in the upper stratosphere, a small fraction

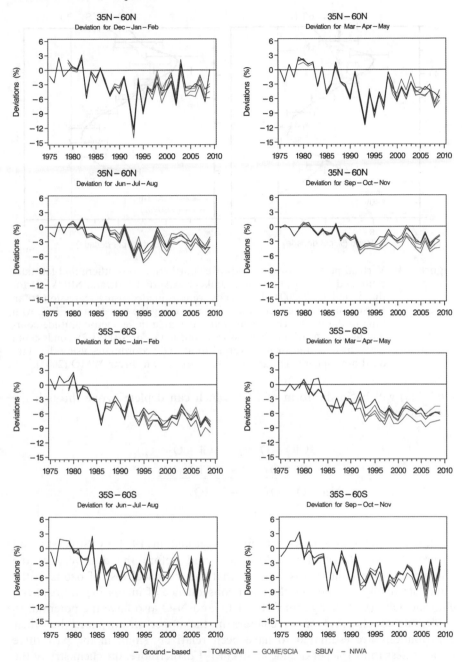

Figure 6.3 Seasonal area-weighted total ozone deviations from the 1964–1980 means, calculated for four seasonal averages, for the latitude bands 35°N–60°N (top) and 35°S–60°S (bottom). Updated from Fioletov *et al.*[9]

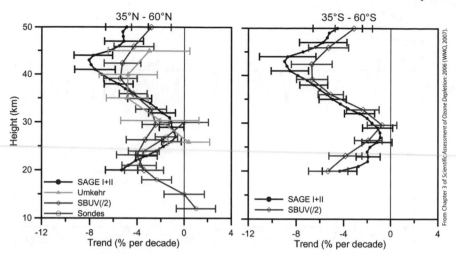

Figure 6.4 Vertical profile of ozone trends over northern and southern mid-latitudes estimated from ozonesondes, Umkehr, SAGE I + II, and SBUV(/2) for the period 1979–2004. The trends were estimated using regression to an EESC curve and converted to %/decade using the variation of EESC with time in the 1980s. The trends were calculated in geometric altitude coordinates for SAGE and in pressure coordinates for SBUV(/2), sondes, and Umkehr data and then converted to altitude coordinates using the standard atmosphere. The 2σ error bars are shown. From WMO (2007).[1]

(10–15%) will be in the form of ClO, which can deplete ozone through the catalytic cycle:

$$\begin{array}{rcl} ClO + O & \rightarrow & Cl + O_2 \\ Cl + O_3 & \rightarrow & ClO + O_2 \\ \hline \text{Net:} \quad O + O_3 & \rightarrow & 2O_2 \end{array}$$

This was the ozone loss cycle first suggested in 1974 by Stolarski and Cicerone[10] and Rowland and Molina,[11] and before the discovery of the Antarctic Ozone Hole it was believed that this chemical loss of ozone in the upper stratosphere would be the main consequence of increasing atmospheric chlorine. Observed changes in CH_4, H_2O, and N_2O also have the potential to contribute to chemical ozone changes in the upper stratosphere, but their contributions are estimated to have been relatively small in the past three decades (less than 1% per decade at 40 km.)[1] Furthermore, the chemistry of the upper stratosphere is strongly dependent upon temperature, so that it is important to consider the feedbacks between ozone depletion and temperature changes in model estimates, and to consider changes in radiatively important gases (especially CO_2 and H_2O). In the upper stratosphere, ozone chemical loss cycles proceed more slowly at colder temperatures and so a cooling of the stratosphere will lead to more ozone (see Chapter 1, Figure 1.2). Climate change, driven by the increase in IR radiatively active gases, will lead (and has already led) to a cooling of the upper stratosphere, thereby accelerating the recovery of upper stratospheric ozone due to the decrease of EESC (Chapters 2 and 9). Eventually, the cooling of the upper stratosphere will lead to higher ozone concentrations than before the onset of halogen-driven ozone depletion, and therefore to a so-called 'super-recovery'.

Lower Stratosphere

In the lower stratosphere, where the chemical lifetime of ozone is comparable to or longer than the transport time scale, both dynamical and chemical processes have the potential to affect ozone and its trends. In particular, changes in chemical loss processes may deplete ozone, but only if they can remove it rapidly enough to compete with the seasonally varying transport that moves ozone through this region of the atmosphere.

In the lower stratosphere, heterogeneous reactions on sulfate aerosols convert NO_y species to HNO_3 and, under cold conditions can activate chlorine (*i.e.* convert chlorine from reservoir species which have no direct impact on ozone to radical species which destroy ozone—see Chapter 4; Bormann *et al.*)[12] The most important heterogeneous reactions are:

$$N_2O_5 + H_2O \quad \rightarrow \quad 2HNO_3$$
$$ClONO_2 + H_2O \quad \rightarrow \quad HNO_3 + HOCl$$
$$ClONO_2 + HCl \quad \rightarrow \quad Cl_2 + HNO_3$$

These reactions also occur in the polar regions on cold sulfate aerosol and other surfaces where they are a key step in causing rapid springtime ozone depletion (Chapter 4). At mid-latitudes the reactions proceed more slowly but they can also cause a repartitioning of chlorine species from the reservoirs to more active forms. The stratospheric aerosol loading varies with time; it is significantly enhanced after major volcanic eruptions which reach the stratosphere. The eruption of El Chichón in 1982 and Mt Pinatubo in 1991 caused significant increases (around two orders of magnitude in optical depth) in stratospheric aerosols.[13] Once in the active form, chlorine can lead to ozone depletion *via* the catalytic cycles which also involve bromine and HO_x species:

$$ClO + BrO \quad \rightarrow \quad Br + Cl + O_2$$
$$Br + O_3 \quad \rightarrow \quad BrO + O_2$$
$$Cl + O_3 \quad \rightarrow \quad ClO + O_2$$
$$\text{Net:} \quad 2O_3 \quad \rightarrow \quad 3O_2$$

$$\begin{array}{rcl}
ClO + HO_2 & \rightarrow & HOCl + O_2 \\
HOCl + h\nu & \rightarrow & OH + ClO \\
Cl + O_3 & \rightarrow & ClO + O_2 \\
OH + O_3 & \rightarrow & HO_2 + O_2 \\
\hline
\text{Net:} \quad 2O_3 & \rightarrow & 3O_2
\end{array}$$

In this way, long-term trends in stratospheric chlorine and bromine can contribute to *in situ* mid-latitude ozone depletion.

As discussed in Chapter 2, halogens reach the stratosphere through the transport of halocarbon source gases from the troposphere. These gases tend to have very long lifetimes in the lower atmosphere, typically many tens of years. However, recently attention has focused on the contribution that transport of short-lived bromine species (species with tropospheric lifetimes of less than six months) may have on stratospheric ozone. At present, short-lived bromine species are believed to contribute about 5 pptv out of a total stratospheric bromine loading of around 21 pptv (Chapter 2; WMO).[2] This additional bromine will enhance mid-latitude ozone depletion.[14,15] Figure 6.5 shows calculations with a 2D model and assumptions of different additional bromine loadings. The impact of the additional bromine is most significant at times of high aerosol loading just after 1982 and 1991, due to the fact that BrO destroys O_3 in a coupled cycle with ClO, and ClO is enhanced by heterogeneous chemistry during these periods.

Effect of Polar Vortex Loss on Mid-latitudes

Ozone depletion processes in polar lower stratosphere can also impact mid-latitudes. This can occur either through the export of ozone-depleted air or by the export of chemically activated air masses, which can continue to destroy ozone in the mid-latitudes. Export of vortex air may occur during the winter/spring, depending on the strength of the polar vortex, and this process will occur earlier in the season for the Arctic where the vortex is weaker and more disturbed. In both hemispheres, when the polar vortex breaks down low-ozone air will be mixed into mid-latitudes—the so-called 'dilution effect'. Given the larger loss in the Antarctic vortex, the dilution effect will be more significant in the southern hemisphere. As polar ozone loss is driven by chlorine and bromine chemistry (see Chapter 5) it is reasonable to consider these processes as ultimately a chemical influence, although transport is clearly important in advecting the low ozone or activated air masses into the mid-latitudes.

Model studies have quantified the effect of Arctic polar processing on mid-latitude ozone amounts. Knudsen and Grooß[16] used a seasonal reverse domain filling (RDF) trajectory calculation to study the dilution of ozone depleted air into northern mid-latitudes. They estimated that approximately 40% of the

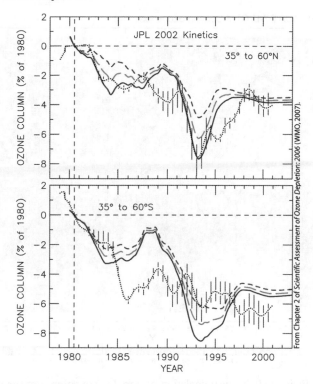

Figure 6.5 Calculated change in column ozone relative to 1980 levels using a 2D model which includes additional stratospheric bromine from very short-lived species (VSLS) of 0 (red), 4 (green) and 8 (blue) pptv for 35°N–60°N (top) and 35°S–60°S (bottom) compared with trends in total column ozone. Based on Salawitch *et al.*[14] and WMO (2007).[1]

6.8% decline in mid-latitude ozone observed by TOMS between 1979 and 1997 could be accounted for by transport of polar ozone depleted air into mid-latitudes. Other studies have used tracers mapped in equivalent latitude (a transformation of coordinates based on the area contained within contours of constant chemical concentration or potential vorticity and which accounts for distortions to the polar vortex), or discriminated between polar and non-polar air in 3D CTMs, to evaluate the connection between polar and mid-latitude ozone loss. From the analysis of several Arctic winters in the late 1990s, Millard *et al.*[17] showed that the contribution to the seasonal mid-latitude ozone loss from high latitudes was strongly dependent on the meteorological conditions and the stability of the polar vortex. They found the largest contribution to mid-latitude ozone loss for 1999/2000 with half the mid-latitude loss originating north of 60°N. A much smaller contribution was found in the winter 1996/1997 characterized by a strong and pole-centered vortex prior to the final breakup. Using a different model and set of tracers, Marchand *et al.*[18] found similar results for the winter 1999/2000. Godin *et al.*[19] used a high-resolution model together with ground-based ozone measurements to study the influence of

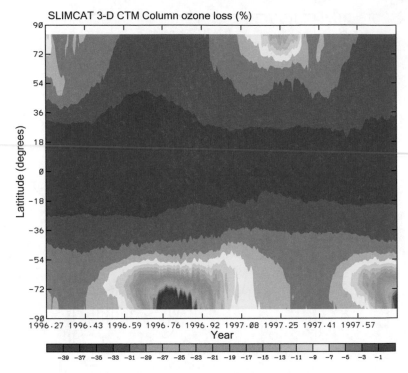

Figure 6.6 Estimated chemical depletion of zonal mean column ozone (%) from the difference between two 3D CTM experiments (with/without chlorine activating heterogeneous reactions) for the mid-1990s. From Chipperfield.[20]

vortex excursions and polar air filaments on mid-latitude ozone amounts during several winters, finding the largest impact during 1999/2000 due to the large Arctic ozone loss that occurred that winter coupled with several vortex excursions at the end of the winter.

The accumulated effect of polar vortex loss on mid-latitudes has been studied by Chipperfield[20] using multiannual simulations of the 1990s using a 3D CTM with full chemistry. They estimated that overall depletion processes related to heterogeneous chlorine activation at high/mid-latitudes results in 2–3% less ozone at 50°N throughout the year, of which 1% was due to processing at mid-latitudes. The effect was larger in the southern hemisphere (5% less O_3 at 50°S) reflecting the large losses occurring in the Antarctic (Figure 6.6).

6.3.2 Dynamical Contributions

Transport of stratospheric ozone is a key factor influencing its seasonal and interannual variability. The large seasonal cycle in column ozone over the mid-latitudes (in both hemispheres), with maximum columns occurring in spring (see Chapter 1 Figure 1.5), is due primarily to enhanced transport from the tropical source region during the winter–spring seasons, and the hemispheric differences in ozone amount are mainly a result of differential transport (larger in the NH). The interannual variability in the winter–spring ozone buildup is greater in the northern hemisphere than in the south, reflecting the greater planetary wave activity in the northern hemisphere stratosphere (Chapter 1). Given the observed large interannual and decadal-scale variability in stratospheric dynamical quantities, it is reasonable to consider a dynamical influence on decadal ozone trends on both regional and hemispheric scales (*e.g.* Hood and Zaff).[21]

Changes in two specific dynamical transport processes can significantly influence mid-latitude ozone trends. These are:

(1) Changes in tropospheric circulation, particularly changes in the frequency of local non-linear synoptic wave forcing events, which lead to the formation of extreme ozone minima ('mini-holes') due to large increases in tropopause height (Steinbrecht *et al.*;[22] Hood *et al.*;[23,24] Reid *et al.*;[25] Orsolini and Limpasuvan;[26] Brönnimann and Hood;[27] Hood and Soukharev;[28] Koch *et al.*;[29] Mangold *et al.*[30]

(2) Interannual and long-term changes in the strength of the stratospheric mean meridional (Brewer-Dobson) circulation, which is responsible for the winter–spring build-up of extratropical ozone (*e.g.* Fusco and Salby;[31] Randel *et al.*;[32] Weber *et al.*;[33] Salby and Callaghan;[34] Hood and Soukharev;[28] Harris *et al.*)[35]

The relationship between tropopause height, mid tropospheric temperature and column ozone is illustrated by Figure 6.7 using data from the Hohenpeissenberg station. For this location Steinbrecht *et al.*[36] found that ∼1/3 of the total ozone trend could be explained by tropopause height changes. Other studies have shown a similar role for tropopause height changes at northern midlatitudes.[37]

Another method to assess the impact of dynamical changes on mid-latitude ozone is to use an off-line 3D chemical transport model forced by analysed winds. In principle these winds contain information about real atmospheric variability. Hadjinicolaou *et al.*[38,39] used this approach to show that dynamical changes enhanced ozone depletion at northern mid-latitudes after the eruption of Mt Pinatubo in 1991. However, this approach needs to be used with caution. Analyzed wind datasets, such as the ECMWF ERA-40, may contain spurious long-term variations (*e.g.* due to assimilation of different datasets) as discussed in Feng *et al.*[15] and Monge-Sanz *et al.*[40] Also, this approach does not give any information about the cause of the dynamical changes in the analyzed winds. Full chemistry off-line CTMs which use these windfields will implicitly contain this dynamical information.

Decadal ozone changes induced by these dynamical processes can be produced from natural variability, from changes in climate resulting from the increase in greenhouse gases such as CO_2, CH_4, N_2O, or from ozone depletion

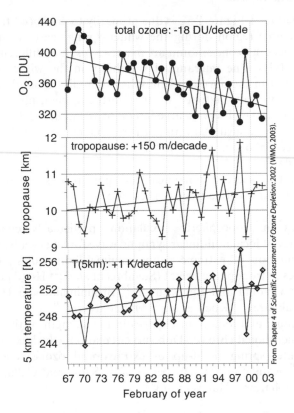

Figure 6.7   February monthly means of total ozone (top), tropopause height (middle) and temperature at 5 km altitude (bottom) at Hohenpeissenberg (47.8°N, 11.0°E). Updated from Steinbrecht *et al.*[36] through February 2002. From WMO (2003).[2]

itself. The importance of knowing the cause depends on the issue being addressed. To quantify the ozone depletion that can be attributed to ozone-depleting substances such as CFCs and halons, the causes of the dynamical influences on ozone changes are unimportant, as long as they are not a result of the chemical ozone depletion itself. However, to make valid predictions of future ozone levels, it is important to understand the fundamental causes of the dynamical influence on ozone changes. Thus, it is important to distinguish between the identification of dynamical contributions to ozone changes, and the explanation of the processes causing the dynamical contribution.

The effects discussed above concern the direct impact of transport on ozone. In addition to these purely dynamical effects, there are chemical and radiative feedbacks between temperature and ozone in the lower stratosphere. These chemical and radiative feedbacks may amplify or skew the effect of natural dynamical variability on ozone. It is thus not possible to cleanly separate 'dynamical' and 'chemical' contributions to ozone changes, attributing a certain fraction to each contribution, because they can be nonlinearly coupled.

Whether the coupling is a first-order or a second-order effect, and what the causality of the relationship is between dynamics and ozone, has to be assessed on a case-by-case basis. Interestingly, Eyring *et al.*[41] found that the impact of greenhouse gas increases and chlorine/bromine changes in northern and southern mid-latitudes were approximately additive.

6.3.3   Other Factors Affecting Mid-latitude Ozone

Other factors may potentially affect the mid-latitude ozone depletion. As shown in Figure 6.1, the 11-year solar cycle exerts a natural decadal variation in column ozone. This column variation has an amplitude of 2–3% from solar minimum to solar maximum over extrapolar latitudes (*e.g.* Hood.)[42]

While halogen species have shown the most significant atmospheric changes over the past few decades, and therefore are the main driver for past chemical ozone depletion, changes in other source gases could potentially affect ozone. N_2O is the source of stratospheric NO_y, which is the main sink of ozone in the mid-stratosphere. H_2O is the souce of stratospheric HO_x species which destroy ozone in the lower and upper stratosphere. Any significant trend in these source gases could, in the future, cause changes to stratospheric ozone (see Chapter 9).

6.3.4   Assessment Model Calculations

Atmospheric numerical models are a mathematical representation of our best current understanding of the physico-chemical processes determining the state of the atmosphere. They contain descriptions of relevant dynamical, radiative and chemical processes. An essential part of model development is testing of the model components by comparison with observation (*e.g.* Eyring *et al.*)[43] This model evaluation is best performed initially at the 'process' level, *e.g.* testing that the model chlorine chemistry has the correct response to changes in aerosol loading compared to observations. When we have confidence in the accuracy of the processes contained in a model, it can be used to simulate the overall behavior of the stratospheric ozone layer. As many processes contribute to ozone depletion, an important test of our understanding is whether a model can reproduce the observed *fingerprint, i.e.* the correct profile, latitudinal, seasonal, and long-term variation of the ozone changes.

The inclusion of all relevant processes in a full 3D model of the stratosphere and troposphere is very computationally demanding and it is only recently that long (multi-decadal) simulations of the development of the ozone layer have been performed with 3D models. The overall ability of models to reproduce the observed changes in stratospheric ozone was most recently evaluated in SPARC (2010)[44] and WMO (2011).[3] The previous WMO (2007)[1] report still largely relied on 2D models, although results from some 3D models (off-line chemical transport models (CTMs) and coupled chemistry-climate models (CCMs) were available). Progress in 3D CCMs, and available computer power,

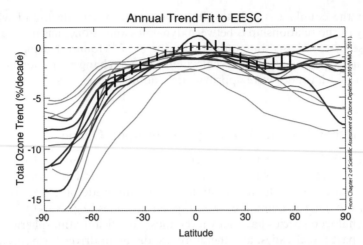

Figure 6.8 Latitude dependence of the annual total ozone trend for 1979–1995 (%/decade) from 13 CCMs used in SPARC (2010) and a merged total ozone dataset. Trends derived from the measurements with a 2σ uncertainty are indicated by the vertical bars. Red lines are from 3 CCMs that rated highest in relevant SPARC (2010) tests. Blue lines are from two models which also performed well these tests. Model trends for 1978–1996 are calculated by fitting to EESC using output from 1978 to the end of the model runs in 2005. From WMO (2011).[3]

has led to the increased coordinated use of these models interpreting ozone changes and they are now the main assessment tool (see SPARC 2010.)[44]

Here we use results from the WMO (2011) Assessment to illustrate how well stratospheric models can reproduce observed past ozone changes. Figure 6.8 shows a comparison between the observed trends in column ozone from 1979–1995, as a function of latitude, with 13 3D models from different research groups. These model simulations used the observed changes in chlorine and bromine source gases (and other greenhouse gases) and used satellite-based observations of aerosol loading. The observations show an annual mean trend near zero in the tropics, and a larger trend in the southern hemisphere mid-latitudes than the north. There is a spread of results between the models, which is one measure of the model uncertainty, but generally the models reproduce this latitudinal behavior. All of the models, however, predict a non-zero trend in the tropics. There is a large spread in modeled trends, especially in the poles where the model variability is large. Detailed process-based model evaluation performed as part of the SPARC CCMVal (2010)[44] report allows us to select the best performing models and thereby reduce this model spread.

Figure 6.9 compares the observed and modeled trends in the ozone profile at northern mid-latitudes for two time periods: 1979–1996 and 2000–2009. For the earlier period, corresponding to the increase in stratospheric halogens (EESC), the 3D chemistry-climate models capture the double peak structure of the observations, with upper stratospheric loss near 40 km, due to gas phase

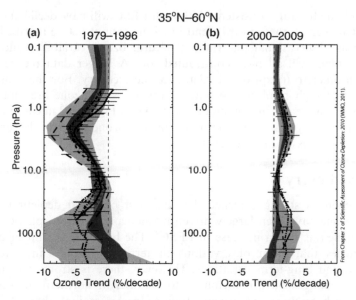

Figure 6.9 Altitude profiles of calculated ozone trend (%/decade) at northern mid-
latitudes from 17 CCMs. (a) Trends for 1979–1996 are obtained by fitting
to EESC. Trends derived from measurements with 2σ uncertainties are
indicated by horizontal lines. The solid line is the trend derived from
SAGE I + SAGE II satellite data. The dashed line is the trend derived
from SAGE-corrected SBUV(/2) data. The dash-dot line is the trend
derived from SBUV(/2) data. The dash-dot-dot line is the trend derived from
sonde data. The long dashed line is the trend derived from Umkehr data.
Red shading is the range of three models with high ratings in all relevant
SPARC (2010) evaluations. Blue shading adds the range of three models that
scored highly in at least one relevant evaluation. (b) Trends for the period
2000–2009. Shaded regions are the same as (a). From WMO (2011).[3]

chemistry, and a lower stratospheric peak near 20 km due to a combination of
aerosol-related chlorine/bromine chemistry and dynamical changes. The range
of model results (shown by the color shading) spans the observations. In
general there is a larger spread in the lower stratosphere where ozone is controlled by both chemical and dynamical processes. Again, the SPARC (2010)
process-based analysis of CCM performance can be used to select the models
which give the most realistic simulation of the processes which control ozone.
These models, indicated by the red and blue shading, show a much smaller
spread and good agreement with the observed trends. This figure therefore
shows that we have a good understanding of the causes of past ozone trends.
Figure 6.9b shows the observed and modeled trends for 2000–2009. This corresponds to the period when ozone 'recovery' is expected. Recovery is used
to describe the increase in ozone due to decreases in chlorine and bromine.
Other processes, such as dynamical changes and stratospheric cooling can also
cause ozone increases and this will mask the signal of recovery due to halogen
decreases. Figure 6.9b shows an upward trend in ozone from observations and

the models are broadly consistent. Although EESC will have declined over this period (Chapter 2), the upward trend in ozone cannot yet be ascribed to this decrease as observational uncertainty; natural ozone variability and stratospheric cooling will all have contributed too. A longer data record will be needed in order to unequivocally detect ozone recovery. For this reason it is essential that high resolution profile observations of ozone are maintained. Similarly, the increases in column ozone shown in Figure 6.2 in certain regions over the past few years cannot yet be ascribed to ozone recovery. Future ozone changes are discussed in Chapter 9.

6.4   Summary

Overall, there is strong evidence that the past observed depletion in mid-latitude ozone have been largely driven by increases in stratospheric chlorine and bromine (*i.e.* by an increase in EESC). The basic fingerprint of modeled loss matches that from observations in terms of profile and latitudinal variations and long-term variations. However, there is strong evidence that dynamical changes have contributed to changes in northern mid-latitudes, which have been more extensively studied. In this region, about 1/3 of the ozone trend over the last few decades can be ascribed to dynamical changes (depending on the period analysed).

Some significant issues of attribution remain however. Models show that large volcanic eruptions should enhance mid-latitude ozone loss and they produce a large signal in both the northern and southern hemispheres. However, observations in the southern hemisphere do not show a clear enhancement of loss in the mid-1990s. It may be that models overestimate the role of mid-latitude aerosols in chlorine activation or that dynamical changes are masking the chemical loss. In a recent study based on the CCMVal (2010) models, Gillett *et al.*,[45] however, found that they did reproduce the observed natural signal (combined impact of solar and volcanic influence). This needs to be investigated further.

In the future, the influence of halogen species will decrease (Chapter 2) and the sign of the chemical loss will reverse. Changes in dynamics, for example caused by climate change, will still exert a strong influence on column ozone. These issues will be discussed further in Chapters 8 and 9. Although stratospheric chlorine and bromine have already started to decrease, it is not yet possible to unequivocally detect an associated increase in ozone (so-called recovery). Ozone increases have been observed over the past decade, but these increases will also have partly been caused by climate changes (*e.g.* stratospheric cooling; dynamical changes). Future analysis of longer data records will be required in order to definitively detect ozone recovery.

Acknowledgements

This chapter draws heavily on the content of recent WMO/UNEP Ozone Assessments. I am grateful to fellow lead authors of the Global Ozone chapters,

Bill Randel (2002), Vitali Fioletov (2006, 2010), and Anne Douglass (2010), along with all the other chapter coauthors. I am grateful to Sophie Godin-Beekman and Nathan Gillett for helpful review comments.

References

1. WMO, *Scientific Assessment of Ozone Depletion*, 2006, Report no 50, WMO Geneva, 2007.
2. WMO, *Scientific Assessment of Ozone Depletion*, 2002, Report no 47, WMO Geneva, 2003.
3. WMO, *Scientific Assessment of Ozone Depletion*, 2010, Report no 52, WMO Geneva, 2011.
4. H. E. Rieder, J. Stahelin, J. A. Mader, T. Peter, M. Ribatet, A. C. Davison, R. Stübi, P. Weihs and F. Holawe, *Atmos. Chem. Phys.*, 2010, **10**, 10033–10045.
5. G. E. Bodeker, H. Shiona and H. Eskes, *Atmos. Chem. Phys.*, 2005, **5**, 2603.
6. M. P. McCormick, J. M. Zawodny, R. E. Viega, J. C. Larsen and P.-H. Wang, *Planet. Space Sci.*, 1989, **37**, 1567–1586.
7. H. U. Dütsch and J. Staehelin, *J. Atmos. Terr. Phys.*, 1992, **54**, 557–569.
8. T. Leblanc and I. S. McDermid, *J. Geophys. Res.*, 2000, **105**, 14 613–14 623.
9. V. E. Fioletov, G. E. Bodeker, A. J. Miller, R. D. McPeters and R. Stolarski, *J. Geophys. Res.*, 2002, **107**, 4647.
10. R. S. Stolarski and R. J. Cicerone, *Canad. J. Chem.*, 1974, **52**, 1610.
11. M. J. Molina and F. S. Rowland, *Nature*, 1974, **249**, 810–812.
12. S. Bormann, S. Solomon, L. Avallone, D. Toohey and D. Baumgardner, *Geophys. Res. Lett.*, 1997, **24**, 2011–2014.
13. S. Solomon, R. W. Portmann, R. R. Garcia, L. W. Thomason, L. R. Poole and M. P. McCormick, *J. Geophys. Res.*, 1996, **101**, 6713–6727.
14. R. J. Salawitch, D. K. Weisenstein, L. J. Kovalenko, C. E. Sioris, P. O. Wennberg, K. Chance, M. K. W. Ko and C. A. McLinden, *Geophys. Res. Lett.*, 2005, **32**, L05811, DOI: 10.1029/2004GL021504.
15. W. Feng, M. P. Chipperfield, M. Dorf, K. Pfeilsticker and P. Ricaud, *Atmos. Chem. Phys.*, 2007, **7**, 2357–2369.
16. B. M. Knudsen and J.-U. Grooß, *J. Geophys. Res.*, 2000, **105**, 6885–6890, DOI: 10.1029/1999JD901076.
17. G. A. Millard, A. M. Lee and J. A. Pyle, *J. Geophys. Res.*, 2003, **108**, 8323, DOI: 10.1029/2001JD000899.
18. M. Marchand, S. Godin, A. Hauchecorne, F. Lefèvre, S. Bekki and M. Chipperfield, *J. Geophys. Res.*, 2003, **108**, 8326, DOI: 10.1029/2001JD000906.
19. S. Godin, M. Marchand, A. Hauchecorne and F. Lefèvre, *J. Geophys. Res.*, 2002, **107**, 8272.
20. M. P. Chipperfield, Multiannual simulations with a three-dimensional chemical transport model, *J. Geophys. Res.*, 1999, **104**, 1781–1805.

21. L. L. Hood and D. A. Zaff, *J. Geophys. Res.*, 1995, **100**, 25791–25800.

22. W. Steinbrecht, H. Claude and U. Köhler, *J. Geophys. Res.*, 1998, **103**, 19183–19192.

23. L. L. Hood, J. P. McCormack and K. Labitzke, *J. Geophys. Res.*, 1997, **102**, 13079–13093.

24. L. L. Hood, S. Rossi and M. Beulen, *J. Geophys. Res.*, 1999, **104**, 24321–24339.

25. S. J. Reid, A. F. Tuck and G. Kildaris, *J. Geophys. Res.*, 2000, **105**, 12169–12180.

26. Y. J. Orsolini and V. Limpasuvan, *Geophys. Res. Lett.*, 2001, **28**, 4099–4102.

27. S. Brönnimann and L. L. Hood, *Geophys. Res. Lett.*, 2003, **30**, 2118, DOI: 10.1029/2003GL018431.

28. L. L. Hood and B. E. Soukharev, *J. Atmos. Sci.*, 2005, **62**, 3724–3740.

29. G. Koch, H. Wernli, C. Schwierz, J. Staehelin and T. Peter, *Geophys. Res. Lett.*, 2005, **32**, L12810, DOI: 10.1029/2004GL022062.

30. A. Mangold, J.-U. Grooß, H. DeBacker, O. Kirner, R. Ruhnke and R. Müller, *Atmos. Chem. Phys.*, 2009, **9**, 6429–6451, DOI:10.5194/acp-9-6429-2009.

31. A. C. Fusco and M. L. Salby, *J. Clim.*, 1999, **12**, 1619–1629.

32. W. J. Randel, F. Wu and R. Stolarski, *J. Meteorol. Soc. Japan*, 2002, **80**(4B), 849–862.

33. M. Weber, S. Dhomse, F. Wittrock, A. Richter, B.-M. Sinnhuber and J. P. Burrows, *Geophys. Res. Lett.*, 2003, **30**, 37.

34. M. L. Salby and P. F. Callaghan, *J. Clim.*, 2004, **17**, 4512–4521.

35. N. R. P. Harris, E. Kyra, J. Staehelin, D. Brunner, S.-B. Andersen, S. Godin-Beekmann, S. Dhomse, P. Hadjinicolaou, G. Hansen, I. Isaksen, A. Jrrar, A. Karpetchko, R. Kivi, B. Knudsen, P. Krizan, J. Lastovicka, J. Maeder, Y. Orsolini, J. A. Pyle, M. Rex, K. Vanicek, M. Weber, I. Wohltmann, P. Zanis and C. Zerefos, *Ann. Geophys.*, 2008, **26**, 1207–1220.

36. W. Steinbrecht, H. Claude, U. Köhler and P. Winkler, *Geophys. Res. Lett.*, 2001, **28**, 1191–1194.

37. P. M. de F. Forster and K. Tourpali, *J. Geophys. Res.*, 2001, **106**, 12 241–12 251.

38. P. Hadjinicolaou, J. A. Pyle, M. P. Chipperfield and J. A. Kettleborough, *Geophys. Res. Lett.*, 1997, **24**, 2993–2996.

39. P. Hadjinicolaou, A. Jrrar, J. A. Pyle and L. Bishop, *Q. J. Roy. Meteorol. Soc.*, 2002, **128**, 1393–1412.

40. B. Monge Sanz, M. P. Chipperfield, A. Simmons and S. Uppala, *Geophys. Res. Lett.*, 2007, **34**, L04801, DOI:10.1029/2006GL028515.

41. V. Eyring, I. Cionni, G. E. Bodeker, A. J. Charlton-Perez, D. E. Kinnison, J. F. Scinocca, D. W. Waugh, H. Akiyoshi, S. Bekki, M. P. Chipperfield, M. Dameris, S. Dhomse, S. M. Frith, H. Garny, A. Gettelman, A. Kubin, U. Langematz, E. Mancini, M. Marchand, T. Nakamura, L. D. Oman, S. Pawson, G. Pitari, D. A. Plummer, E. Rozanov, T. G. Shepherd,

K. Shibata, W. Tian, P. Braesicke, S. C. Hardiman, J. F. Lamarque, O. Morgenstern, J. A. Pyle, D. Smale and Y. Yamashita, *Atmos. Chem. Phys.*, 2010, **10**, 9451–9472.

42. L. L. Hood, Effects of solar UV variability on the stratosphere, *in Solar Variability and its Effect on the Earth's Atmospheric and Climate System*, ed. J. Pap, P. Fox, C. Frohlich, H. Hudson, J. Kuhn, J. McCormack, G. North, W. Sprigg, and S. T. Wu, American Geophysical Union, Washington, D. C., 283–304, 2004.

43. V. Eyring, N. R. P. Harris, M. Rex, T. G. Shepherd, D. W. Fahey, G. T. Amanatidis, J. Austin, M. P. Chipperfield, M. Dameris, P. M. De F. Forster, A. Gettelman, H. F. Graf, T. Nagashima, P. A. Newman, S. Pawson, M. J. Prather, J. A. Pyle, R. J. Salawitch, B. D. Santer and D. W. Waugh, *Bull. Am. Meteorol. Soc.*, 2005, **86**, 1117–1133.

44. SPARC CCMVal (2010), SPARC Report on the Evaluation of Chemistry-Climate Models, V. Eyring, T. G. Shepherd, D. W. Waugh (Eds.), SPARC Report No. 5, WCRP-132, WMO/TD-No. 1526, http://www.atmosp. physics.utoronto.ca/SPARC.

45. N. P. Gillett, H. Akiyoshi, S. Bekki, P. Braesicke, V. Eyring, R. Garcia, A. Yu. Karpechko, C. A. McLinden, O. Morgenstern, D. A. Plummer, J. A. Pyle, E. Rozanov, J. Scinocca and K. Shibata, *Atmos. Chem. Phys.*, 2011, **11**, 599–609.

CHAPTER 7

Impact of Polar Ozone Loss on the Troposphere

N. P. GILLETT*[a] AND S.-W. SON[b]

[a] Canadian Centre for Climate Modelling and Analysis, Environment Canada, University of Victoria, PO Box 3065, STN CSC, Victoria, British Columbia, V8W 3V6, Canada; [b] Department of Atmospheric and Oceanic Sciences, McGill University, 805 Sherbrooke Street West, Montreal, Quebec, H3A 2K6, Canada

7.1 Introduction

This chapter reviews the effects of polar stratospheric ozone loss on the troposphere. Initial interest and research into ozone depletion was largely motivated by its effect on surface ultraviolet (UV) radiation. Ozone is a strong absorber of solar ultraviolet radiation, and hence a reduction in ozone leads to reduced absorption of UV in the stratosphere, and more downwelling UV at the surface. This change in downwelling shortwave (SW) radiation across the tropopause might by itself be expected to have some impact on tropospheric climate: however, ozone also absorbs and emits in the infrared, and its depletion leads to a compensating reduction in downwelling longwave (LW) radiation across the tropopause. Most importantly, due to the reduced absorption of UV, ozone depletion strongly cools the stratosphere, and this cooling leads to a larger reduction in downwelling LW radiation in the upper troposphere and lower stratosphere, which dominates the radiative impact on tropospheric temperature.[1–3] This leads to a small projected global surface cooling associated with stratospheric ozone depletion,[4] though the effect is

Stratospheric Ozone Depletion and Climate Change
Edited by Rolf Müller
© Royal Society of Chemistry 2012
Published by the Royal Society of Chemistry, www.rsc.org

expected to be larger over the Antarctic, where ozone depletion has been much larger.

Stratospheric ozone loss very likely also affects the troposphere by dynamical processes, though the mechanism remains the subject of research. Strong cooling of the polar stratosphere in late spring, induced by stratospheric ozone depletion, strengthens the temperature gradient in the upper troposphere/lower stratosphere (UTLS) in both the meridional and vertical directions, and hence also the stratospheric polar vortex. This change in the circulation of the stratosphere may modulate the propagation of planetary-scale Rossby waves from the troposphere. Changes in the driving of the stratospheric circulation by these planetary waves could then directly influence the tropospheric circulation. An initial tropospheric response may be amplified by changes in baroclinic eddies, in the troposphere.[5] Alternatively, ozone-induced changes in lower stratospheric winds may directly influence baroclinic eddies in the troposphere.

In the Antarctic, there is strong evidence that the ozone depletion observed in springtime over recent decades has led to a strengthening and poleward shift of the westerly wind maximum in the troposphere during spring and summer. This circulation change projects strongly onto the Southern Annular Mode (SAM), the leading mode of tropospheric geopotential height variability in the southern hemisphere (SH),[6] and has been associated with pronounced climate trends over Antarctica.[7–11] It may also be associated with trends in the circulation of the Southern Ocean, which may in turn have influenced ocean-atmosphere carbon fluxes.[12–16] In the Arctic, while a similar relationship between tropospheric circulation and stratospheric ozone depletion might be expected on physical grounds, there is little evidence that the much weaker ozone depletion generally observed there (Chapter 5) has caused significant tropospheric trends.

The extent of polar stratospheric ozone depletion is expected to reduce in the coming decades (Chapter 9). Therefore, the ozone-induced trends in tropospheric climate, which have occurred since the advent of polar ozone depletion, are likely to be reversed in the 21st century. The effects of Antarctic ozone depletion have in some cases added to greenhouse gas-induced effects, for example, in the case of the strengthening and poleward shift of the mid-latitude jet, while in other cases, their effects have been opposed, for example, in the case of the ozone-induced cooling and greenhouse gas-induced warming in the Antarctic interior. Thus as polar ozone recovers through the 21st century while greenhouse gas concentrations continue to increase, the effects of stratospheric ozone on the tropospheric circulation may tend to cancel those of greenhouse gas increases.[10,17–19] By contrast, their effects on Antarctic surface temperature may combine to give enhanced warming.[17,19] This contrasts strongly with the likely polar climate evolution, assuming the emission of ozone-depleting substances had not been regulated, in which case large surface climate effects of ongoing ozone depletion in both hemispheres would likely have occurred[20] (see also Chapter 1).

7.2   Antarctic

7.2.1   Ozone Depletion Effect on the Stratosphere

Antarctic ozone depletion, which exceeds 80% at around 70 hPa in October has had a large cooling influence on the stratosphere.[21] Ozone depletion cools the stratosphere primarily through a reduction in absorption of downwelling SW radiation, with a reduction in absorption of upwelling LW radiation playing only a minor role at high latitudes.[1] The observed cooling trend over recent decades has been largest close to the region of maximum ozone depletion in November in the Antarctic lower stratosphere, where temperatures cooled by ~7 K between 1969 and 1998 (Figure 7.1). Coupled model integrations in which only greenhouse gas changes were prescribed, and not ozone depletion, show only a very weak cooling in this region (Figure 7.1e), whereas simulations including ozone depletion reproduce the observed cooling well (Figure 7.1c). Thus, the spring cooling observed in the Antarctic stratosphere is almost certainly mainly attributable to ozone depletion.

The observed cooling over Antarctica maximizes later in the lower stratosphere around December, perhaps associated with the later maximum in ozone depletion in the lowermost stratosphere due to the downward advection of ozone-depleted air by the Brewer-Dobson circulation.[8] This cooling of the lower stratosphere is associated with a delay in the final warming.[22] The cooling is driven by a reduction in SW heating, offset in part by a reduced emission of LW radiation in response to the cooling. Changes in dynamical heating tend to enhance the cooling in the lowermost stratosphere, prior to the breakup of the vortex, and cancel part of the cooling in December and January—this change is associated with a delay in the dynamical heating associated with the final warming.[23] While the largest cooling has been observed in the Antarctic spring, the Antarctic stratosphere has cooled throughout the year, with the smallest trends observed in winter (Figure 7.1a). The spring cooling of the vortex has directly caused a decrease in Antarctic stratospheric geopotential height at overlying levels, due to an increase in the density of the air, and a strengthening of the polar vortex.[7,8]

7.2.2   Effects of Stratospheric Ozone Depletion on the Troposphere and Ocean

7.2.2.1   *Tropospheric Circulation*

Radiosonde observations of Antarctic geopotential height between 1968 and 1999 show that a significant decrease in tropospheric geopotential height on constant pressure surfaces has occurred in December and January, lagging by one to two months the maximum decrease in geopotential height in the lowermost stratosphere (Figure 7.1b). Based on the seasonality of this response, Thompson and Solomon[7] argued that the tropospheric trends were likely associated with stratospheric ozone depletion. The Antarctic troposphere has

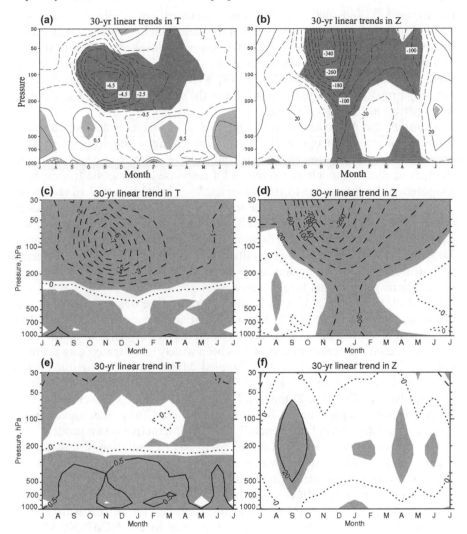

Figure 7.1 30-year (1969–1998) linear trends in temperature (left) and geopotential height (right) averaged over seven radiosonde stations at around 70°S in observations (a,b), 13 CMIP3 coupled models including stratospheric ozone depletion (c, d), and 8 CMIP3 coupled models with no stratospheric ozone depletion (e, f). Trends exceeding one standard deviation of inter-annual variability are shaded in gray. Reproduced from Karpechko *et al.*[28] by permission of the American Geophysical Union.

also exhibited a weak non-significant cooling trend, largest in December, which lags the much larger maximum stratospheric cooling by one to two months.

The delay of one to two months of the maximum tropospheric trends relative to the maximum stratospheric trends is similar to that seen in the downward propagation of anomalies in geopotential height from the stratosphere to the troposphere,[24] and may result from dynamical or radiative processes. Thompson and Solomon[7] suggested that this delay might be associated with the relatively long radiative time scale in the lower stratosphere, which is about a month. The apparent downward propagation could also be due to variations in the climatological strength of the stratospheric polar vortex: planetary waves can only propagate from the troposphere to the stratosphere during a relatively short period in the austral spring.

Although the observations are suggestive of a link between stratospheric ozone depletion and tropospheric circulation changes, they do not prove a causal relationship. However, several studies have prescribed observed ozone changes in climate models and simulated a response in Antarctic tropospheric geopotential height similar to that observed.[8,17–19,25–28] Figure 7.1d shows the mean response to ozone depletion in 13 coupled climate models.[28] As in observations, significant tropospheric geopotential height decreases follow ozone-forced stratospheric geopotential height decreases by around a month. The observations and modeling results taken together provide clear evidence that stratospheric ozone depletion has been an important driver of Antarctic tropospheric geopotential height trends.

Based on the ability of climate models to reproduce observed tropospheric trends as shown in Figure 7.1 (see also the first and second rows of Table 7.1), the impact of stratospheric ozone depletion on the tropospheric circulation and climate, and on the ocean circulation has been widely investigated using climate models. Those models include the Coupled Model Intercomparison Project phase 3 (CMIP3) models[29] and the Chemistry-Climate Model Validation project phase 2 (CCMVal-2) models.[30] The CMIP3 models prescribed stratospheric ozone whereas the CCMVal-2 models used fully interactive stratospheric chemistry but did not generally include interactive ocean models.

The impact of ozone depletion on the zonal-mean circulation is illustrated in Figure 7.2. The CMIP3 models with prescribed ozone depletion (Figure 7.2, left panel) show a strong acceleration of summer zonal winds from the stratosphere all the way down to the surface on the poleward side of the climatological jet. In the troposphere, a weak deceleration is also found on the equatorward side of jet. This pattern of trends is similar to that seen in the NCEP-NCAR and ERA40 reanalyses.[9] A strong acceleration, however, is not simulated in model integrations with no ozone depletion (Figure 7.2, right). This suggests that poleward intensification of the SH-summer jet in the recent past is largely driven by stratospheric ozone depletion. Further analyses of the SH winter when the ozone impact is negligible and of the northern hemisphere (NH), where ozone-forcing is much smaller show no systematic difference between the two sets of CMIP3 models, suggesting that the significant difference in DJF zonal winds found in Figure 7.2 is indeed very likely associated with stratospheric ozone changes.

The strong sensitivity of the westerly jet to stratospheric ozone depletion suggests that stratospheric ozone may affect many aspects of the tropospheric circulation. Table 7.1 shows that lower stratospheric cooling is accompanied by a decrease in tropopause pressure (corresponding to an increase in tropopause

Table 7.1 Summary of SH climate changes in the CMIP3 model integrations. Multi-model mean trends and standard deviations are calculated for models with ozone depletion and those with no ozone depletion for the period 1960–1999. From left to right, columns show trends in October–January (ONDJ) mean polar-cap temperature at 100 hPa integrated south of 64°S with area weight, December–February (DJF) tropopause pressure integrated south of 45°S with area weight, DJF jet-location which is defined by the location of maximum zonal-mean zonal wind at 850 hPa, DJF Hadley-cell boundary location which is identified by the location of zero mass stream function at 500 hPa, and DJF SAM index which is based on the difference in zonal-mean sea level pressure between 40°S and 65°S. Negative trends in the jet and Hadley-cell locations denote poleward shifts over time. For comparison, observed trends from the literature are also indicated where available. The temperature trend is computed at seven radiosonde sites at around 70°S for the time period of 1969–1998,[7] whereas the SAM-index trend is calculated using data from 12 surface stations for the time period of 1958–2000.[75] Reproduced from Son *et al.*[9,19]

Southern hemisphere climate change 1960–1999	ONDJ polar-cap temperature at 100 hPa (K/decade)	DJF tropopause in extratropics (hPa/decade)	DJF jet location (degrees/decade)	DJF Hadley-cell boundary (degree/decade)	DJF SAM index (hPa/decade)
Observations	−1.86				1.05
12 CMIP3 models with ozone depletion	−1.81 ± 1.02	−3.86 ± 1.99	−0.42 ± 0.11	−0.14 ± 0.07	0.95 ± 0.25
8 CMIP3 models with no ozone depletion	−0.11 ± 0.20	−0.95 ± 0.57	−0.19 ± 0.15	−0.09 ± 0.11	0.41 ± 0.63

height) in the extratropics. This is not surprising as strong cooling in the lower stratosphere increases the lapse rate near the tropopause, pushing the location where the temperature lapse rate decreases to $2\,K\,km^{-1}$ (the definition of the tropopause used here) upwards.[31] Although lower stratospheric cooling and upper tropospheric warming caused by increases in greenhouse gases also tend to decrease tropopause pressure, their impacts are weaker. A similar sensitivity to greenhouse gas increases is found in long-term trends of the jet location and the poleward boundary of the SH Hadley cell (Table 7.1). The westerly jet shifts poleward even without ozone depletion, due to increases in greenhouse gas concentration. The poleward shift of the SH-summer jet, however, is substantially strengthened by stratospheric ozone depletion (see also Figure 7.2). Likewise, the expansion of the SH Hadley cell is enhanced by stratospheric ozone depletion. A similar result is also found in the CCMVal-2 model integrations[9,32] and in AGCM integrations.[33,34]

To examine the quantitative relationship between stratospheric ozone depletion and tropospheric circulation changes in more detail, the magnitude of the poleward shift of the jet and the amount of ozone depletion are compared across the ensemble of CCMVal-2 models in Figure 7.3. It is evident that these two quantities are approximately linearly related with each other, indicating that stronger ozone depletion is generally accompanied by a stronger poleward shift in the westerly jet. A similar relationship is also found in the Hadley cell trend.[9] These results suggest that Antarctic ozone depletion affects the entire atmospheric circulation in the SH summer, from the stratosphere to the surface (as further described below), and from the South Pole to the subtropics. These ozone-induced circulations changes are summarized in Figure 7.4.

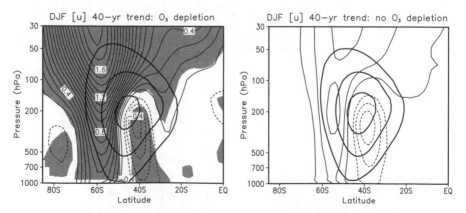

Figure 7.2   Multi-model mean (thick) and linear trend (thin contours) of DJF zonal-mean westerly wind in the CMIP3 integrations over 1960–1999: (left) 12 CMIP3 models with ozone depletion and (right) 8 CMIP models with no ozone depletion. Contour intervals are $10\,ms^{-1}$ starting from $10\,ms^{-1}$ for the climatology and $0.1\,ms^{-1}/decade$ for the trend. Zero lines are omitted and trends exceeding one standard deviation are shaded in gray. Analysis method is described in Son *et al.*[9,10].

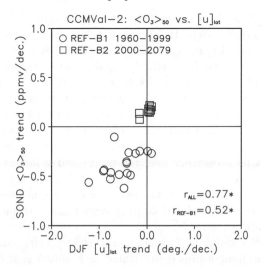

Figure 7.3 Trend relationship between September–December (SOND) mean ozone at 50 hPa integrated south of 64°S and December–February (DJF) mean jet location at 850 hPa as simulated by the CCMVal-2 models. Past and future trends are shown for the periods 1960–1999 and 2000–2079, respectively. Reproduced from Son *et al.*[9] by permission of the American Geophysical Union.

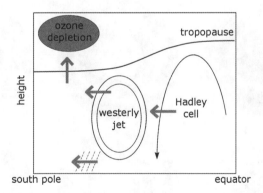

Figure 7.4 Schematic diagram of the impact of stratospheric ozone depletion on Southern Hemisphere summer climate change. Changes in extratropical tropopause height, the location of westerly jet, the poleward boundary of the Hadley cell and precipitation are illustrated. Reproduced from Son *et al.*[9] by permission of the American Geophysical Union.

7.2.2.2 Surface Climate

Based on the tropospheric response to stratospheric ozone depletion described in section 7.2.2.1, one might also expect an influence of ozone depletion on surface climate. Thompson and Solomon[7] showed that circulation changes associated with a trend in the SAM have contributed to a warming of the

Antarctic Peninsula and a cooling of the Antarctic continent in the SH summer and autumn. Warming over the Peninsula has been caused by enhanced surface westerlies (see Figure 7.2), which advect warm air over the oceans to the Peninsula, an increase in northerly winds, and a decrease in the sea ice extent.[35,36] Part of these trends is likely associated with stratospheric ozone depletion, though simulated temperature trends in summer are weaker than those observed,[8] and ozone depletion cannot explain the large winter warming in the Peninsula. Summer/autumn cooling over the continent has been explained by direct radiative cooling associated with ozone depletion[3] and enhanced large-scale subsidence over the pole,[7] and is largely congruent with the trend in the SAM over the period 1968–1999. Nonetheless, this trend is not entirely driven by ozone depletion since the cooling is largest in southern autumn,[35] whereas the simulated cooling response to ozone depletion maximizes in spring and summer.[8]

The response of summer surface climate to stratospheric ozone depletion in the CMIP3 model integrations is illustrated in Figure 7.5. It can be seen that ozone depletion causes a much larger dipolar change of sea level pressure between mid-latitude and subpolar regions (upper left panel) than that

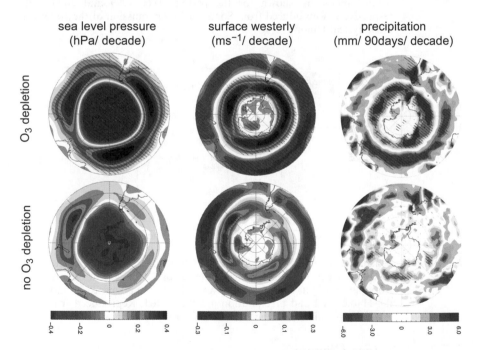

Figure 7.5 DJF surface climate change in the CMIP3 model integrations: (top) models with ozone depletion and (bottom) models with no ozone depletion. From right to left, multi-model mean trends over 1960–1999 are shown for sea level pressure, surface westerly wind and precipitation. Values exceeding one standard deviation are hatched. Adapted from Son *et al.*[19]

associated with greenhouse gas increases alone (lower left panel). The increased enhancement of the meridional pressure gradient is then associated with a larger trend in westerly winds at around 60°S (middle column), consistent with a quasi-barotropic acceleration of the westerly jet shown in Figure 7.2.

Recent modeling studies have shown that stratospheric ozone may also affect global hydrology by modifying the atmospheric general circulation as depicted in Figure 7.4.[19] It is known that expansion of the subtropical dry zone is strongly correlated with the poleward expansion of the Hadley cell in the subtropics.[37] Likewise, the poleward shift of the westerly jet is accompanied by a poleward shift of the storm tracks, which leads to a tripole response in precipitation (Figure 7.5, right), with increases in the southernmost latitudes, decreases in the mid-latitudes, and increases in the subtropics.[33]

Several studies have found a relationship between various aspects of southern hemisphere mid-latitude climate and the SAM in observations. Examples include precipitation over South America,[38] temperature and precipitation over Australia,[39] and precipitation over western South Africa.[40] However, to date, over the extrapolar southern hemisphere as a whole there is not a significant correlation between annual mean trend patterns of surface temperature and precipitation and the regression patterns on the SAM.[41] However, Kang *et al.*[33] find evidence of an observed moistening in austral summer in the southern hemisphere subtropics, which is similar to that simulated in response to stratospheric ozone depletion. Thus, while ozone-induced circulation changes might not dominate temperature and precipitation trends over the extrapolar southern hemisphere, evidence is emerging of a discernible influence on precipitation trends in summer, particularly in the subtropics.

7.2.2.3 Ocean

The Antarctic Circumpolar Current (ACC) is at least partly driven by the surface wind stress associated with the mid-latitude jet. Since the mid-latitude jet is expected to intensify and shift poleward in response to ozone depletion (*e.g.* Figure 7.5 middle), ozone depletion might therefore be expected to have some influence on the ACC.[42] Since the westerly wind stress drives northward Ekman flow away from Antarctica in the Southern Ocean, trends in surface wind stress might also affect the meridional overturning circulation. The resulting ocean circulation changes could then modify upper-ocean temperature and salinity.[43–46]

Sigmond *et al.*[47] find that ozone depletion drives a strengthening of the meridional overturning circulation and a strengthening of the ACC by ∼4 Sv during the last five decades in chemistry-climate model simulations coupled with a dynamical ocean model with relatively course horizontal resolution. In their simulations, ozone depletion drives a cooling of the Southern Ocean south of 50°S, and a weak warming down to ∼1000 m at around 40°S, associated with enhanced Ekman overturning. Spence *et al.*[48] find that projected wind stress changes associated with greenhouse gas increases, which are similar to those due to ozone depletion, drive a broadly similar pattern of subsurface temperature change in the Southern Ocean, even in eddy-permitting simulations with high horizontal resolution. However, observations have shown that the Southern Ocean has been warming and freshening down to depths of over 1000 m south of the ACC at 55–60°S,[49,50] much further south than in the simulations. Taken together, these results suggest that the high-latitude warming of the Southern Ocean[50] is unlikely to have been driven by ozone-induced wind stress changes.

Böning et al.[50] find no evidence of an increased tilt of the isopycnals across the ACC in observations, which is expected in response to an increased surface wind stress based on simulations from coarse-resolution ocean models.[45] This inconsistency may arise from the effects of ocean eddies which are not resolved in coarse-resolution models. Spence et al.[48] find a larger steepening of the isopycnals in response to wind stress changes in a model with weak mesoscale eddy mixing, compared to another model version with high horizontal resolution and stronger mesoscale eddy mixing. Hallberg and Gnanadesikan[51] showed that the response of the Southern Ocean to surface wind stress is very different between eddy-resolving and non-eddy-resolving model integrations. In particular, high-resolution model integrations with explicit representation of ocean eddies[46,51] show that the increase in northward Ekman transport induced by strengthened surface westerly winds steepens isopycnal surfaces on short timescales, but that this response is compensated by an enhanced poleward eddy flux of heat on longer timescales. High resolution simulations also indicate that the response of ocean eddies to surface wind stress changes has a time lag of two to three years.[46,52] However, no simulations of the response to ozone depletion using a coupled model with an eddy-resolving ocean have been carried out. Therefore, while it is reasonable to expect that ozone depletion has contributed to circulation changes in the Southern Ocean, quantitative results remain subject to considerable uncertainty.

It has also been suggested that ozone-induced wind changes may affect atmosphere-ocean fluxes of carbon. The Ekman-driven meridional circulation of the Southern Ocean is associated with outgassing of natural CO_2 at the highest southern latitudes and an uptake of anthropogenic CO_2 at mid-latitudes. Observations show that the Southern Ocean sink of CO_2 has been weakened in the last two decades compared to the strengthening expected based on the increase in atmospheric carbon dioxide.[12] Le Quéré et al.[12] show that observed interannual variations in the sink strength are correlated with inter-annual variations in the SAM, consistent with simulated results.[13–15] Since ozone depletion has caused variations in the strength of the zonal wind stress over the Southern Ocean and variations in the SAM, it may also be linked with a reduction in the strength of the Southern Ocean carbon sink. However, the proposed mechanism linking CO_2 fluxes and variations in westerly wind stress depends on increases in westerly wind stress strengthening the meridional overturning circulation of the Southern Ocean: the role of ocean eddies in influencing carbon fluxes remains to be determined.

7.2.3 Mechanisms of Tropospheric Response to Stratospheric Ozone Depletion

7.2.3.1 Radiative Effects

Since ozone is a strong absorber of solar UV radiation, Antarctic UV-B at the surface can be enhanced by a factor of two or more in the presence of the ozone hole in spring.[53] Although long-term measurements of erythemal UV over Antarctica do not extend back into the pre-ozone-hole period, simulations indicate that erythemal radiation over the Antarctic polar cap has likely increased by a factor of around two since 1960 during spring (Figure 7.6). UV absorption by ozone is also influenced by stratospheric temperature, since the

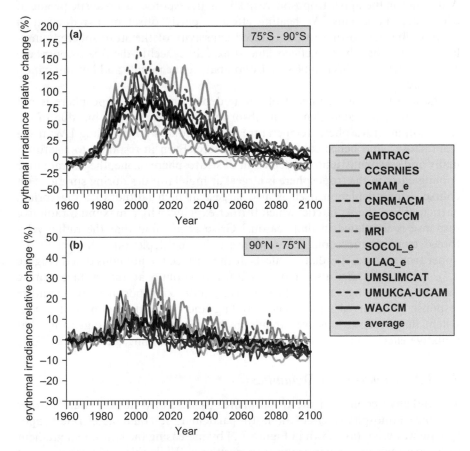

Figure 7.6 (a) Average surface erythemal irradiance change for October–November at 75°S-90°S in the CCMVal-2 simulations (% relative to 1965–1979 mean). (b) As (a) but for March-April at 90°N-75°N. Reproduced from Chapter 10, SPARC CCMVal.[69]

absorption cross-section of ozone is temperature-dependent.[54] Cloudiness and surface albedo can also have large effects on surface UV. It is, however, generally accepted that stratospheric ozone is the dominant driver of springtime trends of Antarctic UV.[54] Since Antarctic ozone depletion is no longer increasing, current levels of Antarctic erythemal UV are simulated to be approximately constant. As ozone recovers through the 21st century, Antarctic erythemal UV is expected to decrease (Figure 7.6).

Ozone depletion also has a radiative impact on tropospheric temperatures. The decreased absorption of downwelling SW in the stratosphere, due to ozone depletion, leads to a SW warming effect in the Antarctic troposphere. Ramanathan and Dickinson[1] argue that this radiative heating effect should be largest at the surface at high latitudes in summer, in response to uniform ozone depletion, but Keeley *et al.*[23] find the largest increase in SW heating over Antarctica in the upper troposphere in spring in response to a realistic profile of ozone depletion. This SW heating effect is partly offset by a reduction in downwelling LW owing to the reduced emissivity of the stratosphere;[1] simulations show that the two effects almost exactly cancel in the free troposphere over Antarctica[3]—this can be seen by comparing the middle and bottom panels of Figure 7.7.

The largest radiative impact of ozone depletion on the troposphere is not through the associated changes in absorptivity and emissivity, but due to the reduction in stratospheric temperature, which reduces downwelling LW at the tropopause,[1] and hence cools the troposphere. Even in response to a latitudinally and seasonally uniform depletion of stratosphere ozone, the net radiative cooling effect in the troposphere is largest at high latitudes during spring.[1,2] Of course, ozone depletion itself is also largest at high latitudes during spring, particularly in the Antarctic, which further explains why a maximum radiative response is simulated in that season.[3] Grise *et al.*[3] find that the radiatively-forced tropospheric cooling in response to ozone depletion is largest in the upper troposphere, if radiative effects at the surface are not considered (Figure 7.7), but Lal *et al.*[2] also find a significant cooling at the surface during the Antarctic spring. Overall, the radiatively-induced tropospheric temperature response to observed stratospheric cooling and ozone depletion is of a consistent magnitude to observed trends in the free troposphere, suggesting that radiative effects may be the dominant cause of the cooling there (Figure 7.7).

7.2.3.2 Zonal-mean Dynamics

The radiative cooling in the troposphere, as described above, could further influence atmospheric circulation. In particular, it could accelerate tropospheric westerlies (as shown in Figure 7.2) by increasing the latitudinal gradient of temperature in the extratropical troposphere. While this is plausible, recent study showed that the polar-cap cooling in the troposphere may not be necessary for the observed wind change. Son *et al.*[9] found a significant acceleration of the westerly jet, quantitatively similar to the one in Figure 7.2, from the CCMVal-2 models, in which tropospheric temperature change over the pole is very weak. This indicates that there are other processes, which at least in part drive the observed wind change in the troposphere.

In the northern hemisphere, stratospheric perturbations associated with stratospheric sudden warmings propagate all the way down through the troposphere, modifying the Northern Annular Mode (NAM) at the surface.[55] Stratospheric anomalies associated with the final warming in the SH[24] are also observed to propagate down to the surface. Several mechanisms have been proposed to explain this effect, and these mechanisms might plausibly also explain the effect of stratospheric ozone depletion on the troposphere. The intraseasonal coupling between the stratosphere and troposphere has often been explained by changes in tropospheric mean meridional circulation in geostrophic and hydrostatic balance with changes in planetary-scale wave drag in the lowermost stratosphere—often called "downward control".[56] Thompson *et al.*[57] argued that this downward control response alone is of a magnitude consistent with observations, even though the structure of the tropospheric anomalies predicted by downward control alone is not consistent with that observed. Other studies,[5,58] however, have found that downward control cannot explain the full tropospheric response to prescribed stratospheric forcing in the absence of a positive feedback on the zonal flow by baroclinic eddies. Recently Son *et al.*[9] further showed that near-surface wind changes simulated by the CCMVal-2 models are not consistent with those predicted by the downward control principle.

Several recent studies have considered the influence of the lower-stratospheric wind on synoptic-scale tropospheric transient eddies.[5,58–62] Chen and Held,[59] and others suggested that ozone-induced wind changes may modify the phase speed of baroclinic waves, changing the location of wave breaking. Ozone depletion accelerates westerlies in the UTLS by increasing the latitudinal gradient of zonal-mean temperature. The enhanced westerlies are then expected to increase the phase speed of baroclinic waves in the upper troposphere. If the zonal wind in the subtropics is not affected by ozone depletion, this increase in the wave phase speed would cause poleward displacement of the critical latitude where wave phase speed is equal to background wind speed, causing eddies to break further poleward. As breaking eddies extract momentum from the mean flow, the critical latitude would shift yet further poleward, and the resulting feedback would displace the entire tropospheric circulation poleward. Although this mechanism is plausible, it is hard to verify in observations or coupled model simulations since changes in baroclinic eddies and zonal-mean flows occur concurrently.[59]

Based on quasi-geostrophic refractive index dynamics,[63] Simpson *et al.*[62] proposed that lower-stratospheric thermal forcing might communicate with the troposphere by changing the potential vorticity gradient, thereby altering the propagation direction of baroclinic waves. Ozone-induced cooling causes the meridional and vertical temperature gradients near the tropopause to increase, effectively modifying the potential vorticity gradient in the UTLS. This could enhance wave propagation upward and equatorward from the extratropics, resulting in a poleward shift in atmospheric general circulation.

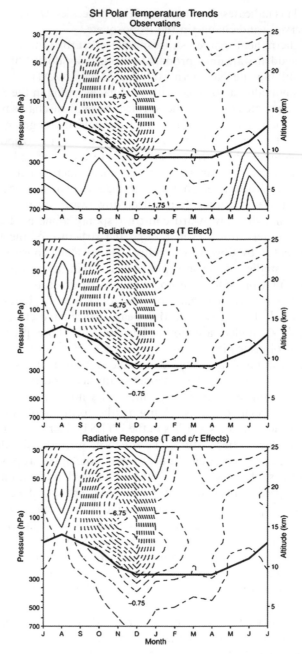

Figure 7.7   Observed polar-cap temperature changes between 1979–1983 and 1997–
2001 based on the NCEP reanalysis (top). The radiative response in the
troposphere to prescribed observed stratospheric temperature changes,
based on a simulation with a fixed dynamical heating model (middle).
The tropospheric radiative response to stratospheric temperature changes
and changes in stratospheric emissivity and transmissivity associated
with stratospheric ozone changes (bottom). Taken from Grise *et al.*[3]
© American Meteorological Society. Reprinted with permission.

This argument has been evaluated in idealized model integrations,[62] but its full applicability to the real world remains to be determined.

It should be noted that the above approaches are all based on zonal-mean linear dynamics where small-amplitude eddies are assumed. It is questionable how well these mechanisms apply to large-amplitude eddies such as those associated with wave breaking. Recent studies have shown that linear theories in fact cannot explain the spatial structure of wave breaking, which is modulated by stratospheric wind. For instance, in baroclinic life-cycle calculations, the wave breaking pattern can change abruptly when the lower-stratospheric wind is gradually altered.[60,61] This change in wave breaking has the opposite effect on tropospheric zonal mean wind compared to that predicted by linear theory, indicating that non-linear wave-mean flow interaction plays a non-negligible role in the baroclinic eddy responses to wind anomalies in the lowermost stratosphere. However, given that ozone-induced circulation change in the Southern Hemisphere has been relatively weak (*e.g.* Figure 7.3), such nonlinear processes have likely played only a minor role.

In summary, while several mechanisms, including radiative effects, downward control, baroclinic eddy-mean flow interaction and non-linear processes, have been proposed to explain the influence of stratospheric ozone depletion on the troposphere, their relative importance remains unclear.

7.2.3.3 Zonal Asymmetry Effects

Son *et al.*[9,10] found that coupled chemistry climate models simulate larger SH circulation trends than climate models in which only zonal mean ozone changes are prescribed. While ozone concentrations vary predominantly with latitude, the stratospheric distribution of ozone is not zonally symmetric, and deviations from zonal symmetry are largest in the Antarctic lower stratosphere in spring, associated with a displacement of the ozone hole away from the pole.[64,65] The mechanisms described so far attempt to explain how zonal mean changes in ozone may influence the zonal mean flow: But several studies have demonstrated that the inclusion of zonal asymmetries in ozone changes in models causes additional cooling of the polar lower stratosphere during spring with an overlying dynamical warming,[64–67] compared to simulations in which zonal mean ozone is specified. This zonal-mean temperature response to the zonal asymmetries in ozone is forced by a change in dynamical heating.[67] The cause of this change in dynamical heating remains unclear, though likely some initial perturbation to the vortex induced by the zonal asymmetries in ozone modulates the mean state, and hence the propagation of planetary waves.

Waugh *et al.*[65] examine the influence of zonal asymmetries in ozone in transient simulations and find that they increase the cooling in the Antarctic lower stratosphere by ~25% (Figure 7.8a and b). Moreover, they find an extension of the effect into the troposphere where the strengthening of the SH mid-latitude jet and the decrease in sea level pressure simulated over Antarctica in summer are both larger in simulations including zonal asymmetries in ozone

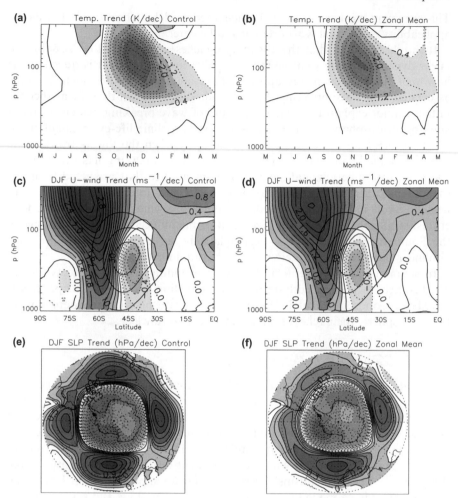

Figure 7.8 Antarctic average temperature trends (a and b), Southern Hemisphere DJF zonal-mean wind trends (c and d), and DJF SLP trends (e and f) over the period 1960–2000 from simulations of the GEOS chemistry climate model with interactive three-dimensional ozone variations (left column) and specified zonal mean ozone changes (right column). Reproduced from Waugh *et al.*[65] by permission of the American Geophysical Union.

(Figure 7.8). These results imply that climate model simulations in which only zonal mean ozone changes are prescribed may not capture the full climate effects of stratospheric ozone depletion or recovery on either the stratosphere or the troposphere.

7.2.4 Future Changes

As stratospheric concentrations of ozone-depleting substances continue their projected decline through the coming decades (Chapter 2), stratospheric ozone concentrations are projected to increase (Chapter 9). Since gas-phase ozone depletion and the formation of polar stratospheric clouds (PSCs) are both sensitive to temperature, and increasing greenhouse gases will tend to cool the stratosphere, ozone recovery will not be an exact reversal of historical ozone depletion, and in particular ozone levels in the Antarctic are expected to return to 1980 levels later than global mean ozone, largely due to a weaker influence of Brewer-Dobson circulation change on ozone in the southern hemisphere.[68] Antarctic spring column ozone is projected to return to 1980 levels by ~ 2050, although there is considerable uncertainty in this date.

As Antarctic ozone recovers, the polar stratosphere is projected to warm in spring, reversing the historical trend, although the rate of warming is projected to be smaller than the rate of cooling over recent decades.[69] In the troposphere, greenhouse gases are expected to cause a strengthening and poleward shift in the mid-latitude jet, but ozone recovery is expected to cancel or reverse this effect in austral summer.[10,11,69] In austral winter, by contrast, when ozone concentrations are unlikely to change very much, an ongoing greenhouse-gas-induced strengthening and poleward shift of the mid-latitude jet is projected.

Ozone depletion has been associated with a cooling trend in Antarctic summer surface temperature.[7,8] As ozone recovers through the 21st century, this contribution to surface temperature trends is expected to reverse, while greenhouse gases are expected to contribute an ongoing warming influence. Thus enhanced warming is projected to occur in the Antarctic in summer,[26] though the projected enhancement in warming due to ozone recovery is modest in size.[19]

7.3 Arctic

Due to the stronger planetary wave driving and higher temperatures in the Arctic stratosphere, PSCs form much less often than in the Antarctic (Chapter 4), ozone depletion is usually much less severe (spring 2011 was an exception with record Arctic column depletion), and hence simulated increases in surface UV are usually much smaller than in the southern hemisphere (Figure 7.6). Moreover, as in the Antarctic, maximum increases in UV are projected for around 2010, with ozone recovery through the 21st century expected to lead to a decrease in surface UV in the Arctic. However, in the NH, increasing green-house gas concentrations are projected to increase the strength of the Brewer-Dobson circulation, increasing the transport of ozone-rich air into the Arctic vortex during winter and spring, and leading to a return of Arctic ozone to 1980 values by ~ 2030[68,70] (see also Chapter 9). Furthermore, Arctic ozone is expected to increase above 1980 levels in subsequent decades, leading to surface UV levels below pre-ozone-hole conditions[70] (Figure 7.6).

While ozone depletion in the Arctic might be expected to have had a similar effect on surface climate as in the Antarctic, Arctic ozone depletion, also at a maximum in the lower stratosphere in spring, has been only on average about 25% of the magnitude of that observed in the Antarctic,[3] although depletion in individual years, such as 2011, has been large. Arctic ozone depletion is thought to have contributed to the cooling observed in the Arctic stratosphere in spring,[21] but the observed cooling is larger than that simulated in response to ozone depletion. While some early studies did simulate an increase in the NAM in response to ozone depletion,[71–73] the simulated change was generally smaller than that observed, and the simulated response to ozone depletion could not explain the large observed trends in mid-winter. Grise *et al.*[3] examined the radiative influence on temperatures of the observed Arctic temperature changes, and found only small tropospheric temperature changes, which were not particularly similar to those observed, indicating that Arctic zonal-mean temperature trends in the troposphere cannot be explained by radiative forcing from the stratosphere.

The tropospheric NAM index in the CCMVal-2 simulations was found to correlate with stratospheric ozone and stratospheric inorganic chlorine in winter and spring.[74] The correlation was higher with ozone, which may be because ozone variations and the tropospheric NAM are both influenced by variations in stratospheric wave driving, but the surface NAM was also significantly correlated with stratospheric inorganic chlorine, which should be less affected by wave driving variations than ozone. Thus, there is some evidence to suggest that stratospheric ozone changes may have contributed somewhat to the winter and spring positive trends in the NAM index, but this effect is likely to have been small compared to internal variability.[11]

7.4 Summary

Antarctic ozone depletion has cooled the Antarctic stratosphere in spring, and strengthened the polar vortex. A decrease in temperature and geopotential height in the Antarctic troposphere has been observed, lagging the stratospheric changes by about a month. A consistent decrease in tropospheric geopotential height in spring and summer is simulated by climate models, which include stratospheric ozone depletion, but such trends are not reproduced in models with greenhouse gas forcing changes only, indicating that the observed Antarctic tropospheric summer circulation trends have likely been largely caused by ozone depletion. Coupled chemistry-climate simulations of the response to stratospheric ozone depletion show a strengthening and poleward shift of the mid-latitude jet in the troposphere, which is proportional to the amount of ozone depletion simulated in the stratosphere. Increases in tropopause height and a poleward shift of the southern boundary of the SH Hadley Cell are also simulated in response to stratospheric ozone depletion: comparable increases in tropopause height and a poleward shift of the Hadley Cell boundary have also been observed.

At the surface, stratospheric ozone depletion has contributed to a strengthening of the mid-latitude westerlies in summer,[8] and to the pattern of summer/autumn warming over the Antarctic Peninsula and cooling over the rest of the continent.[7,8] Simulations of the response to ozone depletion also exhibit an increase in precipitation southward of 55°S, a decrease between 35°S and 55°S, and an increase between 15°S and 35°S in summer: some evidence of such a summer moistening trend in the subtropics has recently been identified in observations.[33] Several studies have suggested that the ozone-depletion-induced strengthening and poleward shift of the westerly wind maximum over the Southern Ocean has caused a strengthening and poleward shift of the Antarctic Circumpolar Current, and a strengthened overturning circulation, but these findings are not fully supported by observations or by simulations with eddy-resolving ocean models. There is some observational and modeling evidence to suggest that the outgassing of natural CO_2 from the high-latitude Southern Ocean may have been enhanced partly by ozone-induced changes in surface winds: however, the role of oceanic eddies in modulating the response remains to be investigated.

Several mechanisms have been proposed to explain the observed tropospheric response to changes in Antarctic stratospheric ozone and, more generally, the tropospheric response to anomalies in stratospheric circulation. Radiative coupling certainly appears to play a role, with the strong stratospheric cooling driving cooling in the free troposphere, and at the surface.[1–3] Adjustment of the flow to changes in wave drag in the stratosphere (so-called downward control) could explain part of the response,[57] but the pattern of tropospheric wind changes is not consistent with this mechanism alone. Several studies suggest that baroclinic eddies are important, either acting to amplify a tropospheric response forced by some other mechanism, or by responding directly to changes in zonal winds in the lowermost stratosphere. Other work demonstrates that the circulation response to ozone depletion through the stratosphere and troposphere is also sensitive to zonal asymmetries in the distribution of stratospheric ozone, which have changed through time with the development of the Antarctic ozone hole.[65]

Since Antarctic ozone depletion is projected to recover in the coming decades, the trends in tropospheric climate forced by ozone depletion over recent decades are likely to be reversed in the 21st century. In some cases ozone recovery and greenhouse gas influences will tend to counteract each other, for example, in their effects on the tropospheric circulation in summer. In other cases, their effects will tend to combine, for example, on surface warming over Antarctica.

Coupled chemistry climate models indicate a weak link between Arctic ozone depletion and NAM trends in the winter and spring.[74] However, the much weaker ozone trends in the Arctic and the much larger interannual variability mean that no such relationship has been identified in the observations.

Acknowledgements

We thank Neil Tandon for helping us generate Figure 7.5, and David Thompson and Martyn Chipperfield for helpful reviews.

References

1. V. Ramanathan and R. E. Dickinson, *J. Atmos. Sci.*, 1979, **36**, 1084–1104.
2. M. Lal, A. K. Jain and M. C. Sinha, *Tellus B*, 1987, **39B**, 326–328.

3. K. M. Grise, D. W. J. Thompson and P. M. Forster, *J. Climate*, 2009, **22**, 4154–4161.
4. P. Forster, V. Ramaswamy, P. Artaxo, T. Bernsten, R. Betts, D. W. Fahey, J. Haywood, J. Lean, D. C. Lowe, G. Myhre, J. Nganga, R. Prinn, G. Raga, M. Schultz and R. V. Dorland, in *The Physical Science Basis. Contribution of Working Group I to the Fourth Assessment Report of the Intergovernmental Panel on Climate Change*, ed. S. Solomon, D. Qin, M. Manning, Z. Chen, M. Marquis, K.B. Averyt, M. Tignor, H.L. Miller, Cambridge University Press, Cambridge, United Kingdom and New York, NY, USA, 2007.
5. Y. C. Song and W. A. Robinson, *J. Atmos. Sci.*, 2004, **61**, 1711–1725.
6. D. W. J. Thompson, J. M. Wallace and G. C. Hegerl, *J. Climate*, 2000, **13**, 1018–1036.
7. D. W. J. Thompson and S. Solomon, *Science*, 2002, **296**, 895–899.
8. N. P. Gillett and D. W. J. Thompson, *Science*, 2003, **302**, 273–275.
9. S. W. Son, E. P. Gerber, J. Perlwitz, L. M. Polvani, N. P. Gillett, K. H. Seo, V. Eyring, T. G. Shepherd, D. Waugh, H. Akiyoshi, J. Austin, A. Baumgaertner, S. Bekki, P. Braesicke, C. Bruhl, N. Butchart, M. P. Chipperfield, D. Cugnet, M. Dameris, S. Dhomse, S. Frith, H. Garny, R. Garcia, S. C. Hardiman, P. Jockel, J. F. Lamarque, E. Mancini, M. Marchand, M. Michou, T. Nakamura, O. Morgenstern, G. Pitari, D. A. Plummer, J. Pyle, E. Rozanov, J. F. Scinocca, K. Shibata, D. Smale, H. Teyssedre, W. Tian and Y. Yamashita, *J. Geophys. Res-Atmos.*, 2010, **115**, DOI:10.1029/2010JD014271.
10. S. W. Son, L. M. Polvani, D. W. Waugh, H. Akiyoshi, R. Garcia, D. Kinnison, S. Pawson, E. Rozanov, T. G. Shepherd and K. Shibata, *Science*, 2008, **320**, 1486–1489.
11. P. M. Forster, D. W. J. Thompson, M. P. Baldwin, M. P. Chipperfield, M. Dameris, J. D. Haigh, D. J. Karoly, P. J. Kushner, W. J. Randel, K. H. Rosenlof, D. J. Seidel and S. Solomon, in *Scientific Assessment of Ozone Depletion: 2010*, World Meteorological Organisation, Geneva, Switzerland, 2011.
12. C. Le Quere, C. Rodenbeck, E. T. Buitenhuis, T. J. Conway, R. Langenfelds, A. Gomez, C. Labuschagne, M. Ramonet, T. Nakazawa, N. Metzl, N. Gillett and M. Heimann, *Science*, 2007, **316**, 1735–1738.
13. A. Lenton and R. J. Matear, *Global Biogeochem. Cy.*, 2007, **21**.
14. N. S. Lovenduski, N. Gruber, S. C. Doney and I. D. Lima, *Global Biogeochem. Cy.*, 2007, **21**, DOI:10.1029/2006GB002900.
15. N. S. Lovenduski, N. Gruber and S. C. Doney, *Global Biogeochem. Cy.*, 2008, **22**, DOI:3010.1029/2007GB003139.
16. A. H. Butler, D. W. J. Thompson and K. R. Gurney, *Global Biogeochem. Cy.*, 2007, **21**, DOI:10.1029/2006GB002796.
17. D. T. Shindell and G. A. Schmidt, *Geophys. Res. Lett.*, 2004, **31**, DOI:10.1029/2004GL020724.
18. J. Perlwitz, S. Pawson, R. L. Fogt, J. E. Nielsen and W. D. Neff, *Geophys. Res. Lett.*, 2008, **35**, DOI:10.1029/2008GL033317.

19. S. W. Son, N. F. Tandon, L. M. Polvani and D. W. Waugh, *Geophys. Res. Lett.*, 2009, **36**, DOI:10.1029/2009GL038671.
20. O. Morgenstern, P. Braesicke, M. M. Hurwitz, F. M. O'Connor, A. C. Bushell, C. E. Johnson and J. A. Pyle, *Geophys. Res. Lett.*, 2008, **35**, DOI:10.1029/2008GL034590.
21. W. J. Randel and F. Wu, *J. Climate*, 1999, **12**, 1467–1479.
22. D. W. Waugh, W. J. Randel, S. Pawson, P. A. Newman and E. R. Nash, *J. Geophys. Res-Atmos.*, 1999, **104**, 27191–27201.
23. S. P. E. Keeley, N. P. Gillett, D. W. J. Thompson, S. Solomon and P. M. Forster, *Geophys. Res. Lett.*, 2007, **34**, DOI:10.1029/2007GL031238.
24. D. W. J. Thompson, M. P. Baldwin and S. Solomon, *J. Atmos. Sci.*, 2005, **62**, 708–715.
25. J. M. Arblaster and G. A. Meehl, *J. Climate*, 2006, **19**, 2896–2905.
26. R. L. Miller, G. A. Schmidt and D. T. Shindell, *J. Geophys. Res-Atmos.*, 2006, **111**, DOI:10.1029/2005JD006323.
27. W. J. Cai and T. Cowan, *J. Climate*, 2007, **20**, 681–693.
28. A. Y. Karpechko, N. P. Gillett, G. J. Marshall and A. A. Scaife, *Geophys. Res. Lett.*, 2008, **35**, DOI:10.1029/2008GL035354.
29. G. A. Meehl, T. F. Stocker, W. D. Collins, P. Friedlingstein, A. T. Gaye, J. M. Gregory, A. Kitoh, R. Knutti, J. M. Murphy, A. Noda, S.C.B. Raper, and A. J. W. a. Z.-C. Z. I.G. Watterson, in *The Physical Science Basis. Contribution of Working Group I to the Fourth Assessment Report of the Intergovernmental Panel on Climate Change*, eds. S. Solomon, D. Qin, M. Manning, Z. Chen, M. Marquis, and M. T. a. H. L. M. K.B. Averyt, Cambridge University Press, Cambridge, United Kingdom and New York, NY, USA, 2007, p. 766.
30. V. Eyring, M. P. Chipperfield, M. A. Giorgetta, D. E. Kinnison, K. Matthes, P. A. Newman, S. Pawson, T. G. Shepherd and D. W. Waugh, *SPARC Newsletter*, 2008, **30**, 20–26.
31. S. W. Son, L. M. Polvani, D. W. Waugh, T. Birner, H. Akiyoshi, R. R. Garcia, A. Gettelman, D. A. Plummer and E. Rozanov, *J. Climate*, 2009, **22**, 429–445.
32. C. Mclandress, T. G. Shepherd, J. F. Scinocca, D. A. Plummer, M. Sigmond, A. I. Jonsson and M. C. Reader, *J. Climate*, 2011, **24**, 1850–1868.
33. S. M. Kang, L. M. Polvani, J. C. Fyfe and M. Sigmond, *Science*, 2011, **332**, 951–954.
34. L. M. Polvani, D. W. Waugh, G. J. P. Correa and S. W. Son, *J. Climate*, 2011, **24**, 795–812.
35. J. Turner, S. R. Colwell, G. J. Marshall, T. A. Lachlan-Cope, A. M. Carleton, P. D. Jones, V. Lagun, P. A. Reid and S. Iagovkina, *Int. J. Climatol.*, 2005, **25**, 279–294.
36. G. J. Marshall, A. Orr, N. P. M. van Lipzig and J. C. King, *J. Climate*, 2006, **19**, 5388–5404.
37. J. Lu, G. Chen and D. M. W. Frierson, *J. Climate*, 2008, **21**, 5835–5851.
38. G. E. Silvestri and C. S. Vera, *Geophys. Res. Lett.*, 2003, **30**, DOI:10.1029/2003GL018277.

39. H. H. Hendon, D. W. J. Thompson and M. C. Wheeler, *J. Climate*, 2007, **20**, 2452–2467.
40. C. J. C. Reason and M. Rouault, *Geophys. Res. Lett.*, 2005, **32**, DOI:10.1029/2005GL022419.
41. N. P. Gillett, T. D. Kell and P. D. Jones, *Geophys. Res. Lett.*, 2006, **33**, DOI:10.1029/2006GL027721.
42. J. R. Toggweiler and J. Russell, *Nature*, 2008, **451**, 286–288.
43. W. Lefebvre, H. Goosse, R. Timmermann and T. Fichefet, *J. Geophys. Res-Oceans.*, 2004, **109**, DOI:10.1029/2004JC002403.
44. A. Sen Gupta and M. H. England, *J. Climate*, 2006, **19**, 4457–4486.
45. J. C. Fyfe, O. A. Saenko, K. Zickfeld, M. Eby and A. J. Weaver, *J. Climate*, 2007, **20**, 5391–5400.
46. J. A. Screen, N. P. Gillett, D. P. Stevens, G. J. Marshall and H. K. Roscoe, *J. Climate*, 2009, **22**, 806–818.
47. M. C. Sigmond, M. C. Reader, J. C. Fyfe and N. P. Gillett, *Geophys. Res. Lett.*, 2011, **38**, DOI:10.1029/2011GL047120.
48. P. Spence, J. C. Fyfe, A. Montenegro and A. J. Weaver, *J. Climate*, 2010, **23**, 5332–5343.
49. S. T. Gille, *Science*, 2002, **295**, 1275–1277.
50. C. W. Boning, A. Dispert, M. Visbeck, S. R. Rintoul and F. U. Schwarzkopf, *Nat. Geosci.*, 2008, **1**, 864–869.
51. R. Hallberg and A. Gnanadesikan, *J. Phys. Oceanogr.*, 2006, **36**, 2232–2252.
52. M. P. Meredith and A. M. Hogg, *Geophys. Res. Lett.*, 2006, **33**, DOI:10.1029/2006GL026499.
53. G. Bernhard, C. R. Booth and J. C. Ehramjian, *J. Geophys. Res-Atmos.*, 2004, **109**, DOI:10.1029/2004JD004937.
54. A. F. Bais, D. Lubin, A. Arola, G. Bernhard, M. Blumthaler, N. Chubarova, C. Erlick, H. P. Gies, N. Krotkov, K. Lantz, B. Mayer, R. L. McKenzie, R. D. Piacentini, G. Seckmeyer, J. R. Slusser and C. S. Zerefos, in *Scientific Assessment of ozone depletion: 2006*, World Meteorological Organisation, Geneva, Switzerland, 2007.
55. M. P. Baldwin and T. J. Dunkerton, *Science*, 2001, **294**, 581–584.
56. P. H. Haynes, C. J. Marks, M. E. Mcintyre, T. G. Shepherd and K. P. Shine, *J. Atmos. Sci.*, 1991, **48**, 651–679.
57. D. W. J. Thompson, J. C. Furtado and T. G. Shepherd, *J. Atmos. Sci.*, 2006, **63**, 2616–2629.
58. P. J. Kushner and L. M. Polvani, *J. Climate*, 2004, **17**, 629–639.
59. G. Chen and I. M. Held, *Geophys. Res. Lett.*, 2007, **34**, DOI:10.1029/2007GL031200.
60. M. A. H. Wittman, A. J. Charlton and L. M. Polvani, *J. Atmos. Sci.*, 2007, **64**, 479–496.
61. T. Kunz, K. Fraedrich and F. Lunkeit, *J. Atmos. Sci.*, 2009, **66**, 2288–2302.
62. I. R. Simpson, M. Blackburn and J. D. Haigh, *J. Atmos. Sci.*, 2009, **66**, 1347–1365.

63. D. G. Andrews, J. R. Holton and C. B. Leovy, *Middle Atmosphere Dynamics*, Academic Press, Orlando; London, 1987.
64. N. P. Gillett, J. F. Scinocca, D. A. Plummer and M. C. Reader, *Geophys. Res. Lett.*, 2009, **36**, DOI:10.1029/2009GL037246.
65. D. W. Waugh, L. Oman, P. A. Newman, R. S. Stolarski, S. Pawson, J. E. Nielsen and J. Perlwitz, *Geophys. Res. Lett.*, 2009, **36**, DOI:10.1029/2009GL040419.
66. F. Sassi, B. A. Boville, D. Kinnison and R. R. Garcia, *Geophys. Res. Lett.*, 2005, **32**, DOI:10.1029/2004GL022131.
67. J. A. Crook, N. P. Gillett and S. P. E. Keeley, *Geophys. Res. Lett.*, 2008, **35**, DOI:10.1029/2007GL032698.
68. S. Bekki, G. E. Bodeker, A. F. Bais, N. Butchart, V. Eyring, D. W. Fahey, D. E. Kinnison, U. Langematz, B. Mayer, R. W. Portmann and E. Rozanov, in *Scientific Assessment of Ozone Depletion: 2010*, World Meteorological Organisation, Geneva, Switzerland, 2011.
69. M. P. Baldwin, N. P. Gillett, P. M. Forster, E. P. Gerber, M. I. Hegglin, A. Y. Karpechko, J. Kim, P. J. Kushner, O. H. Morgenstern, T. Reichler, S.-W. Son and K. Tourpali, in *SPARC Report on the Evaluation of Chemistry-Climate Models, SPARC Report No. 5, WCRP-132, WMO/TD-No 1526*, ed. V. Eyring, T. G. Shepherd and D. W. Waugh, 2010, pp. 379–412.
70. M. I. Hegglin and T. G. Shepherd, *Nat. Geosci.*, 2009, **2**, 687–691.
71. H. F. Graf, I. Kirchner and J. Perlwitz, *J. Geophys. Res-Atmos.*, 1998, **103**, 11251–11261.
72. E. M. Volodin and V. Y. Galin, *J. Climate*, 1999, **12**, 2947–2955.
73. D. T. Shindell, G. A. Schmidt, R. L. Miller and D. Rind, *J. Geophys. Res-Atmos.*, 2001, **106**, 7193–7210.
74. O. Morgenstern, H. Akiyoshi, S. Bekki, P. Braesicke, N. Butchart, M. P. Chipperfield, D. Cugnet, M. Deushi, S. S. Dhomse, R. R. Garcia, A. Gettelman, N. P. Gillett, S. C. Hardiman, J. Jumelet, D. E. Kinnison, J. F. Lamarque, F. Lott, M. Marchand, M. Michou, T. Nakamura, D. Olivié, T. Peter, D. Plummer, J. A. Pyle, E. Rozanov, D. Saint-Martin, J. F. Scinocca, K. Shibata, M. Sigmond, D. Smale, H. Teyssédre, W. Tian, A. Voldoire and Y. Yamashita, *J. Geophys. Res-Atmos.*, 2010, **115**, DOI:10.1029/2009JD013347.
75. G. J. Marshall, *J. Climate*, 2003, **16**, 4134–4143.

CHAPTER 8

Impact of Climate Change on the Stratospheric Ozone Layer

MARTIN DAMERIS[a] AND MARK P. BALDWIN[b]

[a] Deutsches Zentrum für Luft- und Raumfahrt, Institut für Physik der Atmosphäre, Oberpfaffenhofen, 82234 Wessling, Germany; [b] NorthWest Research Associates, Inc., 4118 148th Ave NE, Redmond, WA 98052, U.S.A.

8.1 Introduction

The stratospheric ozone layer shows distinct variability on time-scales ranging from seasonal to decadal. Changes to the ozone layer are driven by both natural processes and human activities. In particular, long-term changes and trends need to be fully explained before robust assessments of the future evolution of the ozone layer can be made. It is necessary to answer questions like:

➤ What has happened in the past?
➤ Why did it happen?
➤ What is the likelihood of it happening again in the future?

Another key question has to be answered in this context:

➤ Are the observed fluctuations and changes driven by natural processes or are they dominated by human activities?

Climate change will affect the future evolution of the ozone layer through changes in temperature, chemical composition, and transport of trace gases and aerosols (*e.g.* Chapter 5 in WMO, 2007;[1] Chapter 4 in WMO, 2011).[2]

Stratospheric Ozone Depletion and Climate Change
Edited by Rolf Müller
© Royal Society of Chemistry 2012
Published by the Royal Society of Chemistry, www.rsc.org

Understanding of cause and effect relationships is often complex because many feedback processes are nonlinear.

The ozone layer is generally assumed to have been nearly unaffected by man-made ozone depleting substances (ODSs; *e.g.* chlorofluorocarbons: CFCs) prior to 1960. By the time concentrations of ODSs have returned to pre-1960 values, which could be the middle of this century (see Chapter 2), concentrations of greenhouse gases (GHGs) will have increased substantially (Chapter 9). An increase of long-lived GHG concentrations (carbon dioxide: CO_2, methane: CH_4, and nitrous oxide: N_2O) in the atmosphere leads to higher tropospheric temperatures (the greenhouse effect). On average, GHGs cool the stratosphere (Subsection 8.2.1–2), which modifies chemical reaction rates (Subsection 8.2.3) for ozone destruction and alters stratospheric circulation (Subsection 8.3.2), both influencing the state of the ozone layer. For those reasons a return to the same ozone layer that existed before 1960 will not occur. The rates of many chemical reactions are temperature dependent, and these reaction rates affect the chemical composition of the atmosphere. There are two types of ozone-destroying chemical reactions that are temperature dependent. Lower stratospheric temperatures lead to a slowing of most gas-phase reactions that destroy ozone (Chapters 1, 3, and 6), but on the other hand yield an intensified depletion of ozone in the lower polar stratosphere due to increased heterogeneous activation of halogens on the surfaces of particles in polar stratospheric clouds (PSCs; see Chapter 4). It is expected that as the stratosphere cools, the slowing of gas-phase reactions will dominate, resulting in a slightly thicker ozone layer (Chapter 9).

Since climate change also influences the dynamics of the troposphere and the stratosphere, and therefore affects the transport of trace gases and particles, dynamically induced temperature changes could locally enhance or weaken the temperature changes caused by radiative processes. So far, estimates of future changes of stratospheric dynamics are uncertain with some numerical models of the atmosphere projecting that stratospheric temperatures will increase in polar regions during the winter and spring seasons, while most models predict a further cooling (Subsection 8.3.2). Hence the net effect of stratospheric radiative, chemical, and dynamical processes and their interactions are poorly understood and quantified.

Due to the expected decrease of ODSs (Chapter 2), the future ozone layer will strongly depend on how increasing concentrations of GHGs affect the temperatures and circulation of the stratosphere. Such a prediction remains a challenge, in large part because the stratospheric circulation depends on how the tropospheric circulation and sea-surface temperatures (SSTs) evolve (Subsection 8.3.3), as well as the details of vertical coupling of the stratosphere and troposphere including the mass exchange and mixing between these atmospheric layers (Section 8.4).

Some aspects of the future evolution of the ozone layer can be projected with numerical models of Earth's atmosphere with reasonable confidence, in particular, with Climate-Chemistry Models (CCMs; see Chapter 9). Climate change is expected, on average, to accelerate the recovery of the ozone layer. A further cooling of the stratosphere will likely result in a slightly thicker ozone layer in the second half of this century ("super-recovery" of the ozone layer). However, such a recovery will not be uniform. Over the polar regions, there is much more uncertainty. Reduced winter temperatures in the lower stratosphere over the poles would be expected to create more PSCs (Chapters 4 and 5), which are needed for the rapid heterogeneous chemical reactions mainly responsible for the observed polar ozone loss in spring, especially in the Antarctic region, creating the most dramatic depletion of ozone, *i.e.* the ozone hole. By itself, this suggests the possibility of increased ozone loss in the northern polar lower stratosphere. However, most numerical models predict that circulation changes due to increasing GHGs will act in the opposite sense during late winter and spring. The net result is not yet clear.

This chapter will give a comprehensive overview about individual physical, dynamical and chemical processes and feedback mechanisms associated with climate change and their implication for ozone depletion, in particular the connection with the expected recovery of the ozone layer, which is discussed in Chapter 9.

8.2 Impact of Enhanced Greenhouse Gas Concentrations on Radiation and Chemistry

GHGs, mainly CO_2, water vapor (H_2O), CH_4, and N_2O, warm the troposphere by absorbing outgoing infrared (IR) radiation from the Earth. The dominant balance in the troposphere is between heating through release of latent heat and radiative cooling by GHGs. In the stratosphere, however, increased GHG concentrations lead to a net cooling as they emit more IR radiation into the upper atmosphere than they absorb. IR emission increases with local temperature. Therefore, the cooling effect increases with altitude, maximizing near the stratopause at around 50 km altitude, where temperatures of the stratosphere are highest. The stratospheric cooling effect of GHGs also varies with latitude, as it depends on the balance between absorption of IR radiation from below and local emission of IR radiation. The net cooling effect of GHGs extends to lower levels at high latitudes, roughly following the tropopause.

Any change in concentrations of radiatively active gases will alter the balance between incoming solar (short-wave) and outgoing terrestrial (long-wave) radiation in the atmosphere. For example, ozone absorbs both short- and long-wave radiation. To determine radiative forcing from stratospheric ozone changes, it is important to distinguish between immediate effects and those after the stratospheric temperature has adjusted, which takes some days (in the lower stratosphere) up to weeks (in the upper stratosphere). Depletion of ozone in the lower stratosphere induces an instantaneous increase in the short-wave solar flux at the tropopause and a slight reduction of the downwelling long-wave radiation. The net instantaneous effect is a positive radiative forcing. However, the decrease in stratospheric ozone causes less absorption of solar and long-wave radiation, yielding a local cooling. After the stratosphere has adjusted, the net effect of stratospheric ozone depletion is a negative radiative forcing (Chapter 1 in IPCC/TEAP, 2005).[3] In contrast, ozone depletion in the middle and upper stratosphere causes a slight positive radiative forcing (Chapter 1 in IPCC/TEAP, 2005). The maximum sensitivity of radiative forcing to stratospheric ozone changes is found in the tropopause region, and the maximum sensitivity of surface temperatures to these ozone changes also peaks near the tropopause.[4]

8.2.1 Past Temperature Changes

There is strong evidence for a large and significant cooling in most of the stratosphere since the 1960s[5–7] (see also Chapter 4 in WMO, 2011).[2] Figure 8.1 shows global average temperature anomalies (90°S–90°N) derived from the RICH radiosonde data set for the time period from 1960 to 2007, spanning a range of altitudes from the upper troposphere (300 hPa) to the middle stratosphere (30 hPa). Corresponding vertical profiles of near-global temperature trends during the period from 1979 to 2007 are shown in Figure 8.2. The radiosonde data sets indicate a warming of the troposphere and reveal an overall long-term cooling of the stratosphere, with trends increasing with altitude. There is reasonable agreement between the lower stratospheric trends derived from satellite data (since 1979) and radiosondes (see Figure 8.3). All data sets suggest that there has been a significant cooling in the stratosphere over the globe for recent decades, including the tropics (see middle part of Figure 8.2).

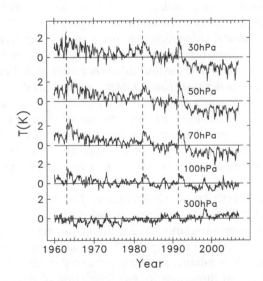

Figure 8.1 Temporal evolution of global average temperature anomalies (K) at pressure levels spanning the upper troposphere to lower stratosphere derived from the radiosonde data set "RICH". The dashed lines denote the major volcanic eruptions of Agung (March 1963), El Chichon (April 1982) and Mt. Pinatubo (June 1991). Figure taken from Randel (2010).

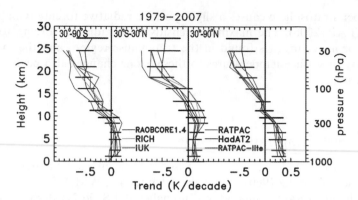

Figure 8.2   Vertical profiles of annual mean temperature trends (K/decade) for 1979–
2007 derived from the separate radiosonde data sets for latitude bands
30°–90°S, 30°N–S, and 30°–90°N. Here the calculated trends are based on
a simple linear approach. Error bars show the two-sigma statistical
uncertainty levels for the RATPAC-lite data. Figure taken from Randel
et al. (2009).

Figure 8.3 shows the time series of global-mean stratospheric temperature
anomalies as derived from satellite data (Microwave Sounding Unit/Strato-
spheric Sounding Unit) weighted over specific vertical levels (black lines). There
is strong evidence for a large and significant cooling in the stratosphere during
the last decades: It is about 0.5 K/decade in the lower stratosphere, whereas in
the upper stratosphere the cooling trend increases to about 1.2 K/decade
(Chapter 4 in WMO, 2011).[2] In addition, Figure 8.3 contains results from
chemistry-climate model simulations (colored lines). The overall development
of stratospheric temperature anomalies is mostly well reproduced by the
majority of CCMs; so far, the obvious difference after 1998 between the SSU26
data series and results derived from CCMs is unexplained.

The stratospheric cooling has not evolved uniformly in recent decades (see
Figures 8.1 and 8.3). A complete interpretation of these changes can only be
given considering natural forcing affecting the behavior of the atmosphere,
including the 11-year activity cycle of the sun, the quasi-biennial oscillation
(QBO) of tropical zonal winds in the lower stratosphere, and large volcanic
eruptions.[8,9]

The 11-year solar activity cycle is documented by obvious fluctuations in the
intensity of solar radiation at different wavelengths. Eleven-year solar ultraviolet (UV) irradiance variations have a direct impact on the radiation and
ozone budget of the middle atmosphere.[10] During years with maximum solar
activity, the solar UV irradiance is clearly enhanced (near 200 nm, which is the
wavelength range most important for the formation of ozone, with the difference between maximum and minimum activity amounts to 6%–8%),[11] which
leads to additional ozone production and heating in the stratosphere and
above. By modifying the meridional temperature gradient, the heating can alter
the propagation of planetary and smaller-scale waves that drive the global

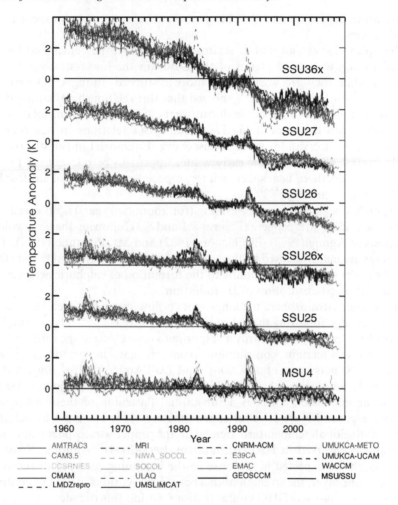

Figure 8.3 Time series of global mean temperature anomalies (K) derived from satellite measurements (Microwave Sounding Unit: MSU; Stratospheric Sounding Unit: SSU; black lines) and chemistry-climate model calculations (colored lines), weighted for MSU/SSU weighting functions. The anomalies are calculated with respect to the period 1980–1994. SSU Channel 27 corresponds to ~34–52 km altitude, channel 36x to ~38–52 km, channel 26 to ~26–46 km, channel 25 to ~20–38 km, channel 26x to ~21–39 km, and channel MSU4 to ~13–22 km. Figure taken from Chapter 4 in WMO (2011).

circulation (see also Section 8.3). Although the direct radiative forcing of the solar cycle in the upper stratosphere is relatively weak, it could lead to a large indirect dynamical response in the lower atmosphere through a modulation of the polar night jet and the Brewer-Dobson circulation.[12] Such dynamical changes can affect the chemical budget of the atmosphere because of the temperature dependence of the chemical reaction rates and transport of chemical species.

The Arctic lower and middle stratosphere tends to be colder and less disturbed during west wind phases of the QBO (in the lower stratosphere near 50 hPa) while they are warmer and more disturbed during QBO east wind phases.[13–15] Further analyses[16,17] showed that this relationship is strong during solar minimum conditions, while during years around maximum solar activity the relationship does not hold. This solar-QBO relationship has remained robust in the observations since its discovery. Equatorial upper stratospheric winds (near 1 hPa) during the early winter appear to be relevant for the evolution of the northern hemisphere winter, especially the timing of stratospheric sudden major warmings.[18–20]

An obvious sign of transient warming (for about two years) is observed in the lower and middle stratosphere (Figures 8.1 and 8.3) following the large volcanic eruptions of Agung (1963), El Chichón (1982) and Mt. Pinatubo (1991). These warmings are mainly caused by the large amount of sulphur dioxide transferred into the lower stratosphere, leading to the formation of sulphuric acid aerosols, and an enhanced absorption of IR radiation.

In the lower stratosphere the long-term cooling manifests itself as more of a step-like change following the volcanic warming events.[9,21,22] The overall lower stratospheric cooling is primarily a response to ozone decreases, with a possible but much less certain contribution from changes in stratospheric water vapor.[23,24] Ramaswamy *et al.* (2006)[8] and Dall'Amico (2010)[9] suggested that the step-like time series behavior is due to a combination of volcanic, solar cycle and ozone influences. There is a substantial flattening of these temperature trends evident in Figure 8.1 after approximately 1995. The latter aspect agrees with small global temperature trends in the upper stratosphere and lower mesosphere observed in HALOE data for 1992–2004.[25] Although some flattening might be expected in response to the beginning recovery of the stratospheric ozone layer, the strength of this behavior is curious in light of continued increases in long-lived GHG concentrations during this decade.

8.2.2 Expected Future Temperature Changes

Long-term changes in radiative forcing over the coming decades are expected to continue to impact global mean temperatures in both the troposphere and stratosphere. As discussed in Subsection 8.2.1, over the past three decades increases in GHG concentrations and the decline in the amount of stratospheric ozone have been the primary forcing mechanisms affecting stratospheric climate.

Global concentrations of GHGs are expected to rise for at least the next half century, although significant uncertainties remain as to the exact rate of increase. It is expected that these changes will be dominated by increases of GHG concentrations and lead, on average, to a cooling of the stratosphere, but there can still be a seasonal warming, particularly at higher latitudes caused by modifications of planetary wave activity. Therefore, the assessment of the future evolution of polar temperatures is uncertain (see Chapter 9). Future changes in stratospheric water vapor are even more difficult to predict, in part because the changes observed over the last four decades are still not fully understood.[5,26,27] While declines in stratospheric ozone also yield a cooling of the stratosphere, within the recent decade (2000–2010) global ozone levels have begun to rise (Chapters 2 and 3 in WMO, 2011;[2] Chapter 9 of this book). Higher ozone levels will increase stratospheric ozone heating, which will at least partially offset the cooling due to increases in GHG concentrations. Because ozone concentrations are so sensitive to the background temperature field, understanding the complex interaction between changing constituent concentrations and temperature requires an evaluation of the coupling between chemistry, radiation and atmospheric dynamics.

8.2.3 Temperature Ozone Feedback

The assessment of the sensitivity of ozone-related chemistry to climate changes is complicated, since accompanying modifications occur in the dynamics, transport and radiation. In particular, temperature changes affect physical, dynamical, and chemical processes influencing the ozone content in the atmosphere in different ways. Moreover, the chemical state of the atmosphere changes as the concentration of trace species and ODSs change. This in turn alters the sensitivity of the stratospheric chemical system to temperature changes.

The sensitivity of ozone chemistry in the upper part of the stratosphere (about 35 to 50 km) to changes in temperature is well explained. There the chemical system is generally under photochemical control and is constrained by gas-phase reaction cycles that are well known. The largest stratospheric cooling, which is associated with increased GHG concentrations has been observed in the upper stratosphere and mesosphere (50 to 100 km). The most important ozone loss cycles in the upper stratosphere (via the catalysts NO_X, ClO_X, and HO_X) are slowing as temperatures decrease[28] (see Chapter 1), leading to higher ozone concentrations. In the lower mesosphere, enhanced ozone concentrations are primarily due to the negative temperature dependence of the reaction $O + O_2 + M \rightarrow O_3 + M$. The situation is more complex in the upper stratosphere and lower mesosphere with different ozone loss cycles having greater influence on ozone concentrations at different altitude ranges (*e.g.* HO_X between about 45 and 60 km; ClO_X between about 45 and 50 km; NO_X between about 30 and 50 km). There the slower loss rates are controlled both by the temperature dependence of the reaction rate constants and by the reduction in the amount of atomic oxygen (change in O_X partitioning). The rate-limiting reactions for all the ozone loss cycles are proportional to the atomic oxygen number density. The atomic oxygen number density, in turn, is also strongly appointed by the reaction $O + O_2 + M \rightarrow O_3 + M$.[29]

The situation is more complicated in the polar lower stratosphere (about 15 to 25 km) in late winter and spring. In addition to the gas-phase ozone loss cycles described above playing a similar role in determining the ozone concentration[30] there is an offset by chlorine- and bromine-containing reservoir species. These chemical substances are activated *via* heterogeneous processes on surfaces of polar stratospheric cloud (PSC) and cold aerosol particles, leading to markedly increased concentrations of ClO_X and BrO_X. This in turn yields significant ozone losses *via* ClO_X and BrO_X catalytic cycles in the presence of sunlight. The rate of chlorine and bromine activation is strongly dependent on stratospheric temperatures, increasing significantly below approximately 195 K. In a polar lower stratosphere with enhanced (over natural levels) concentrations of ODSs, as it is currently observed, chlorine and bromine activation and consequent ozone losses at lower temperatures, counteract any ozone increase through temperature driven reduction in NO_X and HO_X gasphase ozone loss.

Due to systematic differences in the winter and spring season temperatures in the Arctic and Antarctic lower stratosphere, ozone concentrations are developing different in the northern and southern polar stratosphere. In the Antarctic stratosphere temperatures are almost always below the threshold for heterogeneous activation of chlorine and bromine containing species during the winter and in early spring. In the case of elevated ODS concentrations this leads to a significant depletion of ozone and the formation of the Antarctic ozone hole. In contrast, the Arctic lower stratosphere is dynamically much more active and lower stratospheric temperatures are principally higher (about 10 to 15 K). Here temperatures lie close to the threshold value for activation of chlorine and bromine species and ozone depletion *via* heterogeneous chemical reactions are much smaller. Consequently, a significant change in Arctic stratospheric temperatures, for example, a cooling due to climate change would strongly influence springtime ozone concentrations in the Arctic region.

While it is expected that the reduction of ODSs in the next decades will lead to enhanced ozone concentrations, this increase could be affected by modifications in temperature, chemical composition and transport (see also Section 8.3). The future evolution of ozone concentrations is sensitive to changes in both chemical constituents *and* climate. A further cooling of the polar lower stratosphere could delay the recovery of the ozone layer in the Arctic and Antarctic region but on the other hand it could accelerate the ozone recovery in other parts of the stratosphere (see Chapter 9). As mentioned earlier, not only is temperature-dependent chemistry affecting the stratospheric ozone content, but so are dynamical processes. This topic is discussed in the following section.

8.3 Impact of Enhanced Greenhouse Gas Concentrations on Stratospheric Dynamics

Even in the absence of ODSs, climate change can alter the distribution of stratospheric ozone. Climate and circulation changes affect the transport of trace gases and particles within the stratosphere. But climate change also influences the air mass exchange between the troposphere and the stratosphere, for example, the entry of chemical substances into the stratosphere. Therefore, the lifetimes of long-lived chemical substances are determined by the exchange rate of air masses across the tropopause.[31]

Although the stratosphere and troposphere are different in many ways, the atmosphere is continuous, allowing vertical wave propagation and a variety of other dynamical interactions between these regions. A complete description of atmospheric dynamics requires a full understanding of both layers. The dynamical coupling of the stratosphere and troposphere is primarily mediated by the dynamics of atmospheric waves. A variety of such waves originates in the troposphere, propagates upward into the stratosphere and higher up and then dissipates, forming the spatial and temporal structure of stratospheric motions. Moreover, the stratosphere not only shapes its own temporal evolution but also that of the troposphere. Tropospheric and stratospheric dynamics as well as the dynamical coupling of both altitude regions are affected by climate change.

8.3.1 Importance of Atmospheric Waves

The activity of atmospheric waves is divided into three consecutive processes: the generation mechanisms (*i.e.* forcing of waves), the propagation through the atmosphere, and the dissipation of waves primarily due to wave breaking and thermal damping. Outside the tropical region the temperature structure of the stratosphere depends mostly on a balance between diabatic radiative heating and adiabatic heating from induced vertical motion due to dissipation of large-amplitude, planetary-scale Rossby waves.[32] These waves have typical wavelengths of several thousands of kilometres (wave numbers one to three). They either remain stationary or propagate very slowly from the east to the west (*i.e.* quasi-stationary waves) because they are generated due to a combination of overflows of large-scale orographic barriers (like the Rocky Mountains, the Andes, or the Himalaya), meridional temperature gradients, and the Coriolis force (as a consequence of Earth's rotation). Depending on background wind conditions these waves can become unstable somewhere in the stratosphere-mesosphere region (up to 100 km) while propagating upwards.[33] This so-called "wave breaking" leads to a deposition of thermal energy in the upper stratosphere and mesosphere, while planetary waves deposit easterly momentum in the lower stratosphere, which both decelerating the west wind of the stratospheric polar vortex in wintertime (*i.e.* the polar night jet). The weakening of the westerly jets has to be adjusted in the geostrophic balance and causes therefore a small meridional wind component that drives the residual circulation.[34] The stirring of air isentropically across larger distances of the winter stratosphere within a region is known as the "surf zone".[35] This region is bounded by sharp gradients of tracer concentrations in the winter subtropics and at the edge of the polar night jets. The deceleration of the polar vortex (*i.e.* weakening of the zonal wind speed) is accompanied by a warming of the stratosphere at higher latitudes resulting in a thermodynamic imbalance yielding a radiative cooling of the polar stratosphere back towards radiative equilibrium. Consequently, downwelling of polar air masses is enhanced, which must be balanced by a poleward flow of air from lower latitudes. This response pattern describes a meridional circulation which is called the Brewer-Dobson (BD) circulation (see Subsection 8.3.2).

The basic climatology of the extra-tropical stratosphere is mostly understood in terms of large-scale wave dynamics together with the seasonal cycle of radiative forcing. For example, the easterly winds of the summer stratosphere prevent upward propagation of planetary waves.[33] Therefore, in summer stratospheric variability is much smaller than in winter. Dissimilar distribution of the continental land masses between the northern and the southern hemisphere imply asymmetries in the efficiency of planetary wave generation mechanisms. Consequently, in the northern winter stratosphere, planetary wave disturbances are significantly larger than those in southern winter.

In the tropical lower and middle stratosphere between 100 and 10 hPa, the prevailing variability mode is the quasi-biennial oscillation (QBO) of the tropical zonal mean wind field, alternating between westerly and easterly winds with a mean period of approximately 28 months. The alternating wind regimes repeat at intervals that vary from 22 to 34 months (see Baldwin *et al.* (2001)[36] for a review). The peak-to-peak amplitude of the zonal wind speed is about 55 m/s near 20 hPa.[37] Maximum easterlies are generally stronger than westerlies, *i.e.* ~ 35 m/s and ~ 20 m/s, respectively. The QBO signal in temperature amplitude is approximately 8 K. The QBO affects the global stratospheric circulation and, therefore, influences a variety of extra-tropical phenomena including the strength and stability of the polar night jet, and the distribution of ozone and other gases (Baldwin *et al.*, 2001). The QBO is mainly driven by the dissipation of a variety of west- and eastward propagation large-scale equatorial waves.[38,39]

Although stratospheric variability has long been viewed as being caused directly by variability in tropospheric wave sources, it is now widely accepted that the configuration of the stratosphere itself also plays an important role in determining the vertical flux of wave activity from the troposphere because of the strongly inhomogeneous nature of the stratospheric background state, for example, the steep gradient of zonal wind and potential vorticity at the edge of the polar vortex.[40] Given a steady source of planetary waves in the troposphere, any modulation in stratospheric background, for example, gradients of temperature, winds or potential vorticity alter the vertical wave fluxes, giving rise to the possibility of internally driven variability of the stratosphere, as it was already demonstrated in an early numerical model study by Holton and Mass (1976).[41] Some recent idealized modeling studies even suggest that realistic stratospheric variability can also arise in the absence of tropospheric variability.[42,43]

Further, the tropospheric circulation itself is also influenced by the stratospheric configuration. Reflection of stationary planetary wave energy back into the troposphere can occur when the polar vortex exceeds a critical threshold in the lower stratosphere, yielding structural changes of the leading tropospheric variability patterns.[44–46]

Not only is the consideration of large-scale planetary and synoptic wave dynamics very important in determining the climatology of the stratosphere, but also the effects of smaller-scale waves (*e.g.* gravity waves forced by orography or convective events) must be considered for explanations of stratospheric dynamic variability. Any systematic change in processes affecting generation, propagation or dissipation of all waves results in systematic changes of the temperature structure of the stratosphere. The capability of numerical models of the atmosphere to simulate the climatology and space-time changes of stratospheric properties depends critically on the ability to simulate highly nonlinear wave dynamics in a robust way. One of the most challenging aspects of modelling the dynamical coupling of the troposphere and stratosphere is the parameterization of the effects of unresolved waves (in particular gravity waves) and their feedback on the resolved flow. Global atmospheric models mostly have insufficient horizontal and vertical resolution to resolve all necessary characteristics and effects, and therefore they must be prescribed, *i.e.* parameterized in the models. Another issue is that deep convection (an important excitation mechanism for waves that propagate into the stratosphere) is a sub-grid-scale process that must also be parameterized.

8.3.2 The Brewer-Dobson Circulation and Mean Age of Air

The meridional circulation in the stratosphere is called the Brewer-Dobson (BD) circulation, named after Alan Brewer and Gordon Dobson, to honor their fundamental research studies of stratospheric water vapor and ozone measurements.[48,49] The BD circulation is the major driver for transport of stratospheric air masses from tropical to higher latitudes. Its climatology is characterized by rising motion of air in the tropics from the troposphere into the stratosphere and poleward transport there (Figure 8.4). The BD circulation is more pronounced in the winter hemisphere. Due to mass conservation, descending motion of air occurs in stratospheric middle and higher latitudes, mixing stratospheric air back into the troposphere.

One reason could be that the BD circulation should result from solar heating in the tropical region and cooling at higher latitudes, leading to a large circulation cell reaching from the tropics to polar regions as warm tropical air ascends and cold polar air descends. Although this reasoning seems consistent with the observations, the BD circulation mainly results from planetary wave forcing observed in the extra-tropical stratosphere. Thus, the BD circulation has been likened to an enormous wave driven pump.[47]

In particular, the wintertime stratosphere is dominated by planetary (Rossby) waves propagating upward from the troposphere (see Subsection 8.3.1). The existence of the BD circulation is strongly linked to planetary wave activity (Figure 8.4). The BD circulation is different in the southern and northern hemisphere because of hemispheric differences in land–ocean distributions. This leads to more frequent and intense planetary wave activity and stronger BD circulation in the northern winter season. In wintertime, the horizontal mixing in the northern hemisphere often reaches the polar region, whereas in the southern

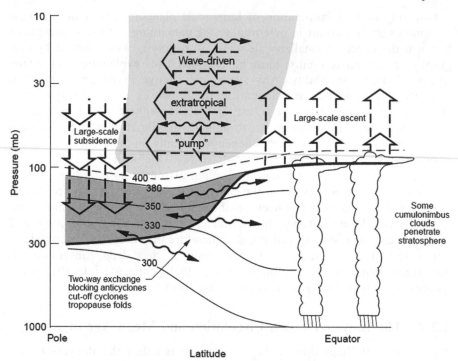

Figure 8.4   Dynamical aspects of stratosphere-troposphere exchange. The tropopause
is shown by the thick black line. Thin lines denote isentropic (*i.e.* constant
potential temperature) surfaces (K). The light shaded area in the stratosphere denotes wave-induced forcing, *i.e.* the extra-tropical pump area,
driving the Brewer-Dobson circulation (see text). The heavy shaded area
indicates the so-called lowermost stratosphere. Figure taken from Holton
et al. (1995).

hemisphere horizontal mixing is most of all confined to lower and middle latitudes and seldom reaches the Antarctic region.

Among other factors, the BD circulation determines the distribution of
ozone (Figure 8.5) and water vapor in the stratosphere, but it also affects the
lifetimes of ozone anthropogenic ODSs, and of some GHGs. It is expected that
climate change will modify processes responsible for the generation of the BD
circulation.

So far observations provide an unclear picture of trends in the strength of the
BD circulation, as intrinsic variability is not well known and effects are difficult
to measure. Analyses of the BD circulation based on chemical measurements
(and estimates of the mean age of air, see below) are often very noisy and
have error bars that exceed the amplitude of the observed trends[51] (see also
Chapter 4 in WMO, 2011).[2]

Recent studies of General Circulation Model (GCM) and Climate-Chemistry
Model (CCM) simulations consistently indicate an acceleration of the BD

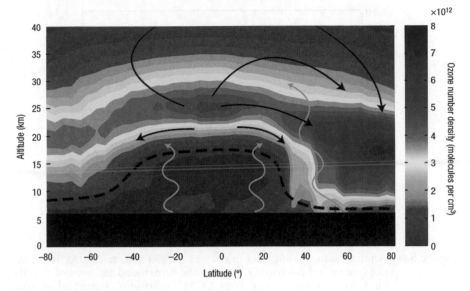

Figure 8.5 Brewer-Dobson circulation and stratospheric ozone. A longitudinally averaged cross-section of the atmosphere shows a schematic of the meridional stratospheric circulation, i.e. the Brewer-Dobson circulation (black arrows), and the ozone distribution (molecules cm^{-3}) as measured by the OSIRIS satellite instrument in March 2004. The circulation is forced by waves propagating up from the troposphere (orange wiggly arrows), especially in the winter hemisphere, and it strongly shapes the distribution of ozone by transporting it from its source region in the tropical upper stratosphere to the high-latitude lower stratosphere. Consequently, ozone number densities are higher at polar latitudes than in the tropics. The dashed line represents the tropopause. Copyright OSIRIS Science Team. Figure taken from Shaw and Shepherd (2008).

circulation in response to increasing greenhouse gas concentrations in future with distinct consequences for the recovery of the ozone layer[52–55] (see also Chapter 4 in SPARC CCMVal, 2010[56] for an overview). As an example, Figure 8.6 presents results derived from a number of CCM simulations showing an almost steady increase of tropical upwelling in the future. The net upward mass flux of air into the tropical stratosphere is taken as a measure of the BD circulation. The implications for a future increase in the BD circulation are substantial. Such an increase would, for example, change the spatial distribution of stratospheric ozone, with increased total ozone at high latitudes and decreased total ozone in the tropics. It would also increase the stratosphere-to-troposphere ozone flux[57] and it would decrease the net age of stratospheric air.[53,58–60]

A stronger BD circulation would tend to warm the extra-tropical regions and cool the tropics. It would have direct implications for injection of tropospheric source gases (*i.e.* original gases transported into the stratosphere and then reacts there), product gases (*i.e.* intermediate or final products produced in the

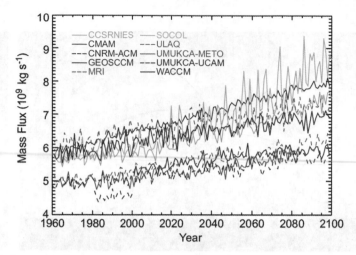

Figure 8.6   Annual mean upward mass flux (kg s⁻¹) at 70 hPa, calculated from residual mean vertical velocity between the turnaround latitudes of the BD circulation; results derived from CCM simulations. Figure taken from Chapter 4 of SPARC CCMVal (2010).

troposphere), and water vapor into the stratosphere and for transport of stratospheric air mass with high ozone concentrations and very low water vapor content into the troposphere. The two-way coupling between troposphere and stratosphere is also demonstrated by the link between ozone depletion and changes in surface climate in the southern hemisphere[61,62] (Chapter 4 in WMO, 2011).[2] Details are discussed in Chapter 7.

Observational investigations of the variability and long-term changes of the BD circulation have not yet provided a clear picture. Nedoluha *et al.* (1998),[63] for example, reported that a slower BD circulation could explain the negative trend in upper stratospheric methane, whereas Waugh *et al.* (2001)[64] argued that a faster BD circulation could explain the observed trend in upper stratospheric chlorine which has been measured by the Halogen Occultation Experiment (HALOE) onboard the Upper Atmosphere Research Satellite (UARS). A weakening of the BD circulation was shown by Salby and Callaghan (2002),[65] consistent with the results presented by Hu and Tung (2003),[66] who found a reduction of planetary wave activity occurring only in late winter (1979–1999) and no obvious change before 1979. They proposed that ozone depletion could be the reason for this reduction due to radiative-dynamical feedback increasing ozone depletion. Conversely, a strengthening of the BD circulation by ozone depletion was suggested by Li *et al.* (2008).[67] An accelerated upwelling in the tropical region was identified in long-term observations by Thompson and Solomon (2005)[68] and Rosenlof and Read (2008).[69] Thompson and Solomon (2009)[7] demonstrated that the contrasting latitudinal structures of recent stratospheric temperature (*i.e.*, stronger cooling in the tropical lower stratosphere than in the extra-tropics) and ozone trends (*i.e.*, enhanced ozone reduction in the tropical lower stratosphere) are consistent with the assumption of an accelerated stratospheric overturning BD circulation.

The observed drop in the tropical lower stratospheric water vapor concentrations after 2001 (see Figure 8.7) is consistent with an enhanced tropical upwelling during that time[27] (see Chapter 4 in WMO, 2011)[2] and a step-like increase in the summed extra-tropical activity of planetary waves from both hemispheres[71] indicating an accelerated in BD circulation in both hemispheres. Other studies suggested that changes in the BD circulation can account for a large fraction of the long-term total ozone decline outside the polar regions (*i.e.* weakening of BD circulation) until the mid-1990s and recent increases in northern middle latitude spring (*i.e.* strengthening of BD circulation).[72–74] Similar results were observed in chemical-transport models (CTMs) using ECMWF reanalysis data.[75]

The mean stratospheric transport time from the tropical lower stratosphere, where tropospheric air enters the stratosphere to any point in the stratosphere

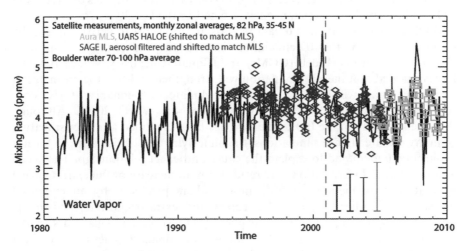

Figure 8.7 Observed changes in stratospheric water vapor mixing ratio (ppmv). Time series of stratospheric water vapor mixing ratio (ppmv) averaged from 70 to 100 hPa near Boulder Colorado (40°N, 105°W) from a balloon-borne frost point hygrometer covering the period 1981 through 2009; satellite measurements are monthly averages, balloon data plotted are from individual flights. Also plotted are zonally averaged satellite measurements in the 35°N-45°N latitude range at 82 hPa from the Aura MLS (turquoise squares), UARS HALOE (blue diamonds) and SAGE II instruments (red diamonds). The SAGE II and HALOE data have been adjusted to match MLS during the overlap period from mid-2004 to the end of 2005, as there are known biases (Lambert *et al.*, 2007). Representative uncertainties are given by the colored bars; for the satellite data sets, these show the uncertainty as indicated by the monthly standard deviations, while for the balloon data set this is the estimated uncertainty provided in the Boulder data files. Figure adapted from Solomon *et al.* (2010).

can in principle be determined by observations of inert trace gases exhibiting a pronounced temporal trend in the troposphere, like sulphur hexafluoride (SF_6), another important GHG used by the electricity industry. The spatial distribution of these transport times, also called mean age of air, is a characterization of the BD circulation. For instance, tropical upwelling is inversely related to mean age of air so that the "age of air" changes as the stratospheric climate changes.[58,76] The global distribution of the mean stratospheric transport time of air can be assessed using observations of the GHG SF_6, for example, from ESA's MIPAS/ENVISAT instrument.[77] So far the available time series derived from satellite measurements are too short to estimate robust trends since interannual variability is very high. Assessments of ascent rates in the lower tropical stratosphere have been provided using different methods including satellite-based observation of water vapor and diabatic heating rates.[78–81] Several CTMs were compared with respect to their transport properties and strategies have been developed to use the meteorological re-analyses for multi-annual simulations and to improve the model performance.[82,83] The multi-annual CTM simulations show a slight acceleration of the BD circulation for the past 30 years, which seems to be in contrast with estimates derived from CO_2 and SF_6 balloon measurements executed in the last 30 years.[51]

While current observational data records are too short to derive statistically significant trends in the BD circulation, several independent studies with both General Circulation Models (GCMs) and Climate-Chemistry Models (CCMs) have indicated that the BD circulation will strengthen and that the mean age of stratospheric air will decrease in a future climate with enhanced GHG concentrations[31,52–55,59,60,67,84–89] (see also Chapter 4 in WMO, 2011).[2] Although this strengthening of the BD circulation has been identified as a robust feature of many climate change simulations, the underlying mechanisms are so far not sufficiently understood to explain the cause and effect relationship. The acceleration of the BD circulation may result from an increase of the extra-tropical generation of planetary waves[59] among others produced by an enhanced temperature gradient between the tropics and extra-tropical regions in the upper troposphere and lower stratosphere (UTLS) region. This is associated with global warming in the models, leading to an enhanced poleward eddy heat flux in the stratosphere.[90] The possible impact of enhanced generation of planetary waves in the (sub)-tropical region caused by higher tropical SSTs and its importance for an intensified upwelling in the tropical UTLS region is discussed for example in Deckert and Dameris (2008).[91] More details are presented in Subsection 8.3.3.

Due to the difficulties in obtaining long-term observations of dynamical parameters in the stratosphere (with the exception of temperature and ozone), a validation of past changes of stratospheric dynamics has not yet been performed in a sufficient way. Li and Waugh (1999),[92] for example, showed that the mean age of air in their two-dimensional model is mainly sensitive to changes in the BD circulation, while changes in mixing show a much weaker effect on mean age. This situation is reversed for tracers with chemical sinks in the stratosphere, for example, nitrous oxide (N_2O) and chlorofluorocarbons

(CFCs). Changes in the intensity of horizontal transport and mixing should influence distributions and correlations of tracers like N_2O and CFCs, observable quantities which can be linked to changes in stratospheric dynamics. Correlations between long-lived tracers in the stratosphere are known to be very robust and are sensitive to mixing processes in the stratosphere[92] (see Plumb, 2007 for a review).[93]

Austin and Li (2006)[58] suggested that a long-term change of mean age of air, on the order of more than half a year, should already have occurred since the mid 1970s. This is in contrast to the findings of Engel *et al.* (2009).[51] As age of air can be derived from observations, this provides a quantity for the validation of modeled changes in the BD circulation. A strengthening of the BD circulation due to increased GHG concentrations would enhance stratosphere-troposphere mass fluxes, which would have important consequences, for example, a reduction of the lifetimes of ODSs, changes in high- and low-latitude temperature, or enhanced downward transport of stratospheric ozone into the troposphere. Such effects influence the future evolution of atmospheric composition, particularly concentrations of ozone and water vapor.

Further consequences will certainly depend on particular future changes in the BD circulation. While the sign of the predicted change in the BD circulation is consistent between atmospheric models, the magnitude and the detailed structure, and therefore the consequences are not. At the moment the differences between models are difficult to interpret because the reasons for the changes of the BD circulation remain unclear. As long as the causes and mechanisms for a possible enhancement of the BD circulation and the magnitude of these changes remain insufficiently explained, the consequences for stratospheric parameters cannot be predicted with the necessary robustness. This uncertainty limits the validity of prognostic studies regarding the future evolution of climate and atmospheric chemical composition, in particular of the stratospheric ozone layer (see Chapter 9).

8.3.3 The Role of Sea Surface Temperatures

With higher GHG concentrations leading to higher tropospheric temperatures, the oceans will absorb heat and sea surface temperatures (SSTs) will tend to increase. SSTs affect the activity of atmospheric waves, *i.e.* generation, propagation and dissipation of planetary as well as gravity waves, and hence the BD circulation.[59,91,94–97] Most of the recent numerical modeling studies that are based on simulations with GCMs and CCMs are imposing SSTs without atmospheric feedback in order to make multi-decadal integrations feasible[52–54,98,99] (see also SPARC CCMVal, 2010).[56] The SST fields are either taken from observations or climate model simulations (*i.e.* coupled Atmosphere-Ocean General Circulation Models, AOGCMs). To a large extent, this practice reproduces zonal-mean hydrological characteristics and interannual variability in stratospheric dynamics of the respective observations and AOGCM simulations.[94,96,100] These investigations indicate that there is an obvious distinction in tropospheric reaction to prescribed tropical and extra-tropical SST anomalies, with implications for stratospheric dynamics (*e.g.* tropical upwelling) and chemistry (*e.g.* polar ozone). The tropospheric reaction to SST anomalies in the tropics is mostly barotropic and deep-convection mediated. The reaction to SST anomalies outside the tropical region is weaker, shallower, and more complicated, because baroclinicity is involved and latent-heat release from convection is much weaker. This makes it more difficult to study atmospheric changes from SST modifications in middle and higher latitudes.[101,102]

In the case of prescribed higher SSTs in tropical regions, numerical simulations with GCMs and CCMs have indicated that deep convection is an important messenger transferring the SST signal into the stratosphere. Currently, there are two mechanisms discussed that may act simultaneously and that both involve SST-related an intensification of tropical deep convection.[103] The first mechanism is based on higher temperatures in the tropical and subtropical upper troposphere resulting from stronger latent-heat release due to intensified deep convection.[59,104] This increases the latitudinal temperature contrast between the tropical upper troposphere and extra-tropical lowermost stratosphere, strengthening locally the zonal wind velocity. The altered wind profile influences planetary wave activity, altering tropical lower stratospheric wave dissipation and hence tropical upwelling and the BD circulation.[97] However, it remains unclear whether a modification of wave generation mechanisms or modifications in wave propagation conditions dominate this impact on the BD circulation and whether waves generated in the tropics or extra-tropics are involved.

The second mechanism focuses on changes in tropical upper tropospheric pressure perturbations that are associated with intensified deep convection due to enhanced SSTs in the tropics. Deckert and Dameris (2008).[91] considered both SST-induced modifications to the convection-related eddy dissipation in the tropical lower stratosphere and the associated implications for the BD circulation. Comparable to a stone hitting a water surface, pressure perturbations excite tropical quasi-stationary planetary waves. They propagate upward as they dissipate but carry enough of the SST signal across the tropical tropopause into the lower stratosphere to affect the tropical upwelling *via* the principle of downward-control.[105] Additionally, Chen *et al.* (2001)[106] and Chen (2001)[107] demonstrated that these eddies can ascend through the tropical easterly winds and cross the tropopause. Rind *et al.* (2002)[108] and Fomichev *et al.* (2007)[89] investigated atmospheric conditions with doubled CO_2 concentrations. They inferred a pattern of amplified eddy dissipation occurring in the tropical lower stratosphere too, appearing to accelerate the upwelling in the tropics by stimulating an anomalous BD cell locally. According to these numerical studies, the strengthening of eddy dissipation is mainly referring to enhanced tropical SSTs, but the importance of deep convective quasi-stationary eddy generation was not yet proved in these investigations.

Both mechanisms discussed above are able to explain observational signs for an accelerated upwelling in the tropics across the tropopause. Satellite as well as radiosonde data indicate that a reduction in temperatures and ozone concentrations have occurred over the past four decades, particularly in the tropical lower stratosphere at all longitudes and during all seasons.[68,109] This finding is consistent with the hypothesis of intensified tropical upwelling. Although studies of stratospheric mass transport trends support this hypothesis, they have large uncertainties.[52,59] Radiative changes as a result of anthropogenic ozone depletion might account for similar modifications of tropical upwelling (Forster *et al.*, 2007).[110] Both convection-related mechanisms fulfil the requirement of enhanced planetary wave breaking at low latitudes. The observed decrease of temperature in the tropical tropopause region in 2001[27] is part of a close relationship between SSTs and lower stratospheric temperatures.[69] There is a clear anticorrelation between SSTs in the western tropical Pacific Ocean—the region on Earth with highest SSTs—and temperatures and ozone and water vapor concentrations in the tropical lower stratosphere. Anomalously high SSTs coincide with low temperatures and ozone concentrations, and *vice versa*. This anticorrelation is unlikely to result from the lifting of air due to convection, since the stratospheric signal occurs at altitudes well beyond the highest-reaching thunderstorms. Both mechanisms discussed above could contribute to this anticorrelation. So far, the response of planetary wave activity to anomalies in convection-related pressure perturbations seems to be a better candidate because it is more immediate than the planetary wave response to anomalies in latitudinal temperature contrast. For example, some of the mentioned numerical sensitivity studies indicated that there are various different latitudes where stratospheric wave breaking is sensitive to SST anomalies.[94,96]

Moreover, it was shown by Brönnimann *et al.* (2006)[111] that wave breaking in the extra-tropical northern stratosphere during winter responds to the El Niño/Southern Oscillation (ENSO) signal in tropical SSTs. Nevertheless, the sensitivity of planetary wave activity to extra-tropical SST anomalies relative to SST anomalies in tropical regions is unknown.[89,112–114] In theory, the tropospheric response to extra-tropical SST changes should influence the life-cycle of planetary waves *via* altered ocean-continent temperature contrast, changes in position and strength of storm tracks, and modified barotropic or baroclinic instability.[32]

8.4 Coupling of the Stratosphere and the Troposphere in a Changing Climate

The net mass exchange between the troposphere and stratosphere is mostly associated with the BD circulation[47,115] with a net upward flux in the tropics balanced by a net downward flux in the extra-tropical regions (Subsection 8.3.2). Air in the tropical lower stratosphere rises slowly (about 0.2 to 0.3 mm/s) and carries ozone-poor air from the troposphere higher up into the stratosphere. There, with increasing altitude, photochemical production of ozone becomes more effective. The upwelling in the tropics is clearly modulated by the seasonal cycle and the tropical QBO phase.[36] When the QBO is in its westerly phase the ascent rate is lower, and there is more time for ozone production, which enhances the tropical total ozone column. However, near the tropopause the picture is more complex, with two-way mixing across the extratropical tropopause at and below synoptic scales, and vertical mixing in the tropical-tropopause layer (TTL) resulting from convective processes. In the subtropics and extra-tropics there is not only transport of chemical species and particles from the troposphere into the lowermost stratosphere occurring through quasi-isentropic motion (associated with synoptic-scale and mesoscale circulations, *e.g.* baroclinic eddies, frontal circulations) but there is also substantial transport of air masses from the stratosphere into troposphere. Quantification of this two-way transport has improved significantly over recent years through analyses of observations and numerical modelling studies (see Stohl *et al.* (2003)[116] for a review). However, significant quantitative uncertainties remain about the role of small-scale circulations, for instance the importance of convective systems in transporting air from the troposphere into the stratosphere and *vice versa*. A further complication is that all related processes are affected by climate change, making it even more difficult to assess future changes of stratosphere-troposphere connections.

8.4.1 Stratosphere-troposphere Coupling

The stratosphere and the troposphere are connected by physical, dynamical and chemical processes. Changes in the concentrations of radiatively active gases in the stratosphere yield significant changes in stratospheric temperature[6,8,23,117] (see Subsection 8.2.1). In addition, especially in winter, the stratosphere is significantly affected by upward propagating tropospheric waves that dissipate in the stratosphere and mesosphere and slow the polar night vortex (Subsection 8.3.1). Statistical analyses of dynamical quantities have demonstrated a strong connection between stratospheric and tropospheric modes of variability.[118–121] For instance, during northern winter, a high correlation exists between the intensity of the stratospheric polar night jet and the North Atlantic Oscillation (NAO) in the middle troposphere,[122] which is a key parameter for weather and climate in Europe. A strong stratospheric polar night jet is associated with a positive phase of the NAO, corresponding to stronger westerlies in the North Atlantic region and positive temperature anomalies over central and northern Eurasia.

Moreover, Thompson and Wallace (1998; 2000)[123,124] identified vertically coherent patterns from the stratosphere down to Earth's surface with more zonally symmetric, quasi-annular anomalies characterized by geopotential anomalies of one sign over the polar cap, offset by anomalies of the opposite sign over lower latitudes. This concept of annular modes applies equally well in either hemisphere, and describes the leading mode of variability from the surface through the stratosphere.[121] This variability pattern is now referred to as the Northern Annular Mode (NAM) and the Southern Annular Mode (SAM). The surface NAM is similar to the NAO pattern, but is geographically broader in scale. For the purposes of stratosphere-troposphere coupling, it makes little difference whether tropospheric variability is described by the NAO or the NAM. While NAM and SAM variability exist in the troposphere throughout the year, the variability extends into the stratosphere during the winter and spring seasons, when stratospheric dynamical variability is enhanced (the stratospheric circulation in the summer is quiescent).

During northern winter, deep positive and negative NAM anomalies are associated with anomalously strong or weak polar zonal wind jets, respectively. Usually they appear first in the upper stratosphere and mesosphere and then propagate downward to the troposphere. There they are seen as anomalies of tropospheric meteorological fields with a time lag of several weeks.[125,126] Thompson *et al.* (2002)[127] found a high correlation between extreme weather events and the strength of the stratospheric polar vortices, hence implying that considering stratospheric anomalies in winter might improve extended-range weather forecasting. Baldwin *et al.* (2003; 2007)[128,129] emphasised the importance of persistent circulation anomalies in the lower stratosphere in winter for the phase of the tropospheric NAM. The annular mode variations are consistent with deep temperature anomalies that result from modulation of the residual circulation. Thus, anomalously strong wave driving (as during a sudden warming) leads to stronger downwelling and warming over the polar cap. The anomalous warming extends into the upper troposphere.

On climate change time scales, changes in tropospheric variability can be associated with stratospheric variability of either natural or anthropogenic origin. For example, Kodera (2002)[130] showed that during the maximum of the 11-year cycle of solar activity, the initial radiative solar signal leads to obvious modifications of the stratospheric zonal mean wind, which are correlated with the NAO index (see above). A similar connection between the stratosphere and troposphere was also found for the SAM index, but only during maximum solar activity.[131] These results, derived from observed data, were supported by investigations performed with GCMs. Matthes *et al.* (2006),[132] for example, detected significant differences in the near-surface geopotential height in the northern hemisphere between minimum and maximum solar activity resembling the signature of the AO, with more positive phases during solar maximum. Moreover, stronger polar night jets during solar maximum are associated with stronger tropospheric cyclone activity in the North Atlantic region and warmer and more humid winters in central Europe and Eurasia.

Temperature anomalies which were identified near Earth's surface following large volcanic eruptions resembled anomalies associated with the positive phase of the NAO.[133,134] Accompanying investigations with data derived from GCM simulations[135,136] showed that dynamical feedback processes initiated by the tropical stratospheric warming after volcanic major eruptions due to enhanced sulphur aerosol loading were responsible for the tropospheric response.[137,138]

In addition to naturally forcing mechanisms of stratosphere-troposphere coupling, man-made contributions are important. The stratosphere is strongly influenced by radiative perturbations, which are caused by reduced stratospheric ozone content and enhanced GHG concentrations including water vapor.[23,24,139] Schwarzkopf and Ramaswamy (2008)[140] analysed multi-decadal climate model simulations with varying (*i.e.* transient) boundary conditions. They found a sustained and significant global, annual mean cooling in the lower and middle stratosphere since about the 1920s, a global temperature change signal developing clearly earlier than in any lower atmospheric region that mostly results from carbon dioxide (CO_2) increases. Particularly since the beginning 1980s, stratospheric ozone depletion has strengthen the cooling in the stratosphere. Forster *et al.* (2007)[110] used a "radiative fixed dynamical heating" model to demonstrate that the effects of tropical ozone decreases at about 70 hPa and lower pressures can lead to significant cooling below which is comparable in magnitude to changes of other radiatively active trace gases (*e.g.* Chapter 5 in WMO, 2007).[1]

Thompson and Solomon (2002)[61] documented a clear increase of the tropospheric southern hemisphere circumpolar circulation since the 1970s with a warming of the Antarctic Peninsula and Patagonia and a marked cooling of the East-Antarctic region and the Antarctic plateau which is obviously associated with a shift to a more positive phase of the SAM. Similar tropospheric signatures were derived from GCM simulations with a prescribed stratospheric polar ozone loss. An induced cooling was shown to lead to more positive phases of the surface NAM and SAM[122,141,142] GCM simulations dealing with the influence of increasing GHG concentrations indicated a positive trend in the NAM in the troposphere.[62,143,144]

However, these studies came to contradictory conclusions about the relevance of the stratospheric contribution. Idealized numerical studies also demonstrated that parts of tropospheric variability can be explained by anomalies propagating downward from the stratosphere into the troposphere and that even climate near the Earth surface is affected.[145–148] So far the mechanisms which are important for the downward coupling of the stratosphere and troposphere remain unclear. There are several mechanisms worth considering: (1) Planetary wave activity (see Subsection 8.3.1): the vertical propagation of planetary waves from the troposphere to the stratosphere is affected by stratospheric dynamical conditions. Perlwitz and Harnik (2003)[149] suggested that wave reflection from the upper stratosphere could influence tropospheric circulation. (2) Planetary scale wave-mean flow interaction: the interaction between planetary waves with the zonal mean flow in the stratosphere may lead to a downward propagation of zonal wind and temperature anomalies that could reach the lower troposphere and Earth surface[137,150,151] (3) Direct responses to variances of potential vorticity: Hartley *et al.* (1998)[152] and Black (2002)[153] showed that changes in lower stratospheric zonal circulation can lead to obvious changes of tropopause height and tropospheric wind speed. (4) "Downward control": in case of adequately long anomalous wave driving, secondary equilibrium circulations develop in the stratosphere extending to the troposphere (Haynes *et al.*, 1991;[105] see also Subsection 8.3.2). (5) Influence of stratospheric conditions on baroclinic instability in the troposphere: for example, Wittmann *et al.* (2004)[154] found that the addition of a stratospheric jet to the tropospheric jet yielded a net near-surface geopotential height anomaly that is strongly similar to the AO. Synoptic-scale tropospheric responses to stratospheric changes were also identified by Charlton *et al.* (2004).[155] Moreover, simulations with simplified GCMs indicated that changes in both planetary wave propagation and planetary wave energy due to tropospheric climate change are important.[112,113] (6) Geostrophic and hydrostatic adjustment of the tropospheric flow to anomalous wave drag[105,156] and anomalous diabatic heating at stratospheric levels.[156]

It must be kept in mind that the mechanisms mentioned above are not independent from each other, and on the other hand it is also not clear so far how they act together to produce the observed stratosphere-troposphere coupling.

On the one hand the investigations by Thompson and Solomon (2002)[61] discussed above implied an active role of the stratosphere in the development of extreme tropospheric weather events. But on the other hand, Polvani and Waugh (2004)[157] pointed out that the stratosphere itself is particularly forced by tropospheric dynamics and rather responds passively or acts as a referrer transferring initial tropospheric anomalies *via* the stratosphere back into the troposphere. Moreover, Fyfe *et al.* (1999)[143] and Gillett *et al.* (2002)[62] demonstrated that effects of increasing GHG concentrations can be simulated adequately in GCMs without stratospheric dynamics. Scaife *et al.* (2005)[158] showed that the IPCC climate projections of the 20th century indeed revealed a positive trend in the NAO—even in models with low stratospheric resolution—however, the magnitude of the observed NAO trend was underestimated by these climate models.

But there are many other numerical model studies showing a different picture. The influence of stratospheric variability on the lower atmosphere seems to be covered qualitatively better by stratosphere-resolving models, *i.e.* GCMs and CCMs containing the complete stratosphere. For instance, the importance of the stratosphere in determining adequately tropospheric climate change patterns was demonstrated by Stenchikov *et al.* (2006)[159] and Miller *et al.* (2006).[160] They concluded that the tropospheric climate response due to large volcanic eruptions was underestimated by ocean-atmosphere GCMs (AOGCMs, *i.e.* climate models), most of which do not resolve the stratosphere. Shindell *et al.* (1999, 2001)[136,144] already emphasised the important role of the stratosphere for simulating realistic tropospheric AO trends in their GCM. They demonstrated that including stratospheric dynamics strongly improved the simulated magnitude of the observed NAO trend between 1960 and 1990. This finding is consistent with investigation of Rind *et al.* (2005a;b)[112,113] who found a larger impact of a stratospheric forcing on the NAO than on the AO.

These results indicate that stratosphere-troposphere coupling may play a specific role for suitable assessments of tropospheric climate change patterns. But so far it is unclear whether the under-representation of the vertical coupling in the climate models is related to the missing stratospheric resolution, or if it is related to reduced stratospheric variability in climate models.

So far there is a clear tendency in the scientific community to support the point of view that stratospheric response to climate change plays a potentially important role for tropospheric climate.[129] Signs of stratospheric response to tropospheric climate change are therefore of major interest. Current projection studies with GCMs, climate models and CCMs, however, do not reveal a coherent picture of the stratospheric change, ranging from a projected increase of the stability of the stratospheric polar night jets in a future climate (due to further radiative cooling) to a projected decrease (due to enhanced tropospheric wave forcing see Subsection 8.3.2). Stratospheric ozone is projected to recover significantly in the first half of the 21st century, leading to a weakening of the polar vortex, while rising carbon dioxide levels are expected to counteract this process (see Chapter 9).

8.4.2   The Tropical and the Extra-tropical Tropopause Layer

The tropopause region both in the tropics and in the extra-tropical regions is of specific importance in understanding the coupling of the troposphere and the stratosphere. It is not only the tropopause itself (*i.e.* defined as the lowest level above which the lapse rate of temperature with height becomes less 2 K/km) which characterize the transition from the troposphere and the stratosphere, but also the closely adjacent layers which must be considered. The status of the tropical tropopause layer (TTL) and the extra-tropical tropopause layer (ExTL) both determine the exchange of air masses from the troposphere into the stratosphere and vice versa. Moreover, anomalies near the tropopause are highly correlated with tropical surface temperature anomalies and with tropopause level ozone anomalies, less so with stratospheric temperature anomalies.[161] Tropopause temperature anomalies are correlated with stratospheric water vapor concentrations.[5,27]

The TTL is usually set as the height region extending from the level of the temperature lapse rate minimum around 11–13 km[162,163] (see Fueglistaler *et al.* (2009)[164] for a review) to the level of highest convective overshoot, slightly above the cold point tropopause (CPT) at about 16–17 km. The very low temperatures (regularly below 200 K) experienced by air propagating upward through this part of the atmosphere play a crucial role for dehydration, and thus for stratospheric humidity. Changes which the TTL has undergone within the last few decades are not fully understood.[161,165]

Therefore, assessments of the future evolution of the TTL are a complex matter. The TTL is environed by a warming troposphere below and a cooling stratosphere above, which makes it difficult to estimate the response of CPT and stratospheric humidity changes. The appreciation of the future evolution of TTL temperatures is complicated by the fact that there is a tropospheric amplification of surface warming.[166] Conversely, a strengthening of the BD circulation, as discussed in Subsection 8.3.2, would imply a lowering of TTL temperatures. For example, Seidel *et al.* (2001)[167] obtained an increase in the height of the CPT of approximately 40 m and a decrease in pressure by about 1 hPa during the years from 1978 to 1997. Furthermore, Seidel *et al.* (2001)[167] and Zhou *et al.* (2001)[168] have detected a cooling of tropical tropopause temperatures of approximately 1 K during this time period, resulting in a decrease in the saturation volume mixing ratio of water vapor of about 0.5 ppmv. These temperature trends in the tropical tropopause region are therefore opposite to tropospheric warming dominating the response of the CPT. The CPT seems to be largely affected by increases in the BD circulation (here mainly by the tropical upwelling) and by increased convection as suggested by Zhou *et al.* (2001)[168] (see also Deckert and Dameris, 2008).[103]

CCMs are able to reproduce the basic dynamical and chemical structures of the TTL.[161,165] Although they are able to simulate the historical trends in tropopause pressure which are obtained from reanalysis data, trends in cold point tropopause temperatures are not consistent across CCMs and reanalyses. The altitude of the tropical tropopause has increased, and the level of main convective outflow appears to have decreased (just as water vapor concentrations) in historical CCM simulations as well as in reanalyses.

Changes in the TTL may not only affect the water vapor content of the stratosphere but also influence the abundance of many other species in the stratosphere. This concerns very short-lived chemical substances (VSLS) mostly of natural origin, such as biogenic bromine compounds that are carried to the stratosphere *via* deep convection followed by transport through the TTL[169,170] (Chapter 2 in WMO, 2007;[1] Chapter 1 in WMO, 2011).[2] Changes in deep convection may further affect the transport of longer-lived chemical substances such as methyl bromide and aerosols produced by biomass burning.[171] Moreover, chemical species may be transported in particulate form across the tropical tropopause, for example, organic sulphur-containing substances.[172]

Little is known about these processes affecting the transport in the TTL, and even less is known about climate induced changes. Given the uncertainties in our understanding of mechanisms in the TTL and of their previous changes, assessments of changes of the future evolution of TTL processes and transport through the TTL are still difficult. A future atmosphere with increasing GHG loadings is expected to develop a warmer troposphere with enhanced deep convection. But for the reasons mentioned above it remains speculative that this will be reflected in a warmer tropopause layer, higher water vapor mixing ratios in the stratosphere with less rapid recovery of ozone.[173] Analyses of the recent decades using observations in combination with CCM results suggest the dominance of other processes, possibly related to changes in the BD circulation.[161,165]

The ExTL is a layer of air adjacent to the local extra-tropical tropopause, which has been interpreted as the result of irreversible mixing of tropospheric air into the lowermost stratosphere[174,175] or as the result of two-way stratosphere-troposphere exchanges.[176,177] It is a global feature with increasing depth towards high latitudes, and has been found to be different for different tracers.[178] The origin of ozone in the ExTL changes markedly with season, with photochemical production dominating in summer and transport from the stratosphere dominating the winter and spring seasons. A general increase of the extra-tropical tropopause height in recent years has been identified by Steinbrecht *et al.* (1998)[179] and Varotsos *et al.* (2004)[180] who related this raising to changes of the ozone column. This long-term change of the tropopause height provides a very sensitive indicator of human effects on climate.[181–183] Changes in the ExTL determine the influence of the stratosphere on the troposphere through: (1) transport of ozone from the stratospheric into the troposphere, (2) UV fluxes,[57] and (3) radiative forcing of the surface climate.[70] Therefore, an accurate representation of dynamical and chemical processes in the upper troposphere and lower stratosphere (UTLS) in CCM is a necessary prerequisite for a robust prediction of the ozone layer and climate change. Hegglin *et al.* (2010)[184] found that the main dynamical and chemical climatological characteristics of the ExTL are generally well represented by most CCMs. Moreover, it is shown that the seasonality in the distribution of lower stratospheric chemical tracers is consistent with the seasonality in the BD circulation.

Possible future changes in BD circulation are all likely to change ozone concentrations, even in the UTLS. So far, limitations of the detailed knowledge on ExTL processes prevent robust assessments of the future evolution of the ExTL state. In particular, the relative contribution of isentropic (quasi-horizontal) and convective (vertical) transport and mixing of tropospheric air into the lowermost stratosphere is not well explained. For example, if the frequency or intensity of mid-latitudinal deep convection would change in a future warmer climate, this would affect the chemical composition of the lowermost stratosphere and, thus, the mid-latitudinal ozone layer. Based on simulations with climate models (AOGCMs) there are indications for decreases in the total number of deep convective events and of extra-tropical storms, but an increase of the mean strength of a single event and in the number of the most intense storms.[185,186] Therefore, more reliable future assessments of implications for ExTL dynamics and chemistry need further investigations.

8.4.3 Expected Future Changes

Numerical model studies indicate that climate change will impact the mass exchange across the tropopause. For instance, Rind *et al.* (2001)[85] estimated a 30% increase in the mass flux due to a doubling of atmospheric CO_2 concentrations, and Butchart and Scaife (2001)[31] estimated that the net upward mass flux above the TTL would increase by about 3% per decade due to climate change. In both studies, the changes in the mass flux resulted from more intensive wave propagation from the troposphere into the stratosphere. Modeling studies of tropospheric ozone[30,187,188] also found that climate change caused a comparable percentage increase in the extra-tropical stratosphere-to-troposphere ozone flux.

For a doubled CO_2 concentration, all 14 climate-change model simulations in Butchart *et al.* (2006)[53] resulted in an increase in the annual mean troposphere-to-stratosphere mass exchange rate, with a mean trend of 11 Gg s^{-1} year^{-1}, or about 2% per decade. The predicted increase occurred throughout the year but was, on average, larger during the boreal winter than during the austral winter.

Butchart and colleagues were unable to conclude whether stratospheric ozone changes or ozone feedbacks had a significant impact on the underlying trend in the mass exchange rate. Other simulations[189] suggest that the trend in tropical upwelling is not constant. Periods (over several years) of enhanced upwelling coincide with periods of significant ozone depletion.

Butchart *et al.* (2010)[53] analyzed the response of stratospheric climate and circulation to increasing amounts of GHG concentrations and ozone recovery in the 21st century. Therefore, simulations of 11 CCMs were investigated using nearly identical forcings and experimental set-ups. Among others they found that on average the annual mean tropical upwelling in the lower stratosphere (at about 70 hPa) increases by almost 2% per decade. 59% of this trend was attributed to parameterised orographic gravity wave drag in the CCMs. They concluded that this is a consequence of the eastward acceleration of the subtropical jets which increases the upward flux of (parameterized) momentum reaching the lower stratosphere in these latitudes.

The majority of CCMs simulate continued decreasing of tropopause altitude and convective outflow pressure by several hPa/decade in the 21st century, along with an approximate 1 K increase per century in cold point tropopause temperature and 0.5–1 ppmv per century increase in water vapor mixing ratio above the tropical tropopause. These changes indicate significant perturbations to TTL processes in a future climate with enhanced GHG concentrations, in particular to deep convective heating and humidity transport.[165]

8.5 Concluding Remarks

It is obvious that understanding of long-term changes of the stratospheric ozone layer is a complex problem which makes robust assessments of its future evolution difficult. On the one hand, the modulation of stratospheric ozone concentrations is driven by natural variability, like solar irradiance and volcanic eruptions, and internal variability of stratospheric circulation on different time-scales affecting the stratospheric thermal structure and the transport of air masses. Ozone production and destruction is controlled by photochemical processes, homogeneous gas-phase reactions and heterogeneous chemistry on surfaces of particles (aerosols, PSCs). It must be considered that the chemical depletion of ozone in the presence of volcanic aerosols or PSCs is of nonlinear nature. On the other hand, the whole story becomes even more complex within a changing climate with enhanced GHG concentrations. Climate change influences net ozone production (*i.e.* sum of ozone destruction and production) both in direct and indirect ways and, therefore, will affect the rate of ozone recovery, which will be different at various altitudes and latitudes. Cooling of the stratosphere due to enhanced GHG concentrations has opposite effects in the upper and lower stratosphere, slowing down the gas-phase ozone loss rate but increasing the heterogeneous ozone loss rate on PSCs. This will accelerate ozone recovery in the upper stratosphere and delay it in the lower stratosphere. Moreover, changes in the stratospheric circulation have the potential to modify the future evolution of the stratospheric ozone layer in the 21st century. For example, it is known that the strength of the BD circulation is directly related to dissipating planetary waves, which are forced in the troposphere, *i.e.* stronger wave forcing coincides with a weaker polar night jet and higher polar temperatures. Furthermore, circulation modes can affect the ozone distribution in the UTLS both directly and indirectly by influencing propagation of planetary waves from the troposphere into the stratosphere. Future changes in the generation of tropospheric waves and circulation modes will influence polar ozone abundance dynamically. Nevertheless, so far there is no consensus from numerical model studies on the sign of this change, making assessments of the rate of ozone recovery uncertain. Generally, a better understanding of stratosphere-troposphere coupling is a key issue for more reliable assessments of future climate change and recovery of the stratospheric ozone layer. Warming and expansion of the tropopause region in future climate could additionally obscure ozone recovery rates as the inverse relation between total ozone and tropopause height seems to hold for long time scales as well.[179,190]

Additionally, future changes of stratospheric water vapor concentrations are uncertain. Chemistry-climate models predict increases of stratospheric water vapor, but confidence in these predictions is low, because these models both have a poor representation of the seasonal cycle in tropical tropopause temperatures (which control global stratospheric water vapor abundances) and cannot reproduce past changes in stratospheric water vapour abundances;[5] (Chapter 4 in WMO, 2011).[2] In a warmer climate, numerical model studies suggest an increase in water vapor outflow to the tropical lower stratosphere. Stratospheric water vapor concentrations may also increase through enhanced methane (CH_4) concentrations. On the other hand, higher CH_4 concentration would remove reactive chlorine, particularly in the upper stratosphere. Numerical modeling studies suggest that increased water vapor concentrations will enhance odd hydrogen (HO_X) in the stratosphere and subsequently increase ozone depletion.[191] Increases in water vapor concentrations in the polar regions would raise the formation of PSCs, potentially increasing springtime ozone depletion. Moreover, if water vapor concentrations would increase in the future, there will be also radiative effects.

An increase in nitrous oxide (N_2O) emission (from extended use of artificial fertilizer) will enhance the amount of stratospheric nitrogen oxides (NO_X). This is expected to reduce ozone in the middle and high stratosphere which would make ozone destruction even worse. N_2O will probably remain the largest ozone-depleting emission for the rest of the century.[192] Also, changes in NO_x and non-methane hydrocarbon emissions are expected to affect the tropospheric concentrations of the hydroxyl radical (OH) and, hence, impact the lifetimes and concentrations of stratospheric trace gases such as CH_4 and organic halogen species.

Future climate change will seriously affect the amount of stratospheric ozone mainly through enhanced GHG concentrations, leading to a cooling of the stratosphere and changes in stratospheric circulation. Beside carbon dioxide changes in stratospheric concentrations of water vapor, methane or nitrous oxide must be also taken into account while influencing ozone chemistry and radiative effects in the stratosphere. Although large uncertainties exist, especially in vertical wave propagation into the stratosphere and stratospheric dynamics, the current consensus view is that the rate of ozone recovery will be accelerated by climate change in most parts of the stratosphere except the polar lower stratosphere in winter and spring (Chapter 9).

Acknowledgements

We thank Rolf Müller for helpful comments on the chapter.

References

1. WMO, *Scientific assessment of ozone depletion: 2006, Global Ozone Research and Monitoring Project-Report No. 50*, Geneva, Switzerland, 2007.
2. WMO, *Scientific assessment of ozone depletion: 2010, Global Ozone Research and Monitoring Project-Report No. 52*, Geneva, Switzerland, 2011.
3. IPCC/TEAP (Intergovernmental Panel on Climate Change/Technology and Economic Assessment Panel), *IPCC/TEAP Special Report on Safeguarding the Ozone Layer and the Global Climate System: Issues Related to Hydrofluorocarbons and Perfluorocarbons. Prepared by Working Groups I and III of the Intergovernmental Panel on Climate Change, and the Technical and Economic Assessment Panel*, ed. B. Metz, L. Kuijpers, S. Solomon, S.O. Andersen, O. Davidson, J. Pons, D. de Jager, T. Kestin, M. Manning, and L. Meyer, 487 pp., Cambridge University Press, Cambridge, U.K. and New York, N.Y., USA, 2005.
4. P. Forster and K. Shine, *J. Geophys. Res.*, 1997, **102**, 10841–10855.
5. W. J. Randel, Variability and trends in stratospheric temperature and water vapor, in: The Stratosphere: Dynamics, Transport and Chemistry, *Geophys. Monogr. Ser.*, **190**, American Geophysical Union, ed. Polvani, Sobel and Waugh, 123–135, doi: 10.1029/2009GM000870, 2010.
6. W. J. Randel, K. P. Shine, J. Austin, J. Barnett, C. Claud, N. P. Gillett, P. Keckhut, U. Langematz, R. Lin, C. Long, C. Mears, A. Miller, J. Nash, D. J. Seidel, D. W. J. Thompson, F. Wu and S. Yoden, *J. Geophys. Res.*, 2009, **114**, DOI: 10.1029/2008JD010421.
7. D. Thompson and S. Solomon, *J. Climate*, 2009, **22**, 1934–1943, DOI: 10.1175/2008JCLI2482.1.
8. V. Ramaswamy, M. Schwarzkopf, W. Randel, B. Santer, B. Soden and G. Stenchikov, *Science*, 2006, **311**, 1138–1141.
9. M. Dall'Amico, L. Gray, K. Rosenlof, A. Scaife, K. Shine and P. Stott, *Clim. Dyn.*, 2010, **34**, DOI: 10.1007/s00382-009-0604-x.

10. J. Haigh, *Nature*, 1994, **370**, 544–546.
11. J. L. Lean, G. J. Rottman, H. L. Kyle, T. N. Woods, J. R. Hickey and L. C. Puga, *J. Geophys. Res.*, 1997, **102**, 29939–29956.
12. K. Kodera and Y. Kuroda, *J. Geophys. Res.*, 2002, **107**, DOI: 10.1029/2002JD002224.
13. J. Holton and H. Tan, *J. Atmos. Sci.*, 1980, **37**, 2200–2208.
14. J. Holton and H. Tan, *J. Meteorol. Soc. Japan*, 1982, **60**, 140–148.
15. M. Dameris and A. Ebel, *Ann. Geophys.*, 1990, **8**, 79–86.
16. K. Labitzke, *Geophys. Res. Lett.*, 1987, **14**, 535–537.
17. K. Labitzke and H. van Loon, *J. Atmos. Terr. Phys.*, 1988, **50**, 197–206.
18. L. Gray, S. Phipps, T. Dunkerton, M. Baldwin, E. Drysdale and M. Allen, *Q. J. R. Meteorol. Soc.*, 2001, **127**, 1985–2004, DOI: 10.1256/smsqj.57606.
19. L. Gray, *Geophys. Res. Lett.*, 2003, **30**, DOI: 10.1029/2002GL016430.
20. L. Gray, S. Crooks, C. Pascoe, S. Sparrow and M. Palmer, *J. Atmos. Sci.*, 2004, **61**, 2777–2796, DOI: 10.1175/JAS-3297.1.
21. S. Pawson, K. Labitzke and S. Leder, *Geophys. Res. Lett.*, 1998, **25**, 2157–2160.
22. D. Seidel and J. Lanzante, *J. Geophys. Res.*, 2004, **109**, DOI: 10.1029/2003JD004414.
23. K. P. Shine, M. S. Bourqui, P. Forster, S. H. E. Hare, U. Langematz, P. Braesicke, V. Grewe, M. Ponater, C. Schnadt, C. A. Smiths, J. D. Haighs, J. Austin, N. Butchart, D. T. Shindell, W. J. Randel, T. Nagashima, R. W. Portmann, S. Solomon, D. J. Seidel, J. Lanzante, S. Klein, V. Ramaswamy and M. D. Schwarzkopf, *Q. J. R. Meteorol. Soc.*, 2003, **129**, 1565–1588, DOI: 10.1256/qj.02.186.
24. U. Langematz, M. Kunze, K. Krüger, K. Labitzke and G. L. Roff, *J. Geophys. Res.*, 2003, **108**, DOI: 10.1029/2002JD002069.
25. E. Remsberg and L. Deaver, *J. Geophys. Res.*, 2005, **110**, DOI: 10.1029/2004JD004905.
26. W. J. Randel, F. Wu, S. J. Oltmans, K. Rosenlof and G. E. Nodoluha, *J. Atmos. Sci.*, 2004, **61**, 2133–2148.
27. W. J. Randel, F. Wu, H. Vömel, G. E. Nedoluha and P. Forster, *J. Geophys. Res.*, 2006, **111**, DOI: 10.1029/2005JD006744.
28. J. Haigh and J. Pyle, *Q. J. R. Meteorol. Soc.*, 1982, **108**, 551–574.
29. A. Jonsson, J. de Grandpré, V. Fomichev, J. McConnell and S. Beagley, *J. Geophys. Res.*, 2004, **109**, DOI: 10.1029/2004JD005093.
30. G. Zeng and J. Pyle, *Geophys. Res. Lett.*, 2003, **30**, DOI: 10.1029/2002GL016708.
31. N. Butchart and A. Scaife, *Nature*, 2001, **410**, 799–802.
32. D. G. Andrews, J. R. Holton and C. B. Leovy, *Middle Atmosphere Dynamics*, Academic Press, San Diego, USA, 1987, p. 489.
33. J. Charney and P. Drazin, *J. Geophys. Res.*, 1961, **66**, 83–109.
34. P. A. Newman, E. R. Nash and J. E. Rosenfield, *J. Geophys. Res.*, 2001, **106**, 19999–20010 DOI: 10.1029/2000JD000061.
35. M. E. McIntyre and T. N. Palmer, *Nature*, 1983, **305**, 593–600.

36. M. Baldwin, L. Gray, T. Dunkerton, K. Hamilton, P. Haynes, W. Randel, J. Holton, M. Alexander, I. Hirota, T. Horinouchi, D. Jones, J. Kinnersley, C. Marquardt, K. Sato and M. Takahashi, *Rev. Geophys.*, 2001, **39**, 179–229.
37. M. Baldwin and L. Gray, *Geophys. Res. Lett.*, 2005, **32**, DOI: 10.1029/2004GL022328.
38. R. Lindzen and J. Holton, *Mon. Wea. Rev.*, 1968, **96**, 385–386.
39. T. Dunkerton, *J. Atmos. Sci.*, 2001, **58**, 7–25.
40. R. Scott, D. Dritschel, L. Polvani and D. W. Waugh, *J. Atmos. Sci.*, 2004, **61**, 904–918.
41. J. Holton and C. Mass, *J. Atmos. Sci.*, 1976, **33**, 2218–2225.
42. R. Scott and L. Polvani, *Geophys. Res. Lett.*, 2004, **31**, DOI: 10.1029/2003GL017965.
43. R. Scott and L. Polvani, *J. Atmos. Sci.*, 2006, **63**, 2758–2776.
44. J. Perlwitz and H.-F. Graf, *Theor. Appl. Climatol.*, 2001, **69**, 149–161.
45. J. Castanheira and H.-F. Graf, *J. Geophys. Res.*, 2003, **108**, DOI: 10.1029/2002JD002754.
46. K. Walter and H.-F. Graf, *Atmos. Chem. Phys.*, 2005, **5**, 239–248.
47. J. R. Holton, P. Haynes, M. E. McIntyre, A. R. Douglass, R. B. Rood and L. Pfister, *Rev. Geophys.*, 1995, **33**, 403–439.
48. A. W. Brewer, *Q. J. R. Meteorol. Soc.*, 1949, **75**, 351–363, DOI: 10.1002/qj.49707532603.
49. G. M. B. Dobson, *Proc. R. Soc. London. A*, 1956, **236**, 187–193.
50. T. A. Shaw and T. G. Shepherd, *Nature Geoscience*, 2008, **1**, 12–13.
51. A. Engel, T. Möbius, H. Bönisch, U. S. R. Heinz, I. Levin, E. Atlas, S. Aoki, S. S. T. Nakazawa, F. Moore, D. Hurst, S. S. J. Elkins, A. Andrews and K. Boering, *Nature Geosci.*, 2009, **2**, 28–31, DOI: 10.1038/ngeo388.
52. N. Butchart, A. A. Scaife, M. Bourqui, J. de Grandpré, S. H. E. Hare, J. Kettleborough, U. Langematz, E. Manzini, F. Sassi, K. Shibata, D. Shindell and M. Sigmond, *Clim. Dyn.*, 2006, **27**, 727–741.
53. N. Butchart, I. Cionni, V. Eyring, T. Shepherd, D. Waugh, H. Akiyoshi, J. Austin, C. Brühl, M. Chipperfield, E. Cordero, M. Dameris, R. Deckert, S. Dhomse, S. Frith, R. Garcia, A. Gettelman, M. Giorgetta, D. Kinnison, F. Li, E. Mancini, C. McLandress, S. Pawson, G. Pitari, D. Plummer, E. Rozanov, F. Sassi, J. Scinocca, K. Shibata, B. Steil and W. Tian, *J. Climate*, 2010, **23**, 5349–5374, DOI: 10.1175/2010JCLI3404.1.
54. V. Eyring, D. W. Waugh, G. E. Bodeker, E. Cordero, H. Akiyoshi, J. Austin, S. R. Beagley, B. A. Boville, P. Braesicke, C. Brühl, N. Butchart, M. P. Chipperfield, M. Dameris, R. Deckert, M. Deushi, S. M. Frith, R. R. Garcia, A. Gettelman, M. A. Giorgetta, D. E. Kinnison, E. Mancini, E. Manzini, D. R. Marsh, S. Matthes, T. Nagashima, P. A. Newman, J. E. Nielsen, S. Pawson, G. Pitari, D. A. Plummer, E. Rozanov, M. Schraner, J. F. Scinocca, K. Semeniuk, T. G. Shepherd, K. Shibata, B. Steil, R. S. Stolarski, W. Tian and M. Yoshiki, *J. Geophys. Res.*, 2007, **112**, DOI: 10.1029/2006JD008332.

55. C. McLandress and T. Shepherd, *J. Climate*, 2009, **22**, 1516–1540, DOI: 10.1175/2008JCLI2679.1.
56. SPARC CCMVal, SPARC Report on the Evaluation of Chemistry–Climate Models, ed. V. Eyring, T.G. Shepherd, D.W. Waugh, SPARC Report No. 5, WCRP-132, WMO/TD-No. 1526, 2010.
57. M. Hegglin and T. Shepherd, *Nature Geosci.*, 2009, **2**, 687–691 DOI: 10.1038/ngeo604.
58. J. Austin and F. Li, *Geophys. Res. Lett.*, 2006, **33**, DOI: 10.1029/2006GL026867.
59. R. R. Garcia and W. J. Randel, *J. Atmos. Sci.*, 2008, **65**, 2731–2739, DOI: 10.1175/2008JAS2712.1.
60. L. Oman, D. Waugh, S. Pawson, R. Stolarski and P. Newman, *J. Geophys. Res.*, 2009, **114**, DOI: 10.1029/2008JD010378.
61. D. Thompson and S. Solomon, *Science*, 2002, **296**, 895–899.
62. N. Gillett, M. Allen, R. McDonald, C. Senior, D. Shindell and G. Schmidt, *J. Geophys. Res.*, 2002, **107**, DOI: 10.1029/2001JD000589.
63. G. E. Nedoluha, D. E. Siskind, J. T. Bacmeister, R. M. Bevilacqua and J. M. Russell, *Geophys. Res. Lett.*, 1998, **25**, 987–990, DOI: 10.1029/98GL00489.
64. D. Waugh, D. Considine and E. Fleming, *Geophys. Res. Lett.*, 2001, **28**, 1187–1190.
65. M. Salby and P. Callaghan, *J. Climate*, 2002, **15**, 3673–3685.
66. Y. Hu and K. Tung, *J. Climate*, 2003, **16**, 3027–3038.
67. F. Li, J. Austin and J. Wilson, *J. Climate*, 2008, **21**, 40–57.
68. D. Thompson and S. Solomon, *J. Climate*, 2005, **18**, 4785–4795.
69. K. Rosenlof and G. Reid, *J. Geophys. Res.*, 2008, **113**, DOI: 10.1029/2007JD009109.
70. S. Solomon, K. Rosenlof, R. Portmann, J. Daniel, S. Davis, T. Sanford and G.-K. Plattner, *Science*, 2010, **327**, 1219–1223, DOI: 10.1126/science.1182488.
71. S. Dhomse, M. Weber and J. Burrows, *Atmos. Chem. Phys.*, 2008, **8**, 471–480.
72. M. L. Salby and P. F. Callahan, *J. Climate*, 2004, **17**, 4512–4521.
73. L. Hood and B. Soukharev, *J. Atmos. Sci.*, 2005, **62**, 3724–3740.
74. S. Dhomse, M. Weber, I. Wohltmann, M. Rex and J. Burrows, *Atmos. Chem. Phys.*, 2006, **6**, 1165–1180.
75. P. Hadjinicolaou, J. A. Pyle and N. R. P. Harris, *Geophys. Res. Lett.*, 2005, **32**, DOI: 10.1029/2005GL022476.
76. R. R. Garcia, D. R. Marsh, D. E. Kinnison, B. A. Boville and F. Sassi, *J. Geophys. Res.*, 2007, **112**, DOI: 10.1029/2006JD007485.
77. G. P. Stiller, T. von Clarmann, M. Höpfner, N. Glatthor, U. Grabowski, S. Kellmann, A. Kleinert, A. Linden, M. Milz, T. Reddmann, T. Steck, H. Fischer, B. Funke, M. Lopez-Puertas and A. Engel, *Atmos. Chem. Phys.*, 2008, **8**, 677–695.
78. T. Hall and D. Waugh, *J. Geophys. Res.*, 1997, **102**, 8991–9001.

79. M. Nivano, K. Yamazaki and M. Shiotani, *J. Geophys. Res.*, 2003, **108**, 4794, DOI: 10.1029/2003JD003871.
80. S. M.R., A. Douglass, R. Stolarski, S. Pawson, S. Strahan and W. Read, *J. Geophys. Res.*, 2008, **113**, DOI: 10.1029/2008JD010221.
81. K. Krüger, S. Tegtmeier and M. Rex, *Atmos. Chem. Phys.*, 2008, **8**, 813–823.
82. B. Monge-Sanz, M. P. Chipperfield, A. J. Simmons and S. M. Uppala, *Geophys. Res. Lett.*, 2007, **34**, DOI: 10.1029/2006GL028515.
83. I. Wohltmann and M. Rex, *Atmos. Chem. Phys.*, 2008, **8**, 265–272.
84. D. Rind, D. Shindell, P. Lonergan and N. K. Balachandran, *J. Climate*, 1998, **11**, 876–894.
85. D. Rind, M. Chandler, P. Lonergan and J. Lerner, *J. Geophys. Res.*, 2001, **106**, 20195–20212.
86. M. Sigmond, P. C. Siegmund, E. Manzini and H. Kelder, *J. Climate*, 2004, **17**, 2352–2367.
87. S. Eichelberger and D. Hartmann, *Geophys. Res. Lett.*, 2005, **32**, DOI: 0.1029/2005GL022924.
88. X. Jiang, S. Eichelberger, D. Hartmann and Y. Yung, *J. Atmos. Sci.*, 2007, **64**, 2751–2755.
89. V. Fomichev, A. Jonsson, J. de Grandpré, S. B. C. McLandress, K. Semeniuk and T. Shepherd, *J. Climate*, 2007, **20**, 1121–1144.
90. A. Haklander, P. Siegmund, M. Sigmond and H. Kelder, *Geophys. Res. Lett.*, 2008, **35**, DOI: 10.1029/2007GL033054.
91. R. Deckert and M. Dameris, *Geophys. Res. Lett.*, 2008, **35**, DOI: 10.1029/2008GL033719.
92. S. Li and D. Waugh, *J. Geophys. Res.*, 1999, **104**, 30559–30569.
93. R. A. Plumb, *Rev. Geophys.*, 2007, **45**, DOI: 10.1029/2005RG000179.
94. J. Austin and R. Wilson, *J. Geophys. Res.*, 2010, **115**, DOI: 10.1029/2009JD013292.
95. N. Calvo, R. R. Garcia, W. J. Randel and D. Marsh, *J. Atmos. Sci.*, 2010, **67**, 2331–2340, DOI: 10.1175/2010JAS3433.1.
96. H. Garny, M. Dameris and A. Stenke, *Atmos. Chem. Phys.*, 2009, **9**, 6017–6031.
97. H. Garny, M. Dameris, W. Randel, G. Bodeker and R. Deckert, *J. Atmos. Sci.*, 2011, **68**, 1214–1233.
98. J. Austin, D. Shindell, S. R. Beagley, C. Brühl, M. Dameris, E. Manzini, T. Nagashima, P. Newman, S. Pawson, G. Pitari, E. Rozanov, C. Schnadt and T. G. Shepherd, *Atmos. Chem. Phys.*, 2003, **3**, 1–27.
99. V. Eyring, N. Butchart, D. W. Waugh, H. Akiyoshi, J. Austin, S. Bekki, G. E. Bodeker, B. A. Boville, C. Brühl, M. P. Chipperfield, E. Cordero, M. Dameris, M. Deushi, V. E. Fioletov, S. M. Frith, R. R. Garcia, A. Gettelman, M. A. Giorgetta, V. Grewe, L. Jourdain, D. E. Kinnison, E. Mancini, E. Manzini, M. Marchand, D. R. Marsh, T. Nagashima, E. Nielsen, P. A. Newman, S. Pawson, G. Pitari, D. A. Plummer, E. Rozanov, M. Schraner, T. G. Shepherd, K. Shibata, R. S. Stolarski, H. Struthers, W. Tian and M. Yoshiki, *J. Geophys. Res.*, 2006, **111**, DOI: 10.1029/2006JD007327.

100. P. Braesicke and J. Pyle, *Q. J. R. Meteorol. Soc.*, 2004, **130**, 2033–2045, DOI: 10.1256/qj.03.183.
101. N. Hall, J. Derome and H. Lin, *J. Climate*, 2001, **14**, 2035–2053.
102. N. Hall, H. Lin and J. Derome, *J. Climate*, 2001, **14**, 2696–2709.
103. R. Deckert and M. Dameris, *Science*, 2008, **322**, 53–55, DOI: 10.1126/science.1163709.
104. W. J. Randel, R. R. Garcia and F. Wu, *J. Atmos. Sci.*, 2008, **65**, 3584–3595.
105. P. H. Haynes, C. J. Marks, M. E. McIntyre, T. G. Shepherd and K. P. Shine, *J. Atmos. Sci.*, 1991, **48**, 651–678.
106. P. Chen, M. Hoerling and R. Dole, *J. Atmos. Sci.*, 2001, **58**, 1827–1835.
107. P. Chen, *J. Atmos. Sci.*, 2001, **58**, 1585–1594.
108. D. Rind, J. Lerner, J. Perlwitz, C. McLinden and M. Prather, *J. Geophys. Res.*, 2002, **107**, DOI: 10.1029/2002JD002483.
109. W. Randel and A. Thompson, *J. Geophys. Res.*, 2011, **116**, DOI: 10.1029/2010JD015195.
110. P. Forster, G. Bodeker, R. Schofield, S. Solomon and D. Thompson, *Geophys. Res. Lett.*, 2007, **34**, DOI: 10.1029/2007GL031994.
111. S. Brönnimann, M. Schraner, B. Müller, A. Fischer, D. Brunner, E. Rozanov and T. Egorova, *Atmos. Chem. Phys.*, 2006, **6**, 4669–4685.
112. D. Rind, J. Perlwitz and P. Lonergan, *J. Geophys. Res.*, 2005, **110**, DOI: 10.1029/2004JD005103.
113. D. Rind, J. Perlwitz, P. Lonergan and J. Lerner, *J. Geophys. Res.*, 2005, **110**, DOI: 10.1029/2004JD005686.
114. M. Olsen, M. Schoeberl and J. Nielsen, *J. Geophys. Res.*, 2007, **112**, DOI: 10.1029/2006JD008012.
115. T. Shepherd, *J. Meteor. Soc. Japan*, 2002, **80**, 769–792.
116. A. Stohl, P. Bonasoni, P. Cristofanelli, W. Collins, J. Feichter, A. Frank, C. Forster, E. Gerasopoulos, H. Gaeggeler, P. James, T. Kentarchos, S. Kreipl, H. Kromp-Kolb, B. Krueger, C. Land, J. Meloen, A. Papayannis, A. Priller, P. Seibert, M. Sprenger, G. Roelofs, E. Scheel, C. Schnabel, P. Siegmund, L. Tobler, T. Trickl, H. Wernli, V. Wirth, P. Zanis and C. Zerefos, *J. Geophys. Res.*, 2003, **108**, DOI: 10.1029/2002JD002490.
117. V. Ramaswamy, M. L. Chanin, J. Angell, J. Barnett, D. Gaffen, M. Gelman, P. Keckhut, Y. Koshelkov, K. Labitzke, J. J. R. Lin, A. O'Neill, J. Nash, W. Randel, R. Rood, K. Shine, M. Shiotani, and R. Swinbank, Stratospheric temperature trends: Observations and model simulations, *Rev. Geophys.*, **39**(1), 71–122, 2001.
118. M. Baldwin, X. Cheng and T. Dunkerton, *Geophys. Res. Lett.*, 1994, **21**, 1141–1144.
119. J. Perlwitz and H.-F. Graf, *J. Climate*, 1995, **8**, 2281–2295.
120. K. Kodera, M. Chiba, H. Koide, A. Kitoh and Y. Nikaidou, *J. Meteorol. Soc. Japan*, 1996, **74**, 365–382.
121. M. P. Baldwin and D. W. J. Thompson, *Q. J. R. Meteorol. Soc.*, 2009, **135**, 1661–1672, DOI: 10.1002/qj.479.

122. C. Schnadt and M. Dameris, *Geophys. Res. Lett.*, 2003, **30**, 1487, DOI: 10.1029/2003GL017006.
123. D. W. J. Thompson and J. M. Wallace, *Geophys. Res. Lett.*, 1998, **25**, 1297–1300, DOI: 10.1029/98GL00950.
124. D. Thompson and J. Wallace, *J. Climate*, 2000, **13**, 1000–1016.
125. M. P. Baldwin and T. J. Dunkerton, *J. Geophys. Res.*, 1999, **104**, 30937–30946.
126. M. Baldwin and T. Dunkerton, *Science*, 2001, 581–584.
127. D. W. J. Thompson, Lee, S. and Baldwin, M. P. Atmospheric processes governing the Northern Hemisphere annular mode/North Atlantic Oscillation, AGU monograph on "The North Atlantic Oscillation", 293, 85–89, 2002.
128. M. Baldwin, T. Hirooka, A. O'Neill and S. Yoden, *SPARC Newsletter*, 2003, **1**, 24–26.
129. M. P. Baldwin, M. Dameris and T. G. Shepherd, *Science*, 2007, **316**, 1576–1577, DOI: 10.1126/science.1144303.
130. K. Kodera, *Geophys. Res. Lett.*, 2002, **29**, DOI: 10.1029/2001GL014557.
131. Y. Kuroda and K. Kodera, *Geophys. Res. Lett.*, 2005, **32**, DOI: 10.1029/2005GL022516.
132. K. Matthes, Y. Kuroda, K. Kodera and U. Langematz, *J. Geophys. Res.*, 2006, **111**, DOI: 10.1029/2005JD006283.
133. A. Robock and J. Mao, *Geophys. Res. Lett.*, 1992, **19**, 2405–2408.
134. P. M. Kelly, P. Jones and J. Pengqun, *Int. J. Climatol.*, 1996, **16**, 537–550.
135. I. Kirchner, G. Stenchikov, H.-F. Graf, A. Robock and J. Antuna, *J. Geophys. Res.*, 1999, **104**, 19039–19055.
136. D. T. Shindell, *Geophys. Res. Lett.*, 2001, **28**, 1551–1554, DOI: 10.1029/1999GL011197.
137. K. Kodera, *J. Geophys. Res.*, 1994, **99**, 1273–1282, DOI: 10.1029/93JD02731.
138. K. Walter and H.-F. Graf, *Q. J. R. Meteorol. Soc.*, 2006, **132**, 467–483, DOI: 10.1256/qj.05.25.
139. S. Rosier and K. Shine, *Geophys. Res. Lett.*, 2000, **27**, 2617–2620.
140. M. Schwarzkopf and V. Ramaswamy, *Geophys. Res. Lett.*, 2008, **35**, DOI: 10.1029/2007GL032489.
141. I. Kindem and B. Christiansen, *Geophys. Res. Lett.*, 2001, **28**, 1547–1550.
142. N. P. Gillett and D. W. J. Thompson, *Science*, 2003, **302**, 273–275, DOI: 10.1126/science.1087440.
143. J. Fyfe, G. Boer and G. Flato, *Geophys. Res. Lett.*, 1999, **26**, 1601–1604.
144. D. T. Shindell, R. L. Miller, G. A. Schmidt and L. Pandolfo, *Nature*, 1999, **399**, 452–455, DOI: 10.1038/20905.
145. B. Christiansen, *J. Geophys. Res.*, 2000, **105**, 29461–29474.
146. L. Polvani and P. Kushner, *Geophys. Res. Lett.*, 2002, **29**, DOI: 10.1029/2001GL014284.
147. W. A. Norton, *Geophys. Res. Lett.*, 2003, **30**, DOI: 10.1029/2003GL016958.

148. M. Taguchi, *J. Atmos. Sci.*, 2003, **60**, 1835–1846.
149. J. Perlwitz and N. Harnik, *J. Climate*, 2003, **16**, 3011–3026.
150. A. Ebel, M. Dameris and H. Jakobs, *Ann. Geophys.*, 1988, **6**, 501–512.
151. B. Christiansen, *J. Atmos. Sci.*, 1999, **56**, 1858–1872.
152. D. Hartley, J. Villarin, R. Black and C. Davis, *Nature*, 1998, **391**, 471–474.
153. R. X. Black, *J. Climate*, 2002, **15**, 268–277.
154. M. Wittmann, L. Polvani, R. Scott and A. Charlton, *Geophys. Res. Lett.*, 2004, **31**, DOI: 10.1029/2004GL020503.
155. A. J. Charlton, A. O'Neill, W. A. Lahoz and P. Berrisford, *J. Atmos. Sci.*, 2004, **62**, 590–602.
156. D. Thompson, J. Furtado and T. Shepherd, *J. Atmos. Sci.*, 2006, **63**, 2616–2629.
157. L. Polvani and D. Waugh, *J. Climate*, 2004, **17**, 3548–3554.
158. A. A. Scaife, J. R. Knight, G. K. Vallis and C. K. Folland, *Geophys. Res. Lett.*, 2005, **32**, DOI: 10.1029/2005GL023226.
159. G. Stenchikov, K. Hamilton, R. Stouffer, A. Robock, V. Ramaswamy, B. Santer and H.-F. Graf, *J. Geophys. Res.*, 2006, **111**, DOI: 10.1029/2005JD006286.
160. R. Miller, G. Schmidt and D. Shindell, *J. Geophys. Res.*, 2006, **111**, DOI: 10.1029/2005JD006323.
161. A. Gettelman, T. Birner, V. Eyring, H. Akiyoshi, D. Plummer, M. Dameris, S. Bekki, F. Lefèvre, F. Lott, C. Brühl, K. Shibata, E. Rozanov, E. Mancini, G. Pitari, H. Struthers, W. Tian and D. Kinnison, *Atmos. Chem. Phys.*, 2009, **9**, 1621–1637.
162. S. C. Sherwood and A. E. Dessler, *J. Atmos. Sci.*, 2001, **58**, 765–779.
163. A. Gettelman and P. Forster, *J. Meteorol. Soc. Japan*, 2002, **80**, 911–924.
164. S. Fueglistaler, A. Dessler, T. Dunkerton, I. Folkins, Q. Fu and P. Mote, *Rev. Geophys.*, 2009, **47**, DOI: 10.1029/2008RG000267.
165. A. Gettelman, M. I. Hegglin, S.-W. Son, J. Kim, M. Fujiwara, T. Birner, S. Kremser, M. Rex, J. A. Anel, H. Akiyoshi, J. Austin, S. Bekki, P. Braesike, C. Brühl, N. Butchart, M. Chipperfield, M. Dameris, S. Dhomse, H. Garny, S. Hardiman, P. Jöckel, D. E. Kinnison, J.-F. Lamarque, E. Mancini, M. Marchand, M. Michou, O. Morgenstern, S. Pawson, G. Pitari, D. Plummer, J. Pyle, E. Rozanov, J. Scinocca, T. Shepherd, K. Shibata, D. Smale, H. Teyssèdre and W. Tian, *J. Geophys. Res.*, 2010, **115**, DOI: 10.1029/2009JD013638.
166. B. Santer, T. Wigley, C. Mears, S. K. F.J. Wentz, D. Seidel, K. Taylor, M. W. P.W. Thorne, P. Gleckler, W. C. J.S. Boyle, K. Dixon, C. Doutriaux, Q. F. M. Free, J. Hansen, G. Jones, R. Ruedy, J. L. T.R. Karl, G. Meehl, V. Ramaswamy, G. Russell and G. Schmidt, *Science*, 2005, **309**, 1551–1556, DOI: 10.1126/science.1114867.
167. D. J. Seidel, R. J. Ross, J. K. Angell and G. C. Reid, *J. Geophys. Res.*, 2001, **106**, 7857–7878.
168. X. L. Zhou, M. A. Geller and M. H. Zhang, *J. Geophys. Res.*, 2001, **106**, 1511–1522.

169. R. J. Salawitch, D. K. Weisenstein, L. J. Kovalenko, C. E. Sioris, P. O. Wennberg, K. Chance, M. K. W. Ko and C. A. McLinden, *Geophys. Res. Lett.*, 2005, **32**, DOI: 10.1029/2004GL021504.

170. B.-M. Sinnhuber and I. Folkins, *Atmos. Chem. Phys.*, 2006, **6**, 4755–4761.

171. M. Andreae and P. Merlet, *Global Biogeochem. Cycles*, 2001, **15**, 955–966.

172. J. Notholt, B. Luo, S. Fueglistaler, D. Weisenstein, M. Rex, M. Lawrence, H. Bingemer, I. Wohltmann, T. Corti, T. Warneke, R. von Kuhlmann and T. Peter, *Geophys. Res. Lett.*, 2005, **32**, DOI: 10.1029/2004GL022159.

173. A. Stenke and V. Grewe, *Atmos. Chem. Phys.*, 2005, **5**, 1257–1272.

174. P. Hoor, C. Gurk, D. Brunner, M. I. Hegglin, H. Wernli and H. Fischer, *Atmos. Chem. Phys.*, 2004, **4**, 1427–1442.

175. M. Hegglin, D. Brunner, T. Peter, V. W. J. Staehelin, P. Hoor and H. Fischer, *Geophys. Res. Lett.*, 2005, **32**, DOI: doi:10.1029/2005GL022495.

176. L. L. Pan, W. J. Randel, B. L. Gary, M. J. Mahoney and E. J. Hintsa, *J. Geophys. Res.*, 2004, **109**, DOI: 10.1029/2004JD004982.

177. S. A. Bischoff, P. O. Canziani and A. E. Yuchechen, *Int. J. Climatol.*, 2007, **27**, 189–209 DOI: 10.1002/joc.1385.

178. M. Hegglin, C. Boone, G. Manney and K. Walker, *J. Geophys. Res.*, 2009, **114**, DOI: 10.1029/2008JD009984.

179. W. Steinbrecht, H. Claude, U. Köhler and K. Hoinka, *J. Geophys. Res.*, 1998, **103**, 19183–19192.

180. C. Varotsos, C. Cartalis, A. Vlamakis, C. Tzanis and I. Keramitsoglou, *J. Climate*, 2004, **17**, 3843–3854.

181. R. Sausen and B. Santer, *Meteorol. Z.*, 2003, **12**, 131–136.

182. B. Santer, M. Wehner, T. Wigley, G. M. R. Sausen, K. Taylor, C. Ammann, W. W. J. Arblaster, J. Boyle and W. Brüggemann, *Science*, 2003, **301**, 479–483 DOI: 10.1126/science.1084123.

183. B. Santer, T. Wigley, A. Simmons, P. Kållberg, G. Kelly, S. Uppala, C. Ammann, J. Boyle, W. Brüggemann, C. Doutriaux, M. Fiorino, C. Mears, G. Meehl, R. Sausen, K. Taylor, W. Washington, M. Wehner and F. Wentz, *J. Geophys. Res.*, 2004, **109**, DOI: 10.1029/2004JD005075.

184. M. Hegglin, A. Gettelman, P. Hoor, R. Krichevsky, G. Manney, L. Pan, S.-W. Son, G. Stiller, S. Tilmes, K. Walker, V. Eyring, T. Shepherd, D. Waugh, H. Akiyoshi, J. Anel, J. Austin, A. Baumgaertner, S. Bekki, P. Braesicke, C. Brühl, N. Butchart, M. Chipperfield, M. Dameris, S. Dhomse, S. Frith, H. Garny, S. Hardiman, P. Jöckel, D. Kinnison, J.-F. Lamarque, E. Mancini, M. Michou, O. Morgenstern, T. Nakamura, D. Olivié, S. Pawson, G. Pitari, D. Plummer, J. Pyle, E. Rozanov, J. Scinocca, K. Shibata, D. Smale, H. Teyssèdre, W. Tian and Y. Yamashita, *J. Geophys. Res.*, 2010, **115**, DOI: 10.1029/2010JD013884.

185. S. Brinkop, *Meteorol. Z.*, 2002, **11**, 323–333.

186. S. Lambert and J. Fyfe, *Clim. Dyn.*, 2006, **26**, 713–728.

187. W. Collins, R. Derwent, B. Garnier, C. Johnson, M. Sanderson and D. Stevenson, *J. Geophys. Res.*, 2003, **108**, DOI: 10.1029/2002JD002617.

188. K. Sudo, M. Takahashi and H. Akimoto, *Geophys. Res. Lett.*, 2003, **30**, 2256, DOI: 10.1029/2003GL018526.

189. J. Austin, J. Wilson, F. Li and H. Vömel, *J. Atmos. Sci.*, 2007, **64**, 905–921, DOI: 10.1175/JAS3866.1.

190. D. Seidel, Q. Fu, W. Randel and T. Reichler, *Nature Geosci.*, 2008, 21–24, DOI: 10.1038/nego.2007.38.

191. B. Vogel, T. Feck and J.-U. Grooß, *J. Geophys. Res.*, 2011, **116**, D05301 DOI: 10.1029/2010JD014234.

192. A. R. Ravishankara, J. S. Daniel and R. W. Portmann, *Science*, 2009, **326**, 123–125, DOI: 10.1126/science.1176985.

CHAPTER 9
Stratospheric Ozone in the 21st Century

D. W. WAUGH,*[a] V. EYRING[b] AND D. E. KINNISON[c]

[a] Department of Earth and Planetary Sciences, Johns Hopkins University, Baltimore, Maryland, USA; [b] Deutsches Zentrum für Luft- und Raumfahrt, Institut für Physik der Atmosphäre, Oberpfaffenhofen, Germany; [c] National Center for Atmospheric Research, Boulder, CO, USA

9.1 Introduction

The stratospheric ozone layer has been depleted by anthropogenic emissions of halogenated species over the last decades of the 20th century. Observations show that tropospheric halogen loading is now decreasing, which reflects the controls of ozone-depleting substances (ODSs) by the Montreal Protocol and its Amendments and Adjustments (see Chapter 2).[1,2] The total abundance of ODSs in the troposphere peaked around 1993 and has slowly declined since then. This slow decline is expected to continue over the 21st century (21C), and ODS are expected to be back to 1980 levels around 2040, and to around 1970 levels by the end of the century (see Chapter 2). Atmospheric concentrations of greenhouse gases (GHGs) have also increased and are expected to further increase in the future, with consequences for the ozone layer.[3] As a result of climate change, the ozone layer will not return to precisely its unperturbed state when the abundance of halogens returns to background levels. Furthermore, climate change complicates the attribution of ozone recovery to the decline of ODSs.

To project the future evolution of stratospheric ozone and attribute its change in response to the different forcings, numerical models are required that

Stratospheric Ozone Depletion and Climate Change
Edited by Rolf Müller
© Royal Society of Chemistry 2012
Published by the Royal Society of Chemistry, www.rsc.org

can adequately represent the chemistry and dynamics of the ozone layer, along with the energetics and natural variability of the atmosphere. The coupling of stratospheric chemistry with climate models has led to a new generation of models far more complex than those available when the Montreal Protocol was signed over 20 years ago. Such models, known as Chemistry-Climate Models (CCMs), are three-dimensional atmospheric circulation models with fully coupled chemistry, *i.e.* where chemical reactions drive changes in atmospheric composition which in turn change the atmospheric radiative balance and hence dynamics. CCMs are key tools for the detection, attribution and projection of the response of stratospheric ozone to ODSs and other factors, and allow questions about future stratospheric ozone and solar ultraviolet (UV) radiation levels to be studied. In particular, by including an explicit representation of tropospheric climate change, they make it possible to address the coupling between climate change and ozone depletion/recovery in a comprehensive manner.

Over the past decade there have been several international projects evaluating stratospheric CCMs, and related General Circulation Models (GCMs), most of which have been organized under the auspices of the WCRP's (World Climate Research Programme) SPARC (Stratospheric Processes and their Role in Climate) project. For example, the GCM-Reality Intercomparison Project (GRIPS) and the Chemistry-Climate Model Validation (CCMVal) Activity.[4,5] These multi-model projects have contributed directly to the assessment of CCMs during the preparation of the World Meteorological Organization/ United Nations Environment Programme (WMO/UNEP) Scientific Assessments of Ozone Depletion.[5–10]

This chapter discusses projections of the evolution of stratospheric ozone during the 21st century. We first describe the CCMs and simulations that have been used in the last decade to project stratospheric ozone (Section 9.2). Section 9.3 briefly reviews the major factors that are affecting the ozone projections which are discussed in Section 9.4. The uncertainties and open questions in the evolution of ozone (O_3) in the 21C are discussed in Section 9.5, and a summary is in Section 9.6.

9.2 Models and Simulations

9.2.1 Chemistry-climate Models

CCMs consist of coupled modules that calculate the dynamical fields (temperatures and winds), radiation (heating and cooling rates), and chemistry, see Figure 9.1. At each time step, the simulated concentrations of the radiatively active gases are used in the calculations of the net heating rates so that a change in the abundance of radiatively active gases feeds back on atmospheric dynamics fields (*e.g.* winds and temperature). Similarly, changes in dynamics feed back on the chemical composition.

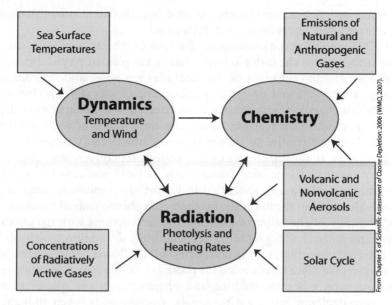

Figure 9.1   Schematic of a Chemistry-Climate Model (CCM). The core of a CCM
(oval symbols) consists of a general circulation model (GCM) that
includes calculation of the heating and cooling rates and a detailed
chemistry module. They are interactively coupled. Photolysis rates are
calculated online or are determined from a lookup table. Arrows indicate
the direction of effect. Rectangular boxes denote external impacts. From
Figure 5.1 of WMO (2007).[2]

The dynamics (*i.e.* the temporal evolution of wind, temperature and pressure,
or other prognostic variables) in state-of-the-art CCMs is determined by sol-
ving the "primitive" equations. The basic dynamical state of the atmosphere
within which transport takes place depends on a number of physical processes.
These include the propagation of Rossby and gravity waves, wave-mean-flow
interaction, and the diabatic circulation. Correct reproduction of the climatological mean state of the stratosphere by CCMs, including inter-hemispheric
differences, inter-annual and intra-seasonal variability, is important but not
sufficient: the basic dynamical mechanisms must be well represented in the
underlying GCMs on which the CCMs are based if future changes are to be
modeled credibly. A major issue with GCMs of the middle atmosphere is the
treatment of gravity waves.[4] In addition, prescribed sea surface temperatures
(SSTs) and sea ice concentrations (SICs) hinder the feedback between chemistry-climate interactions, so there is a need for a range of simulations looking
at all aspects of the atmosphere-ocean system.

Radiative calculations are used in CCMs to derive photolysis rates and
heating rates. Photolysis rates in the stratosphere affect the abundance of many
chemical constituents that in turn control radiatively active constituents, such
as O_3, nitrous oxide (N_2O), methane (CH_4), and chlorofluorocarbons (CFCs).

These radiatively active constituents are used in radiative heating calculations and therefore affect temperature and dynamics.

All CCMs used to make projections of ozone in the 21C include a comprehensive stratospheric chemistry scheme that is coupled to physical processes through the radiation calculations. This includes gas-phase and heterogeneous chemistry on aerosols and on polar stratospheric clouds (PSCs). One of the ways in which chemistry and dynamics are coupled is the temperature dependence of many chemical reaction rates. The importance of local control of ozone by chemistry relative to transport varies substantially between various times and places. In the upper stratosphere transport plays a role by controlling the concentrations of long-lived tracers such as inorganic chlorine, but photochemical timescales are so short that transport has a minimal direct impact on ozone. However, in the lower stratosphere, the photochemical timescales are longer (typically of the order of months) and interactions with dynamics are complex and more challenging to model accurately. In addition, aerosols and PSCs play an important role in chemistry of the lower stratosphere, since reactions can take place within or on the particles. In this region, heterogeneous reactions convert inorganic chlorine and bromine reservoir species to more active ozone-depleting species (Chapter 4). Consequently, even thought the photochemical lifetime of ozone is typically many months in the lower stratosphere, rapid chemical loss of ozone occurs when temperatures are cold, aerosols or PSCs exist, and sunlight is available (Chapter 5).

Transport in the stratosphere involves both meridional overturning (the so-called "Brewer-Dobson" circulation), and mixing. The most important aspects are the vertical (diabatic) mean motion and the horizontal mixing. Horizontal mixing is highly inhomogeneous, with transport barriers in the subtropics and at the edge of the wintertime polar vortex; mixing is most intense in the wintertime "surf zone", *i.e.* the region surrounding the polar vortex, and is comparatively weak in the summertime extratropics. Accurate representation of this structure in CCMs is important for the ozone distribution itself, as well as for the distribution of chemical families and species that affect ozone chemistry, *e.g.* Cl_y, total inorganic nitrogen (NO_y), total inorganic bromine (Br_y) water vapor (H_2O), and CH_4.

9.2.2 Simulations

CCMs have been used to perform several different types of simulations. Transient simulations consider observed or projected changes in concentrations of radiatively active gases and other boundary conditions (*e.g.*, emissions), whereas time-slice simulations are applied to study the internal variability of a CCM under fixed conditions, *e.g.*, GHG concentrations and SSTs, to estimate the significance of specific changes. Transient simulations are preferred for studying past and projecting future ozone changes because in these simulations, ozone responds interactively to the gradual secular trends in GHGs, ODSs, and other boundary conditions. The CCM simulations are commonly separated into "past" (or "historical") transient simulations that are forced by observations of ODSs, GHGs, and SSTs, and are carried out to see how well the models can reproduce the past behavior of stratospheric ozone, and "future" transient simulations that are forced by trace gas projections and modeled SSTs and are carried out to make projections for the future evolution of stratospheric ozone. In addition, sensitivity or idealized simulations are performed where one or more of the forcing fields are held fixed or vary in an unrealistic manner to isolate the role of particular factors in driving changes in stratospheric O_3.

In recent years, the community has defined reference simulations, with a set of anthropogenic and natural forcings, to encourage consistency and comparison between simulations by different modeling groups.[11-13]

The *past reference simulation*, is defined as a transient run from 1960 to the present and is designed to reproduce the well-observed period of the last 30 years during which ozone depletion is well recorded.[14] This simulation examines the role of natural variability and other atmospheric changes important for ozone balance and trends. All forcings in this simulation are taken from observations. This transient simulation includes all anthropogenic and natural forcings based on changes in trace gases, solar variability, volcanic eruptions, quasi-biennial oscillation (QBO), and SSTs/SICs.[13]

The corresponding *future reference simulation* is a transient simulation from the past into the future (ideally 1960 to 2100), whose objective is to produce best estimates of future ozone-climate change up to 2100 under specific assumptions about GHG increases and decreases in halogen emissions in this period. GHG concentrations (N_2O, CH_4, and CO_2) in this reference simulation are prescribed following the Intergovernmental Panel on Climate Change Special Report on Emission Scenarios (IPCC SRES) "A1B" GHG scenario and surface mixing ratios of ODSs are based on the adjusted halogen scenario A1 from the 2006 WMO/UNEP Assessment, which includes the earlier phaseout of hydrochlorofluorocarbons (HCFCs) that was agreed to by the Parties to the Montreal Protocol in 2007, see Figure 9.2.[2,3] The future reference simulations typically include only anthropogenic forcings, and external natural forcings such as solar variability, and volcanic eruptions are not considered, as they cannot be known in advance.

The CCMVal reference simulations have been performed by most CCM groups in support of the 2006 and 2010 WMO/UNEP Assessments.[2,14] The first round of CCMVal (CCMVal-1) included 13 CCMs, whereas 18 CCMs participated in the most recent second round of CCMVal (CCMVal-2), see Table 9.1.[11] In addition, several different types of *sensitivity simulations* have been performed by a small subgroup of CCMs. For example, simulations with fixed halogens have been performed to study the effect of halogens on stratospheric ozone (and climate) in a changing climate' and "no greenhouse-gas induced climate change" simulations have been performed to address the coupling of ozone depletion/recovery and climate change.[15-17] The SRES A1B GHG scenario is only one of several scenarios for the possible evolution of GHGs, and future simulations have also been performed using a different GHG scenarios to assess the dependence of the future ozone evolution on the GHG scenario.[10,18]

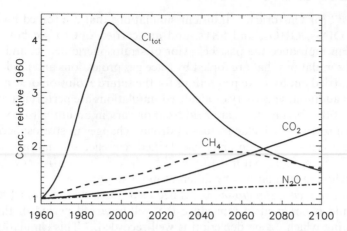

Figure 9.2 Time series of the surface concentrations of total chlorine from the WMO (2007) scenario, and GHGs from the IPCC SRES A1B scenario.[2,3] Concentrations are shown relative to their 1960 concentrations (820 ppt for Cl_{tot}, 1265 ppb for CH_4, 316 ppm for CO_2, and 291 ppb for N_2O).

Table 9.1 CCMs that are used for ozone projections in the CCMVal-2 intercomparison of SPARC CCMVal (2010): name of the model, group and references for model documentations.

	CCM	Group and Location	References
1	AMTRAC3	GFDL, USA	Austin and Wilson (2010)[46]
2	CAM3.5	NCAR, USA	Lamarque et al. (2008)[47]
3	CCSRNIES	NIES, Tokyo, Japan	Akiyoshi et al. (2009)[48]
4	CMAM	MSC, University of Toronto, York Univ., Canada	Scinocca et al. (2008);[49] deGrandpre et al. (2000)[50]
5	CNRM-ACM	Meteo-France; France	Déqué (2007);[51] Teyssèdre et al. (2007)[52]
6	E39CA	DLR, Germany	Stenke et al. (2009);[53] Garny et al. (2008)[54]
7	EMAC	MPI Mainz, Germany	Jöckel et al. (2006)[55]
8	GEOSCCM	NASA/GSFC, USA	Pawson et al. (2008)[56]
9	LMDZrepro	IPSL, France	Jourdain et al. (2008)[57]
10	MRI	MRI, Japan	Shibata and Deushi (2008a,b)[58,59]
11	NIWA-SOCOL	NIWA, NZ	Schraner et al. (2008);[60] Egorova et al. (2005)[61]
12	SOCOL	PMOD/WRC and ETHZ, Switzerland	Schraner et al. (2008);[60] Egorova et al. (2005)[61]
13	ULAQ	University of L'Aquila, Italy	Pitari et al. (2002);[62] Eyring et al. (2006; 2007)[8,9]
14	UMETRAC	NIWA, NZ	Austin and Butchart (2003)[63]
15	UMSLIMCAT	University of Leeds, UK	Tian and Chipperfield (2005);[64] Tian et al. (2006)[65]
16	UMUKCA-METO	MetOffice, UK	Morgenstern et al. (2008, 2009)[66,67]
17	UMUKCA-UCAM	University of Cambridge, UK	Morgenstern et al. (2008, 2009)[66,67]
18	WACCM	NCAR, USA	Garcia et al. (2007)[68]

9.2.3 Evaluation

Confidence in, and guidance in interpreting, CCM projections of future changes in atmospheric composition can be gained by first ensuring that the CCMs are able to reproduce past observations. Limitations and deficiencies in the models can be revealed through intermodel comparisons and through comparisons with observations. As well as evaluating the simulations of the state of the atmosphere, it is also important to evaluate the representation in the models of key processes that control the distribution of stratospheric ozone.[5] Also, with the increasing number of CCMs and the large spread in ozone projections, there is a need for multi-model comparison in addition to single model studies.

Over the last decade there have been several multi-model comparisons that have evaluated different aspects of the CCMs. Austin *et al.* evaluated a mixture of time-slice and transient simulations from eight CCMs.[6] They focused on diagnostics to evaluate the representation of dynamics in polar regions and found that many of the participating CCMs indicated a significant cold bias in high latitudes, the so-called "cold pole problem", particularly in the southern hemisphere during winter and spring. They concluded that the main uncertainties of CCMs at that time stemmed from the performance of the underlying GCM. Cold biases have been found to exist in the stratosphere in many CCMs, consistent with that previously found for models without chemistry.[4]

The 13 CCMs that participated in CCMVal-1 were evaluated and considered in the 2006 WMO/UNEP Assessment.[8,19] In contrast to previous studies, the CCM simulations were all transient simulations and had almost identical forcings (*e.g.*, SSTs, GHGs, and ODSs). This eliminated many of the uncertainties in the conclusions of the earlier assessments that resulted from the differences in experimental setup of individual models. Also, and perhaps most importantly, this study was the first multi-CCM assessment to evaluate the representation of transport and distributions of important trace gases. It was shown that there were substantial quantitative differences in the simulated stratospheric Cl_y, with the October mean Antarctic Cl_y peak value varying from less than 2 ppb to over 3.5 ppb in the participating CCMs. These large differences in Cl_y among the CCMs have been found to be key to diagnosing the intermodel differences in simulated ozone recovery, in particular in the Antarctic, (see further discussion below).[9] Several other studies evaluated and analyzed different aspects in the CCMVal-1 simulations. For example, Gettelman *et al.* showed that the CCMs were able to reproduce the basic structure of the Tropical Tropopause Layer (TTL) but differences were found in cold point tropopause temperatures trends.[20] Austin *et al.* found that the mean model response is about 2.5% in ozone and 0.8 K in temperature during a typical solar cycle, which is at the lower end of the observed ranges of peak responses.[21]

A much more extensive evaluation of CCMs was performed as part of the 2010 SPARC CCMVal Report.[11] This report analyzed simulations from the 18 CCMs that participated in CCMVal-2. All 13 CCMs in CCMVal-1 participated again, but partly with updated and improved or new model versions; in addition five new models submitted output to the CCMVal archive.[8] The SPARC CCMVal report included evaluation of a much larger set of processes than in previous evaluations, with an evaluation of dynamical, radiative, chemical, and transport processes, as well as upper troposphere/lower stratosphere and stratosphere-troposphere coupling.[8] The report also included the application of observationally based performance metrics to quantify the ability of models to reproduce key processes. Overall, the performance of CCMVal-2 models is similar to those in CCMVal-1. There are some diagnostics for which there is improvement (*e.g.*, Cl_y) but for other diagnostics the general model performance is worse and the spread of models is larger (*e.g.*, 100 hPa temperature and water vapor).

9.3 Changes in Major Factors Affecting Stratospheric Ozone

As discussed in the Introduction, the increase in stratospheric chlorine and bromine over the last few decades was the dominant cause of decreases in O_3 over this period. However, as discussed in Chapter 8, climate change is likely to have an increasing role in O_3 changes over the next century. Therefore, before discussing the projected ozone evolution, we examine the changes in the factors that control the distribution of O_3 as projected in the CCM simulations.

9.3.1 Stratospheric Halogens

The abundance of surface total chlorine (Cl_{tot}) and total bromine (Br_{tot}) peaked around 1993 and has slowly declined since then (see Figure 9.3).[14] This slow decline is expected to continue over the 21C, and ODSs are expected to be back to 1980 levels in the 2030s and be below 1970 levels by the end of this century. The stratospheric chlorine and bromine loading is expected to evolve in a similar manner, although with a transport-related time delay. A commonly used variable for the effect of halogens on stratospheric ozone is the Effective Equivalent Stratospheric Chlorine (EESC).[22,23] The EESC is an empirical estimate of stratospheric reactive chlorine and bromine, based on measurements and projections of surface ODS, observational estimates of the transit times between the troposphere and stratosphere and the fractional release rates of different ODS. Fractional release is the fraction of inorganic halogen released from halocarbons at a given location and time. Calculations of EESC appropriate for mid-latitude lower stratosphere (mean age of air ~3 years) indicate a return of EESC to 1980 values around 2040, while calculations for polar regions (mean age of air ~5.5 years) show a later return dates around 2065. There is a ~25 year difference in return dates for polar and mid-latitude EESC, even though there is only a ~2.5 year difference in mean age, because of the rapid growth of EESC around 1980 and slow decay around 2050.

The CCMs simulate Cl_y and Br_y within the stratosphere, and these concentrations can be used, rather than EESC, to examine variations in halogen

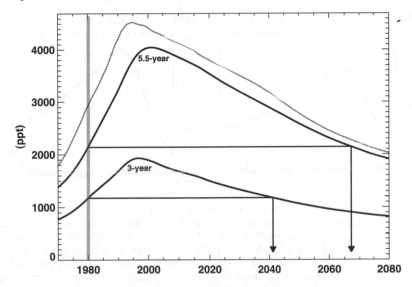

Figure 9.3 Evolution of surface $Cl_{tot} + 60Br_{tot}$ (gray curve) and EESC for mean age-of-air values of 3 and 5.5 years (black curves) for WMO (2002) ODS scenarios. The gray vertical line indicates the reference year of 1980. The black horizontal and vertical lines indicate the recovery date of EESC to 1980 values. From Figure 5 of Newman *et al.* (2007).[23]

that effect ozone (Chapter 2). As discussed in Section 9.2, the CCM simulations use a common scenario for the surface concentrations of ODSs when making projections. However, although the general evolution of Cl_y and Br_y is similar between CCMs, there are significant variations in the peak values of Cl_y and Br_y and the timing of the return to historical levels. For example, in the polar lower stratosphere, the simulated peak value of Cl_y in the individual CCMVal-2 models varies between 2.2 and 3.3 ppb, and the year when Cl_y returns to its 1980 value varies between 2040 and 2080 (see Figure 9.13 in the SPARC CCMVal report).[11] These two aspects are generally related, with CCMs with lower peak Cl_y values projecting an earlier return of Cl_y to pre-1980 values. The resulting evolution of Cl_y in the CCMVal-2 multi-model mean along with its uncertainty is shown in Figure 9.4.

As discussed in Section 9.2.3, the observed peak Cl_y in polar regions is around 3.3 ppb and most CCMs underestimate this peak value. This bias and the fact that models with lower peak return to 1980 values earlier needs to be considered when interpreting the CCM projections of ozone.

9.3.2 Temperature

As discussed in Chapter 8, changes in stratospheric temperatures can impact ozone loss by changing the rate of chemical reactions and the formation of PSCs. The stratosphere has cooled over the last four decades, and cooling is expected to continue through the 21C.[9,11,24] The cooling over the past few

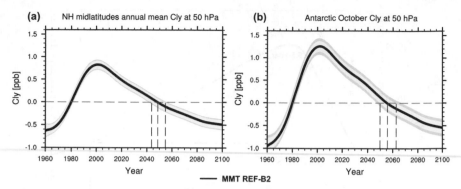

Figure 9.4 1980 baseline-adjusted multi-model trend estimates of annually averaged inorganic chlorine (Cl_y) at 50 hPa (ppb) for **(a)** annual northern midlatitude mean 35–60°N and **(b)** Antarctica (60°S–90°S) in October. The red vertical dashed line indicates the year when the multi-model trend in Cl_y returns to 1980 values and the blue vertical dashed lines indicate the uncertainty in these return dates. Multi-model mean derived from Figures 3.6 and 3.11 of WMO (2011).[14]

decades can be attributed to contributions from both increasing CO_2 and decreasing O_3 (the latter caused by increasing ODSs).[25,26] The impact of increasing CO_2 is expected to be dominant through the 21C, and CCMs show cooling in extra-polar regions throughout the 21C. The cooling rate increases with altitude, and the projected cooling in the upper stratosphere in the 21C is around 1 K/decade (Figure 9.5). This cooling in the middle and upper stratosphere will decrease the rate of the gas-phase chemical reactions that destroy O_3 (Chapter 1, 6), and cause O_3 to return to historical values earlier than expected just from changes in stratospheric halogens.

In the polar lower stratosphere, temperature trends could alter O_3 by a different mechanism; namely, a cooling could lead to increase in the temporal and spatial extent of PSCs. This would increase the occurrence of heterogeneous chemical reactions that lead to ozone depletion. There are large seasonal variations in polar temperature trends, but the winter/early spring trends are most relevant for understanding polar ozone depletion. CCM projections are not uniform on whether there is a cooling or warming in either polar region, but nearly all indicate that any trends will likely be small.[9,24] Understanding and quantifying trends in polar lower stratospheric temperatures during late winter/ spring is complicated by the large year-to-year variability (especially in the northern hemisphere) which tends to interfere with any long-term trend detection. The lack of large winter trends could be due to increased downwelling and diabatic warming, compensating the radiative cooling due to increasing GHGs.

9.3.3 Transport

Another factor that could influence the 21C evolution of stratospheric O_3 is a change in stratospheric circulation. Changes in stratospheric dynamics alter the

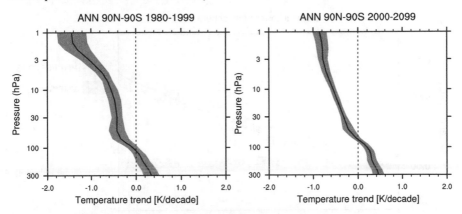

Figure 9.5 Multi-model annual mean 90°S–90°N temperature trend from 1980 to 1999 (left panel), and from 2000 to 2099 (right panel). The black line shows the CCMVal-2 multi-model mean and the shaded region shows ±1 standard deviation about the mean. Multi-model mean derived from Figure 4.4 of *SPARC CCMVal* (2010).[11]

stratospheric circulation, which then impacts the O_3 distribution by altering the direct transport of O_3 among regions, as well as by altering the distribution of Cl_y and Br_y, and other species involved in O_3 destruction. A robust result of modeling studies is that an increase in GHGs leads to an increase in the tropical vertical velocities (upwelling).[24,27,28] This is illustrated in Figure 9.6, which shows the multi-model mean and variance for projections of the tropical upwelling over the 21C. When GHGs and SSTs are increased, there is a significant increase in the upwelling ("REF"), whereas in a simulation with no climate change with GHG and SSTs fixed at 1960 levels, there is no change in the upwelling ("fGHG").[16] The increase in tropical upwelling leads to reduced transport time scales and a decrease in the mean age of air in the stratosphere.[24] The CCM simulations also indicate that there has been an increase in the tropical upwelling and a decrease in mean age of air over the past few decades.[17,29–31] However, this decrease has not been confirmed by observations, and a study by Engel *et al.* provided evidence that there has been a very small increase in the mean age of air over the last three decades.[32] There are, however, large uncertainties in the estimates of mean age from observations, and the cause of the discrepancy between the model and observations is unknown.[33,34]

The impact on O_3 of this acceleration of the stratospheric circulation will vary between regions. For example, an increase in upwelling will decrease tropical ozone below the peak in O_3 concentrations (below ~ 10 hPa) but will increase tropical ozone above this peak. An increase in the meridional circulation will also likely increase O_3 in middle latitudes due to increased transport from tropical source region. We also expect the impact of changes in transport to be larger in the lower stratosphere, where photochemical timescales are longer, than in the upper stratosphere where ozone is under photochemical control.

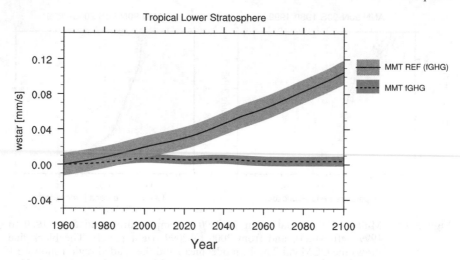

Figure 9.6   Multi-model mean and 95% confidence interval of the 1960 baseline-adjusted annual mean tropical upwelling mass flux between 20°S and 20°N at 70 hPa from the CCMVal-2 reference simulations (MMT REF, solid black line and grey shaded area) and the fixed greenhouse gas simulations (MMT fGHG, black dashed line and blue shaded area). From Figure 5 of Eyring *et al.* (2010).[16]

9.3.4   Other Factors

The long-term evolution of O_3 could also be altered by changes in nitrogen and hydrogen species that are involved in ozone destruction (see Chapter 8).

Increases in tropospheric N_2O are expected to occur in the 21C, leading to an increase in stratospheric NO_y. This would in turn be expected to lead to a decrease in ozone in the middle and upper stratosphere due to increased O_3 destruction by nitrogen oxides (NO_x). However, the percentage decrease in ozone is expected to be much smaller than the percentage N_2O increase. The increase in NO_y is smaller than N_2O due to temperature decreases in the upper stratosphere.[35] For example, the surface concentration of N_2O under scenario A1B increases by around 16% between 1980 and 2050, but the CCMs project that over the same period the stratospheric NO_y generally increases by around 10% or less.

Although the increase in NO_y is small over the 21C and not a dominant factor when GHGs follow the SRES A1B scenario, this is not the case for all IPCC scenarios.[3] For example, for the SRES A2 scenario there is a larger increase in N_2O, and changes related to NO_y make a significant contribution to changes in upper stratospheric O_3 (*e.g.*, in the GEOS CCM the decrease in O_3 at 3 hPa related to NO_y is $\sim 1/3$ the increase related to Cl_y decreases).[36]

The evolution of ozone in the 21C could also be affected by changes in stratospheric water vapor. An increase in water vapor would increase hydrogen oxide (HO_X), and lead to increased ozone loss in the extra-polar lower and upper stratosphere, where HO_X dominates ozone loss. In addition to changing

HO_X, an increase in water vapor would affect PSC formation and hetero-geneous reactions in the CCMs, which could lead to increased springtime polar ozone loss (see Chapter 8).

There are two mechanisms that could cause long-term increases in strato-spheric H_2O: (i) increases in CH_4, which will lead to an increase in H_2O, due to increased production from CH_4 oxidation, and (ii) a warming of the tropical cold-point temperature (which controls the stratospheric entry value of H_2O). Surface concentrations of CH_4 are projected to increase over 21C, although there are large variations between GHG scenarios. Most CCMs indicate a warming of the tropical tropopause in the future, which would cause an additional increase in stratospheric H_2O, by increasing the concentrations entering the stratosphere.[9] However, the increase in stratospheric H_2O due to the warming of tropical tropopause is generally smaller than contribution from a CH_4.[37] Overall, the stratospheric global-mean water vapor trends simulated by the CCMs are small (for the A1B scenario), and are not likely to be a major cause of changes in stratospheric O_3. This might change, however, if methane increases, for example to the melting of permafrost.

9.4   Projections of the Behavior of Ozone

The impact of the different climate factors discussed above on ozone varies between regions, both with latitude and altitude. As a consequence, the ozone evolution varies between regions. This is illustrated in Figure 9.7, which shows the multi-model mean change in O_3 between 2000 and 2100 from the CCMVal-2 models.[18]

9.4.1   Tropical Ozone

As shown in Figure 9.7a, the change in tropical ozone between 2000 and 2100 is very different above and below ~15 hPa. In the upper stratosphere ozone is projected to increase, whereas a decrease is projected in lower stratospheric O_3.[9,16] This contrast is clearly seen in the solid black curves in Figure 9.8, which shows the multi-model mean evolution of (a) upper and (b) lower stratospheric O_3 from the CCMVal-2 reference simulations.[16]

The different evolution of ozone in the tropical upper and lower stratosphere is due to the different role of climate change (and the different role of the mechanisms discussed in Section 9.3) in the two regions. The increase in tro-pical upper stratosphere O_3 is due mainly to decreases in halogen levels, which reduces the O_3 loss due to catalytic chlorine and bromine reactions, and cooling due to increased GHGs, which slows the chemical reactions that destroy O_3, (see Section 9.3 and Chapter 1). These two mechanisms make roughly equal contributions to the O_3 increase over the 21C (for the A1B GHG sce-nario).[9,15,36,38] This can be seen in Figure 9.8a by comparing the dashed (orange) and dotted (blue) curves (shaded regions), which show the projected changes in O_3 due to climate change and ODSs, respectively. The change due to

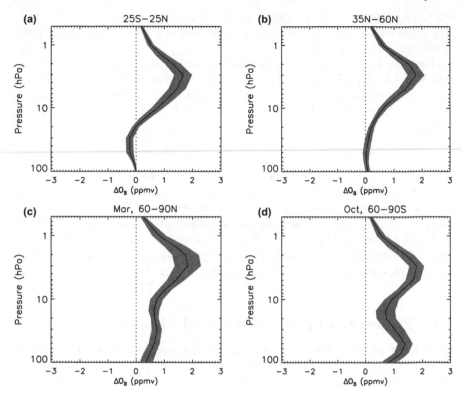

Figure 9.7 Multi-model mean of 2000–2100 ozone changes for different regions. The black line shows the CCMVal-2 multi-model mean and the shaded region shows ±1 standard deviation about the mean. Based on analysis in Oman *et al.* (2010).[18]

ODSs dominates over the latter part of the 20th century, but there is a similar increase in both terms over the 21C.

In the tropical lower stratosphere, the major mechanism causing long-term decreases in O_3 is the increase in tropical upwelling. As discussed in Section 9.3.3, a robust result in CCMs is an increase in tropical upwelling through the 21C. A future increase in upwelling in the tropics would result in a faster transit of air through the tropical lower stratosphere from an enhanced Brewer-Dobson circulation, which would lead to less time for production of ozone and hence lower ozone levels in this region.[14]

The fact that climate change is expected to increase O_3 in the tropical upper stratosphere, but to decrease O_3 in the tropical lower stratosphere means that if, and when, O_3 returns to historical values (*e.g.*, to values of O_3 in 1960 or 1980) varies between these altitudes. In the upper stratosphere, O_3 is projected to return to historical values several decades before upper stratospheric Cl_y and Br_y (and equivalent stratospheric chlorine, ESC; see Chapter 2) return to their historical values. For example, O_3 returns to 1960 values over 70 years before

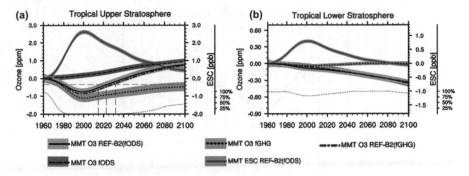

Figure 9.8 Tropical (25°S–25°N) annual mean 1960 baseline-adjusted ozone projections and 95% confidence for the (a) upper and (b) lower stratosphere. The multi-model trend (MMT) is shown for the CCMVal-2 reference run (REF-B2; black curves and grey shaded area), fixed ODS runs (fODS, black dotted line and orange shaded area), and fixed GHG runs (fGHG, black dashed line and blue shaded area). Also shown is the multi-model trend plus 95% confidence interval for Equivalent Stratospheric Chlorine (ESC), displayed with the red solid line and light red shaded area. See Figure 2 of Eyring *et al.* (2010a) for more details.[16]

Cl_y and Br_y return to their 1960 values (see Figure 9.8a). In the tropical lower stratosphere, O_3 may never return to historical values, even when anthropogenic ODSs have been removed from the atmosphere. For example, the multi-model mean O_3 decreases steadily from 1960 to 2100, even though Cl_y and Br_y return to 1960 values by the middle of the 21C, see Figure 9.8b.

The evolution of tropical column ozone depends on the balance between the increase in upper stratospheric concentrations and the decrease in lower stratospheric values, and as a result the projected changes are small, see Figure 9.9. There is no consensus between CCMs on whether tropical column ozone will return to pre-1980 values, with some models showing O_3 increasing slightly above 1980 values by the second half of the 21C, with O_3 remaining below 1980 values through the 21C. There is, however, a consensus that tropical column ozone will not return to 1960 values, *i.e.* even when stratospheric halogens return to historical (pre-1960) values, tropical column ozone will remain below its historical values.

9.4.2 Mid-latitude Ozone

The projected evolution of mid-latitude middle and upper stratosphere O_3 is very similar to that in the tropics (Figure 9.7b), with cooling in the upper stratosphere causing O_3 to return to historical values before Cl_y and Br_y. Furthermore, the magnitude of changes in O_3 and dates for returning to historical values are very similar to the tropics, as is the spread between CCMs. In the lower stratosphere, the evolution of mid-latitude O_3 differs somewhat from that in the tropics. In the subtropics O_3 decreases but at higher latitudes

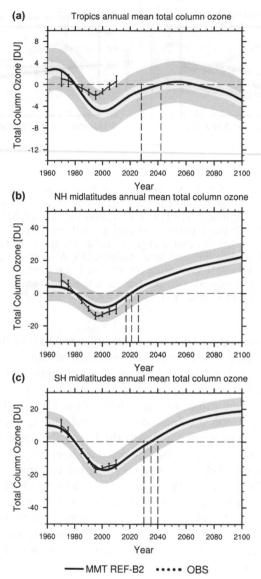

Figure 9.9 1980 baseline-adjusted multi-model trend estimates of annually averaged total column ozone (DU) for the tropics (25°S–25°N, upper panel) and mid-latitudes (middle panel: 35°N–60°N, lower panel: 35°S–60°S) (thick dark gray line) with 95% confidence and 95% prediction intervals appearing as light- and dark-gray shaded regions, respectively, about the trend (note the different vertical scale among the panels). The red vertical dashed line indicates the year when the multi-model trend in total column ozone returns to 1980 values and the blue vertical dashed lines indicate the uncertainty in these return dates. The black dotted lines show observed total column ozone, where a linear least squares regression model was used to remove the effects of the quasi-biennial oscillation, solar cycle, El Niño-Southern Oscillation, and volcanoes from four observational data sets. Multi-model mean derived from Figure 3.6 of WMO (2011).[14]

(and averaged over middle latitudes) there is an increase in lower stratospheric O_3. As in the tropics, changes in transport play an important role in these O_3 changes. However, in mid-latitudes the increase in the meridional circulation leads to an increase rather than a decrease in lower stratospheric O_3.[38,39]

Because ozone averaged over mid-latitudes increases in the upper and lower stratosphere over the 21C, a similar evolution is projected for mid-latitude total column ozone (see Figure 9.9). The evolution of mid-latitude column ozone is similar among the CCMs, with a broad minimum around 2000, which is followed by a slow increase back to and above 1980 values. In all CCMs, the return of O_3 to 1980 values occurs before that of Cl_y and Br_y. However, as can be seen by the multi-model mean standard deviation that is shown in Figure 9.9, there is a spread in the magnitude of the changes and time of return to 1980 values. This spread is closely linked to the spread in simulated Cl_y.[7,9]

In most CCMs there are interhemispheric differences in the evolution of column ozone. The qualitative evolution is the same but there is a difference in magnitude of anomalies and in the date of return to historical values. The anomalies are larger in the SH (because of spreading of ozone hole air into mid-latitudes) and the return of mid-latitude column O_3 to 1980 values occurs later in the SH. The difference in the date of return to 1980 values appears to be due to interhemispheric difference in changes in transport. The increase in strato-spheric circulation driven by climate change transports more O_3 into NH mid-latitude lower stratosphere than SH.[38]

9.4.3 Springtime Polar Ozone

The largest ozone depletion is observed in the polar lower stratosphere during spring, especially in the Antarctic (see Chapter 5). As a result, a major focus of model simulations is the projected evolution of polar lower stratospheric ozone during spring.

Antarctic
All models project a qualitatively similar evolution for Antarctic (60 °–90 °S) ozone in spring, with a broad minimum around 2000, followed by a very slow increase and a return to 1980 values sometime around the middle of the century, see Figure 9.10b. There are, however, as in the extrapolar regions, significant quantitative differences among the models, including a wide spread in the minimum values around 2000 and dates when ozone returns to 1980 (or 1960) values.[7,9]

Several different ozone indices have been used to quantify variations in Antarctic ozone, including the polar cap average (Figure 9.10), ozone mass, daily minimum ozone, and area of the ozone hole (area of ozone less than 220 DU).[40] There are some differences in the evolution of these different diagnostics, in particular over the 2000 to 2030 period, where the rate of change varies. However, all diagnostics show the same broad evolution with a return to 1980 values around the middle of century, and a wide spread among the models in quantitative details.

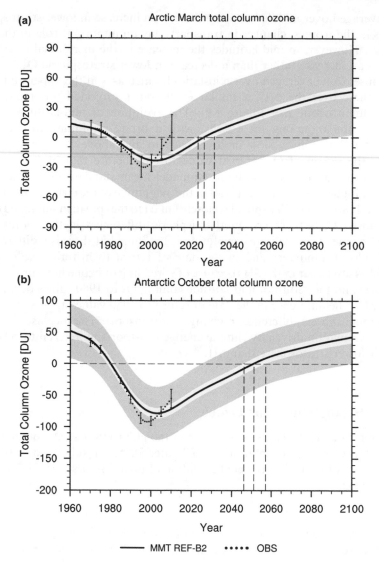

Figure 9.10   As in Figure 9.9, but for the latitude range 60°N–90°N in March (upper row) and the latitude range 60°S–90°S in October (lower row). The red vertical dashed line indicates the year when CCMVal-2 multi-model trend in total column ozone (DU) returns to 1980 values and the blue vertical dashed lines indicate the uncertainty in these return dates. Note the different vertical scale among the panels. Multi-model mean derived from Figure 3.10 of WMO (2011).[14]

The evolution of Antarctic spring ozone is dominated by the changes in Cl_y and Br_y, and changes in climate (temperature and transport) are not, in general, a major factor. This can be seen by the very close correspondence of the evolution of ozone and Cl_y (or ESC),[7,9,16] The spread in Antarctic ozone projections are, as a result, primarily due to differences in simulated Cl_y (and Br_y) among the models. Models that simulate a smaller peak Cl_y have an earlier return of Cl_y to 1980 values, and generally also have smaller ozone depletion and earlier return of ozone to 1980 values. This relationship suggests that the low bias in Cl_y in most models results in an early bias in the projected return of ozone to 1980 values, *i.e.*, the return to 1980 values will likely occur later than indicated by the multi-model mean shown in Figure 9.10b.

Projections of the recovery of the Antarctic ozone hole have also been made using parametric models based on estimates of EESC and analyzed polar temperatures.[41] These calculations indicate that the ozone hole area will remain constant until around 2015, and then decrease to zero around 2070. This recovery date is later than that simulated by most models (see Figure 9b of Eyring *et al.* (2007)), but this difference is consistent with a bias in the models dynamics and transport.[9]

Arctic

Dynamical effects play a much larger role in the evolution of springtime Arctic ozone than in the Antarctic, and as a consequence there is large interannual variability. This interannual variability is much larger than the long-term changes, and time series need to be filtered (smoothed) to see long-term trends. However, long-term evolution of the filtered Arctic ozone is qualitatively the same as in other regions: there is a broad minimum around 2000 with slow increase over the first half of the 21C, see Figure 9.10a.

Although there is qualitative agreement among the models, there are large quantitative variations in the simulated changes in Arctic ozone, with some models showing only a small or even no change in ozone, while others show a large response to changes in halogens. There are substantial variations among the models in the date when ozone returns 1980 values (2020 to 2060), *e.g.*, see Figure 16 of Austin *et al.* (2010).[7]

An early study by Shindell *et al.* projected a substantial increase in Arctic ozone depletion, and the development of an Arctic ozone hole, because of climate change.[42] However, subsequent studies have not reproduced this result. Even though there is a large spread among the CCMs, in both CCMVal-1 and CCMVal-2, none of the CCMs predict large Arctic ozone decreases in the future.

Models project that Arctic ozone will return to 1980 values before Antarctic ozone, with the difference varying from only a few years in some models to over 25 years in others. The evolution of Cl_y and Br_y is similar in both polar regions, but because changes in temperature and transport play a significant role in the Arctic, the evolution of Arctic O_3 does not follow that of Cl_y and Br_y as closely as Antarctic ozone. In particular, acceleration of the Brewer-Dobson circulation and increases in polar temperatures cause an earlier return than that expected just because of changes in halogens.[43] These dynamical changes vary substantially among models, and as a consequence so do the differences in return dates of Arctic and Antarctic ozone.

9.5   Summary and Concluding Remarks

The evolution of stratospheric ozone in the 21st century will depend not only on changes (expected decreases) in the abundance of stratospheric halogens but also on changes due to increases in well-mixed greenhouse gases. The latter will cool the stratosphere, increase the abundance of nitrogen and hydrogen species involved in ozone destruction, and alter transport within the stratosphere. The impact of the greenhouse-gas-induced change on ozone varies between regions, and as a consequence the ozone evolution will vary between regions.

In the upper stratosphere the projected ozone evolution is very similar in the tropics and middle latitudes, with ozone increasing back to 1960 values in the first 2–3 decades of the 21st century. This rapid increase in ozone is due to both decreases in ODSs and cooling due to increased GHGs. The evolution in the mid-latitude lower stratosphere is similar to that in the upper stratosphere, although climate-change induced increases in the circulation play more of a role than cooling. In the tropical lower stratosphere the evolution is, however, very different: here ozone decreases throughout the 21st century due to climate-change-induced increases in the tropical upwelling. In the Antarctic, the models consistently show a broad minimum near year 2000 followed by a slower return to 1980 values than in middle latitudes (with the return delayed until the middle of the century). In the Arctic, the interannual variability in ozone is larger than long-term trends and the models are less consistent in their representation of ozone recovery. However, in most models Arctic ozone returns to 1980 values before Antarctic ozone.

Although there is generally qualitative agreement among models in the evolution of ozone, there are some substantial quantitative differences. In particular, there is a wide spread in projected ozone values at specified periods and in the dates when ozone returns to historical values. In many cases, the differences among model projections can be related to differences in the simulated Cl$_y$, and improving the transport of Cl$_y$, which will reduce the uncertainty in ozone projections, is a remaining major challenge.

It is also important to note that the ozone projections depend on scenarios for surface concentrations of ODSs and GHGs, and the majority of the projections have considered very similar ODS scenarios and the same GHG scenario. However, a recent examination of CCM projections for six different GHG scenarios found that lower GHG emissions result in: (i) smaller reductions in ozone in the tropical lower stratosphere (due to smaller increases in tropical upwelling), and (ii) smaller increases in upper stratospheric ozone globally (due to less severe stratospheric cooling).[10] Largest differences among the six GHG scenarios were found over northern mid-latitudes (~ 20 DU by 2100) and in the Arctic (~ 40 DU by 2100) with divergence mainly in the second half of the 21st century. The results suggest that effects of GHG emissions on future stratospheric ozone should be considered in climate change mitigation policy and ozone projections should be assessed under more than a single GHG scenario.

Future assessments should also consider the uncertainty in how the models force the organic halogen lower boundary condition. Currently, projections of future organic halogen loadings are based on projected emission rates and an estimate of the global atmospheric lifetime of each organic halogen. These factors are then used to create time-dependent volume mixing ratio lower boundary conditions, that are then used to force the CCMs. However, the destruction of each halogen in the CCMs is dependent on the tropical upwelling, meridional mixing, and chemical loss rates (*e.g.*, photolysis rates), and the CCM-derived halogen lifetimes can be very different from the lifetimes assumed for the given projection scenario. By forcing all CCMs to use fixed mixing ratio lower boundary conditions, the flux into the tropical lower stratosphere is fixed in all the models. This minimizes the spread in model-derived ozone return dates.[44] If models were forced with flux lower boundary conditions, the simulated ODS would be consistent with the simulated loss of ODSs and any changes in the model circulation would also feedback on these loss rates. Models with a more realistic circulation (*e.g.*, representation of mean age) would also more accurately represent the fractional release of inorganic halogens from their parent organic. This approach would give a more accurate representation of ozone depletion and recovery.

Another consideration for future simulations is inclusion of interactive ocean and sea ice modules in the CCMs. In all but one of the CCM simulations discussed above, the sea surface temperatures and sea ice concentrations were prescribed, and the important coupling between the atmosphere and oceans/cyrosphere are not represented. The one exception is the Canadian Middle Atmosphere Model (CMAM), in which the atmospheric model is coupled to an ocean/sea ice model.[45] Inclusion of these couplings in the CCMs leads to a more complete representation of the climate system and feedbacks, which could be particularly important for simulations of stratospheric polar ozone and its impact on tropospheric climate.

Acknowledgements

We acknowledge the Chemistry-Climate Model Validation (CCMVal) Activity of the World Climate Research Programme's (WCRP) Stratospheric Processes and their Role in Climate (SPARC) project for organizing and coordinating the model data analysis, and the British Atmospheric Data Centre (BADC) for collecting and archiving the CCMVal model output. We thank Anne Douglass for a careful review of this chapter and Irene Cionni and Luke Oman for preparing some of the figures. This research was supported by the US National Science Foundation and by the German Aerospace Center (DLR).

References

1. S. A. Montzka, J. H. Butler, B. D. Hall, D. J. Mondeel and J. W. Elkins, *Geophys. Res. Lett.*, **30**, 1826, doi:10.1029/2003GL017745, 2003.

2. World Meteorological Organization (WMO)/United Nations Environment Programme (UNEP), Scientific Assessment of Ozone Depletion: 2006, World Meteorological Organization, Global Ozone Research and Monitoring Project, Report No. 50, Geneva, Switzerland, 2007.

3. Intergovernmental Panel on Climate Change (IPCC) (2000), *Special Report on Emissions Scenarios: A Special Report of Working Group III of the Intergovernmental Panel on Climate Change*, 599 pp., Cambridge Univ. Press, Cambridge, U. K.

4. S. Pawson, K. Kodera, K. Hamilton, T. G. Shepherd, S. R. Beagley, B. A. Boville, J. D. Farrara, T. D. A. Fairlie, A. Kitoh, W. A. Lahoz, U. Langematz, E. Manzini, D. H. Rind, A. A. Scaife, K. Shibata, P. Simon, R. Swinbank, L. Takacs, R. J. Wilson, J. A. Al-Saadi, M. Amodei, M. Chiba, L. Coy, J. de Grandpré, R. S. Eckman, M. Fiorino, W. L. Grose, H. Koide, J. N. Koshyk, D. Li, J. Lerner, J. D. Mahlman, N. A. McFarlane, C. R. Mechoso, A. Molod, A. O'Neill, R. B. Pierce, W. J. Randel, R. B. Rood and F. Wu, *Bull. Am. Meteorol. Soc.*, 2000, **81**, 781–796.

5. V. Eyring, N. R. P. Harris, M. Rex, T. G. Shepherd, D. W. Fahey, G. T. Amanatidis, J. Austin, M. P. Chipperfield, M. Dameris, P. M. De, F. Forster, A. Gettelman, H. F. Graf, T. Nagashima, P. A. Newman, S. Pawson, M. J. Prather, J. A. Pyle, R. J. Salawitch, B. D. Santer and D. W. Waugh, *Bull. Am. Meteorol. Soc.*, 2005a, **86**, 1117–1133.

6. J. Austin, D. Shindell, S. R. Beagley, C. Brühl, M. Dameris, E. Manzini, T. Nagashima, P. Newman, S. Pawson, G. Pitari, E. Rozanov, C. Schnadt and T. G. Shepherd, *Atmos. Chem. Phys.*, 2003, **3**, 1–27.

7. J. Austin, J. Scinocca, D. Plummer, L. Oman, D. Waugh, H. Akiyoshi, S. Bekki, P. Braesicke, N. Butchart, M. P. Chipperfield, D. Cugnet, M. Dameris, S. Dhomse, V. Eyring, S. Frith, R. Garcia, H. Garny, A. Gettelman, S. C. Hardiman, D. Kinnison, J. F. Lamarque, E. Mancini, M. Marchand, M. Michou, O. Morgenstern, T. Nakamura, S. Pawson, G. Pitari, J. Pyle, E. Rozanov, T. G. Shepherd, K. Shibata, H. Teyssedre, R. J. Wilson and Y. Yamashita, *J. Geophys. Res.*, doi:10.1029/2010JD013857, 2010.

8. V. Eyring, N. Butchart, D. W. Waugh, H. Akiyoshi, J. Austin, S. Bekki, G. E. Bodeker, B. A. Boville, C. Brühl, M. P. Chipperfield, E. Cordero, M. Dameris, M. Deushi, V. E. Fioletov, S. M. Frith, R. R. Garcia, A. Gettelman, M. A. Giorgetta, V. Grewe, L. Jourdain, D. E. Kinnison, E. Mancini, E. Manzini, M. Marchand, D. R. Marsh, T. Nagashima, P. A. Newman, J. E. Nielsen, S. Pawson, G. Pitari, D. A. Plummer, E. Rozanov, M. Schraner, T. G. Shepherd, K. Shibata, R. S. Stolarski, H. Struthers, W. Tian and M. Yoshiki, *J. Geophys. Res.*, **111**, D22308, doi:10.1029/2006JD007327, 2006.

9. V. Eyring, D. W. Waugh, G. E. Bodeker, E. Cordero, H. Akiyoshi, J. Austin, S. R. Beagley, B. Boville, P. Braesicke, C. Brühl, N. Butchart, M. P. Chipperfield, M. Dameris, R. Deckert, M. Deushi, S. M. Frith, R. R. Garcia, A. Gettelman, M. Giorgetta, D. E. Kinnison, E. Mancini,

E. Manzini, D. R. Marsh, S. Matthes, T. Nagashima, P. A. Newman, J. E. Nielsen, S. Pawson, G. Pitari, D. A. Plummer, E. Rozanov, M. Schraner, J. F. Scinocca, K. Semeniuk, T. G. Shepherd, K. Shibata, B. Steil, R. Stolarski, W. Tian and M. Yoshiki, *J. Geophys. Res.*, 112, D16303, doi:10.1029/2006JD008332, 2007.

10. V. Eyring, I. Cionni, J. F. Lamarque, H. Akiyoshi, G. E. Bodeker, A. J. Charlton-Perez, S. M. Frith, A. Gettelman, D. E. Kinnison, T. Nakamura, L. D. Oman, S. Pawson and Y. Yamashita, *Geophys. Res. Lett.*, 37, L16807, doi:10.1029/2010GL044443, 2010.

11. SPARC CCMVal (2010), SPARC Report on the Evaluation of Chemistry-Climate Models, V. Eyring, T. G. Shepherd, D. W. Waugh (Eds.), SPARC Report No. 5, WCRP-132, WMO/TD-No. 1526, http://www.atmosp. physics.utoronto.ca/SPARC.

12. V. Eyring, D. E. Kinnison and T. G. Shepherd, *SPARC Newsletter No. 25*, 2005b, 11–17.

13. V. Eyring, M. P. Chipperfield, M. A. Giorgetta, D. E. Kinnison, E. Manzini, K. Matthes, P. A. Newman, S. Pawson, T. G. Shepherd and D. W. Waugh, *SPARC Newsletter No. 30*, p. 20–26, 2008.

14. World Meteorological Organization (WMO)/United Nations Environment Programme (UNEP), Scientific Assessment of Ozone Depletion: 2010, World Meteorological Organization, Global Ozone Research and Monitoring Project, Report No. 52, Geneva, Switzerland, 2011.

15. D. W. Waugh, L. Oman, S. R. Kawa, R. S. Stolarski, S. Pawson, A. R. Douglass, P. A. Newman and J. E. Nielsen, *Geophys. Res. Lett.*, 36, L03805, doi:10.1029/2008GL036223, 2009.

16. V. Eyring, I. Cionni, G. E. Bodeker, A. J. Charlton-Perez, D. E. Kinnison, J. F. Scinocca, D. W. Waugh, H. Akiyoshi, S. Bekki, M. P. Chipperfield, M. Dameris, S. Dhomse, S. M. Frith, H. Garny, A. Gettelman, A. Kubin, U. Langematz, E. Mancini, M. Marchand, T. Nakamura, L. D. Oman, S. Pawson, G. Pitari, D. A. Plummer, E. Rozanov, T. G. Shepherd, K. Shibata, W. Tian, P. Braesicke, S. C. Hardiman, J. F. Lamarque, O. Morgenstern, D. Smale, J. A. Pyle and Y. Yamashita, *Atmos. Chem. Phys.*, 10, 9451-9472, doi:10.5194/acp-10-9451-2010, 2010.

17. R. R. Garcia and W. J. Randel, *J. Atmos. Sci.*, 2008, 65, 2731–2739.

18. L. D. Oman, D. A. Plummer, D. W. Waugh, J. Austin, J. F. Scinocca, A. R. Douglass, R. J. Salawitch, T. Canty, H. Akiyoshi, S. Bekki, P. Braesicke, N. Butchart, M. P. Chipperfield, D. Cugnet, S. Dhomse, V. Eyring, S. Frith, S. C. Hardiman, D. E. Kinnison, J.-F. Lamarque, E. Mancini, M. Marchand, M. Michou, O. Morgenstern, T. Nakamura, J. E. Nielsen, D. Olivie, G. Pitari, J. Pyle, E. Rozanov, T. G. Shepherd, K. Shibata, R. S. Stolarski, H. Teyssedre, W. Tian, Y. Yamashita and J. R. Ziemke, *J. Geophys. Res.*, 115, D24306, doi:10.1029/2010JD014362, 2010.

19. D. W. Waugh and V. Eyring, *Atmos. Chem. Phys.*, 2008, 8, 5699–5713.

20. A. Gettelman, T. Birner, V. Eyring, H. Akiyoshi, S. Bekki, C. Brühl, M. Dameris, D. E. Kinnison, F. Lefevre, F. Lott, E. Mancini, G. Pitari,

D. A. Plummer, E. Rozanov, K. Shibata, A. Stenke, H. Struthers and W. Tian, *Atmos. Chem. Phys.*, 2009, **9**, 1621–1637.

21. J. Austin, K. Tourpali, E. Rozanov, H. Akiyoshi, S. Bekki, G. Bodeker, C. Brühl, N. Butchart, M. Chipperfield, M. Deushi, V. I. Fomichev, M. A. Giorgetta, L. Gray, K. Kodera, F. Lott, E. Manzini, D. Marsh, K. Matthes, T. Nagashima, K. Shibata, R. S. Stolarski, H. Struthers and W. Tian, *J. Geophys. Res.*, 2008, **113**, D11306.

22. J. S. Daniel, S. Solomon, R. W. Portmann and R. R. Garcia, *J. Geophys. Res.*, 1999, **104**, 23,871-23,880.

23. P. A. Newman, J. S. Daniel, D. W. Waugh and E. R. Nash, *Atmos. Chem. Phys.*, 2007, **7**, 4537–4552.

24. N. Butchart, I. Cionni, V. Eyring, T. G. Shepherd, D. W. Waugh, H. Akiyoshi, J. Austin, C. Bruhl, M. P. Chipperfield, E. Cordero, M. Dameris, R. Deckert, S. Dhomse, S. M. Frith, R. R. Garcia, A. Gettelman, M. A. Giorgetta, D. E. Kinnison, F. Li, E. Mancini, C. McLandress, S. Pawson, G. Pitari, D. A. Plummer, E. Rozanov, F. Sassi, J. F. Scinocca, K. Shibata, B. Steil and W. Tian, *J. Climate*, **23**, 5349–5374, doi: 10.1175/2010JCLI3404.1, 2010.

25. T. G. Shepherd and A. I. Jonsson, *Atmos. Chem. Phys.*, 2008, **8**, 1435–1444.

26. R. S. Stolarski, A. R. Douglass, P. A. Newman, S. Pawson and M. R. Schoeberl, *J. Climate*, 2010, **23**, 28–42.

27. N. Butchart, A. A. Scaife, M. Bourqui, J. de Grandpré, S. H. E. Hare, J. Kettleborough, U. Langematz, E. Manzini, F. Sassi, K. Shibata, D. Shindell and M. Sigmond, *Clim. Dyn.*, 2006, **27**, 727–741.

28. N. Butchart and A. A. Scaife, *Nature*, 2001, **410**, 799–802.

29. J. Austin, J. Wilson, F. Li and H. Vomel, *J. Atmos. Sci.*, 2007, **64**, 905–921.

30. L. Oman, D. W. Waugh, S. Pawson, R. S. Stolarski and P. A. Newman, *J. Geophys. Res.*, **114**, D03105, doi:10.1029/2008JD010378, 2009.

31. C. McLandress and T. G. Shepherd, *J. Clim*, 2009, **22**, 1516–1540.

32. A. Engel, T. Möbius, H. Bönisch, U. Schmidt, R. Heinz, I. Levin, E. Atlas, S. Aoki, T. Nakazawa, S. Sugawara, F. Moore, D. Hurst, J. Elkins, S. Schauffler, A. Andrews and K. Boering, *Nature Geosciences*, **2**, 28–31, 2009.

33. D. W. Waugh, *Nature Geosciences*, 2009, **2**, 14–16.

34. R. R. Garcia, W. J. Randel and D. E. Kinnison, *J. Atm. Sci.*, 2011, **68**, 139–154.

35. J. E. Rosenfield and A. R. Douglass, *Geophys. Res. Lett.*, 1998, **25**, 4381–4384.

36. L. Oman, D. W. Waugh, S. R. Kawa, R. S. Stolarski, A. R. Douglass and P. A. Newman, *J. Geophys. Res.*, **115**, D05303, doi:10.1029/2009JD012397, 2010.

37. L. Oman, D. W. Waugh, S. Pawson, R. S. Stolarski and J. E. Nielsen, *J. Atmos. Sci.*, 2008, **65**, 3278–3291.

38. T. G. Shepherd, *Atmos. Ocean*, 2008, **46**, 117–138.

39. F. Li, R. S. Stolarski and P. A. Newman, *Atmos. Chem. Phys.*, 2009, **9**, 2207–2213.

40. G. E. Bodeker, H. Shiona and H. Eskes, *Atmos. Chem. Phys.*, 2005, **5**, 2603–2615.
41. P. A. Newman, E. R. Nash, S. R. Kawa, S. A. Montzka and S. M. Schauffler, *Geophys. Res. Lett.*, **33**, L12814, doi:10.1029/2005GL025232, 2006.
42. D. T. Shindell, D. Rind and P. Lonergan, *Nature*, 1998, **392**, 589–592.
43. J. Austin and F. Li, *Geophys. Res. Lett.*, **33**, L17807, doi:10.1029/2006GL026867, 2006.
44. A. R. Douglass, R. S. Stolarski, M. R. Schoeberl, C. H. Jackman, M. L. Gupta, P. A. Newman, J. E. Nielsen and E. L. Fleming, *J. Geophys. Res.*, **113**, D14309, doi:10.1029/2007JD009575, 2008.
45. C. McLandress, A. I. Jonsson, D. A. Plummer, M. C. Reader, J. F. Scinocca and T. G. Shepherd, *J. Climate*, **23**, 5002–5020, doi: 10.1175/2010JCLI3586.1, 2010.
46. J. Austin and R. J. Wilson, *J. Geophys. Res.*, **115**, D18303, doi:10.1029/2009JD013292, 2010.
47. J.-F. Lamarque, D. E. Kinnison, P. G. Hess and F. M. Vitt, *J. Geophys. Res.*, **113**, D12301, doi:10.1029/2007JD009277, 2008.
48. H. Akiyoshi, L. B. Zhou, Y. Yamashita, K. Sakamoto, M. Yoshiki, T. Nagashima, M. Takahashi, J. Kurokawa, M. Takigawa and T. Imamura, *J. Geophys. Res.*, **114**, D03103, doi:10.1029/2007JD009261, 2009.
49. J. F. Scinocca, N. A. McFarlane, M. Lazare, J. Li and D. Plummer, *Atmos. Chem. Phys.*, 2008, **8**, 7055–7074.
50. J. de Grandpre, S. R. Beagley, V. I. Fomichev, E. Griffioen, J. C. McConnell, A. S. Medvedev and T. G. Shepherd, *J. Geophys. Res.*, 2000, **105**, 26475–26491.
51. M. Déqué, *Global and Planetary Change*, 2007, **57**, 16–26.
52. H. Teyssèdre, M. Michou, H. L. Clark, B. Josse, F. Karcher, D. Olivie, V.-H. Peuch, D. Saint-Martin, D. Cariolle, J.-L. Attié, P. Nédélec, P. Ricaud, V. Thouret, R. J. van der A, A. Volz-Thomas and F. Cheroux, *Atmos. Chem. Phys.*, 2007, **7**, 5815–5860.
53. A. Stenke, M. Dameris, V. Grewe and H. Garny, *Atmos. Chem. Phys.*, 2009, **9**, 5489–5504.
54. H. Garny, M. Dameris and A. Stenke, *Atmos. Chem. Phys.*, 2009, **9**, 6017–6031.
55. P. Jöckel, H. Tost, A. Pozzer, C. Brühl, J. Buchholz, L. Ganzeveld, P. Hoor, A. Kerkweg, M. G. Lawrence, R. Sander, B. Steil, G. Stiller, M. Tanarhte, D. Taraborrelli, J. van Aardenne and J. Lelieveld, *Atmos. Chem. Phys.*, 2006, **6**, 5067–5104.
56. S. Pawson, R. S. Stolarski, A. R. Douglass, P. A. Newman, J. E. Nielsen, S. M. Frith and M. L. Gupta, *J. Geophys. Res.*, **113**, D12103, doi:10.1029/2007JD009511, 2008.
57. L. Jourdain, S. Bekki, F. Lott and F. Lefèvre, *Annales Geophysicae*, 2008, **26**, 1391–1413.
58. K. Shibata and M. Deushi, *Annales Geophysicae*, 2008, **26**, 1299–1326.

59. K. Shibata and M. Deushi, Simulation of the stratospheric circulation and ozone during the recent past (1980–2004) with the MRI chemistry-climate model, CGER's Supercomputer Monograph Report Vol. 13, National Institute for Environmental Studies, Japan, 154 pp, 2008.

60. M. Schraner, E. Rozanov, C. Schnadt-Poberaj, P. Kenzelmann, A. Fischer, V. Zubov, B. P. Luo, C. Hoyle, T. Egorova, S. Fueglistaler, S. Brönnimann, W. Schmutz and T. Peter, *Atmos. Chem. Phys.*, **8**, 5957–5974, doi:10.5194/acp-8-5957-2008, 2008.

61. T. Egorova, E. Rozanov, V. Zubov, E. Manzini, W. Schmutz and T. Peter, *Atmos. Chem. Phys.*, 2005, **5**, 1557–1576.

62. G. Pitari, E. Mancini, V. Rizi and D. T. Shindell, *J. Atmos. Sci.*, 2002, **59**.

63. J. Austin and N. Butchart, *Q. J. R. Meteorol. Soc.*, 2003, **129**, 3225–3249.

64. W. Tian and M. P. Chipperfield, *Q. J. R. Meteor. Soc.*, 2005, **131**, 281–303.

65. W. Tian M. P. Chipperfield, L. J. Gray and J. M. Zawodny, *J. Geophys. Res.*, **111**, D20301, doi:10.1029/2005JD006871, 2006.

66. O. Morgenstern, P. Braesicke, M. M. Hurwitz, F. M. O'Connor, A. C. Bushell, C. E. Johnson and J. A. Pyle, *Geophys. Res. Lett.*, **35**, L16811, doi:10.1029/2008GL034590, 2008.

67. O. Morgenstern, P. Braesicke, F. M. O'Connor, A. C. Bushell, C. E. Johnson, S. M. Osprey and J. A. Pyle, *Geosci. Model Dev.*, 2009, **1**, 43–57.

68. R. R. Garcia, D. R. Marsh, D. E. Kinnison, B. A. Boville and F. Sassi, *J. Geophys. Res.*, **112**, D09301, doi:10.1029/2006JD007485, 2007.

CHAPTER 10

Impact of Geo-engineering on Stratospheric Ozone and Climate

SIMONE TILMES* AND ROLANDO R. GARCIA

National Center for Atmospheric Research, Boulder, CO, USA

10.1 Motivation for Proposed Geo-engineering Approaches

Geo-engineering is commonly understood as the active manipulation of the Earth's system to change weather or climate conditions. The idea of weather modification and geo-engineering of the Earth was raised as far back as 1830 when J. P. Espy suggested producing convective updrafts by lighting huge fires in order to change rain intensity and frequency of occurrence.[1] Later, in 1960, N. Rusin and L. Flit[2] proposed finding a way to remove ice from the Arctic "to make it more suitable for life", or to light up the night sky of the world by injecting tiny white particle into space. Already at that time, N. Rusin and L. Flit warned that large-scale environmental modification could cause undesirable side effects. With the increasing concern that efforts to reduce global greenhouse gas emission will not be sufficient to prevent Earth's climate from severe or unforeseeable impacts on life on Earth, the discussion about geo-engineering has attracted increasing attention in recent years. A modern definition of geo-engineering, also called "climate engineering", is "the deliberate large-scale manipulation of the planetary environment to counteract anthropogenic climate change".[3] It is commonly thought of as a short-term intervention to reduce the worst impacts of greenhouse warming on the climate system while the replacement of fossil fuel with renewable energy sources is

Stratospheric Ozone Depletion and Climate Change
Edited by Rolf Müller
© Royal Society of Chemistry 2012
Published by the Royal Society of Chemistry, www.rsc.org

getting underway. Geo-engineering includes two approaches: carbon dioxide removal from the atmosphere (CDR) and solar radiation management (SRM). CDR would reduce atmospheric concentrations of CO_2 by actively sequestering atmospheric CO_2, whereas SRM would reduce shortwave radiation before it reaches Earth, or would increase the reflectivity of the Earth. Very little is known about the potential impact on ozone of CDR (for example, fertilizing the ocean to promote uptake of CO_2 by phytoplankton and industrial approaches) and, therefore, CDR is not discussed further in this chapter. One idea of reducing shortwave radiation to enhance the planetary albedo can be seen as analogous to conditions after a large volcanic eruption, where the presence of enhanced aerosol loading in the stratosphere results in a cooling of the Earth's surface, a long-recognized effect.[4] Concern has been raised, however, that SRM using sulfate aerosol particles to counteract global warming will likely impact stratospheric ozone.[5] Other SRM approaches including marine cloud brightening, reflectors between the Earth and the Sun, and enhancing the reflectivity of Earth's surface are not further discussed. The impact of those approaches on the climate system might be significant, but studies have not explored their impacts on ozone.

In 1974, Budyko[6] was the first to suggest producing an artificial aerosol layer in the atmosphere to reflect sunlight and therefore cool the planet. He also addressed the concern about possible changes in weather conditions as a result of this approach, and acknowledged that such effects could not be foreseen at that time. After this, the question of climate engineering attracted little interest until about 20 years ago, when concerns about climate change were becoming widespread. For example, geo-engineering received the attention of the US National Research Council panel on the policy implications of global warming (US National Academy of Sciences 1992). In 2006, a paper by Paul Crutzen[7] stimulated a lively debate on geo-engineering as a possible last resort to counteract global warming, followed by Tom Wigley,[8] who discussed potential scenarios to combine SRM and mitigation scenarios, and others.[3]

One of the most discussed SRM schemes is the injection of small aerosol particles into the atmosphere to reflect sunlight to enhance the Earth's albedo.[7,9] This approach has received favorable consideration because its effects can be estimated by considering natural events in the past: the injection of aerosol particles into the stratosphere as a result of volcanic eruptions. A decrease of global temperatures by $0.5 \, K$ was observed after the last major eruption of Mount Pinatubo in 1991. This approach is judged to be relatively inexpensive compared to other proposed SRM techniques.[10,11] However, these costs estimates are rather uncertain and many risks and side effects have yet to be studied in detail.[12] Most engineering issues are far from being resolved and to date it is not clear how much cooling could be achieved using sulfate particles to enhance the albedo of the planet, or if other particles would be better suited for this purpose.[13,14]

The injection of particles into the stratosphere raises major concerns about the impacts on ozone and climate.[5] Changes in the Earth's albedo and, therefore, the reduction of shortwave radiation reaching the Earth's surface as a result of enhanced reflecting particles in the stratosphere may result in an offset of the global radiative forcing due to greenhouse gases. However, differences in the forcing of short and longwave radiation result in local changes in temperature and precipitation patterns. For example, changes in the hydrological cycle as a result of volcanic eruptions have been suggested, and the decrease of shortwave radiation is expected to increase the prevalence of droughts.[15–18]

The enhanced burden of liquid sulfate aerosols facilitates heterogeneous activation of halogen compounds, and there is clear evidence that the ozone layer is strongly influenced by those.[19–21] Even though volcanic eruptions can be seen as an analog for geo-engineering, the long-term presence of enhanced sulfate aerosol levels in the stratosphere can be significantly different from sporadic volcanic eruptions with regard to its impact on climate and ozone.[19,22]

In Section 10.2, the impact of enhanced stratospheric aerosols on temperatures, precipitation and stratospheric ozone is outlined as observed one year after the eruption of Mt. Pinatubo, when the stratospheric halogen loading was at a high level (Chapter 2). Climate model simulations and empirical projections of past conditions into the future have been performed to study the impact of potential geo-engineering on future ozone and climate, as summarized in Section 10.3. The potential impact of geo-engineering on tropospheric ozone is not discussed due to the lack of scientific studies. Also, we do not discuss any existing political, ethical, technical, or legal concerns about geo-engineering in general, or specifically for the aerosol approach, although this has been done elsewhere.[3]

10.2 Impact of Major Volcanic Eruptions on Climate and Ozone

Large volcanic eruptions typically inject different gases, such as H_2O, N_2 and CO_2, into the stratosphere. Since these species are relatively abundant in the stratosphere, the injection of additional amounts during volcanic eruptions does not have a significant influence on the climate. However, most important for the climate system is the emission of sulphur species in the form of SO_2 and sometimes H_2S, which are then further oxidized to gaseous H_2SO_4, increasing the net sulfur abundance in the stratosphere by up to two orders of magnitude after a large volcanic eruption.[23] The last two major volcanic eruptions in the northern hemisphere (NH) were El Chichón in 1982, which injected 7 Tg of SO_2 (~ 3.5 Mt S) into the stratosphere, and Mount Pinatubo in 1991, which injected 20 Tg of SO_2 (~ 10 Mt S) into the stratosphere.[24]

During the last two large eruptions, the large amount of injected sulfur resulted in enhanced surface area densities (SAD) of liquid sulfate aerosols in the stratosphere. SAD values for volcanically quiescent conditions are less than $2\,\mu m^2\ cm^{-3}$ in the lower stratosphere. During background conditions the partial pressures of sulfur gases remain relatively low and the particles are quite small,[25] with typical sizes that can be described by a lognormal distribution with an effective radius of $0.17\,\mu m$. Six weeks after volcanic eruptions, when sulfur species concentrations become much higher, the particles grow to much larger sizes and the effective radius was estimated to be 0.43 microns, with some uncertainty.[26–28] After the volcanic eruption of El Chichón in 1982, enhanced SAD values were observed, and somewhat elevated amounts were sustained in the following years by the eruptions of Negra in 1979 and Ulawan in late 1980.[29] The largest increase in recent decades in liquid sulfate aerosol surface area density (SAD), up to $50 \, \mu m^2 \, cm^{-3}$ at altitudes between 15 and 25 km, was observed in 1991 after the eruption of Mount Pinatubo, as derived from 1 μm extinction measurements from the SAM II and SAGE I and II satellites (SPARC 1996).

10.2.1 Impact on Climate and Stratospheric Dynamics

Large particles, as observed after volcanic eruptions, scatter and absorb in the shortwave part of the solar spectrum. This effect resulted in a decrease of global mean temperatures for a few years after the eruption of Mount Pinatubo.[30] Effects of the Mt. Pinatubo eruption on the hydrological cycle were also found,[14] as a substantial decrease in precipitation over land and a record decrease in runoff and river discharge into the ocean from October 1991 to September 1992.

Besides the impact on the shortwave part of the solar spectrum volcanic aerosols also absorb in the infrared and trap some outgoing energy, which results in increasing temperatures in the lower stratosphere.[22,26] The increase of sulfate aerosols in the stratosphere after the eruption of Mount Pinatubo has shown a clear imprint on stratospheric temperatures as shown in Figure 10.1. Tropical temperatures were 1–2 K higher compared to the years with low sulfur burden in the stratosphere. The resulting increase of the temperature gradient between low and high latitudes resulted in a strengthening of the stratospheric winter polar vortex.[31]

10.2.2 Impact on Stratospheric Ozone

The increase of halogen compounds in the stratosphere has led to major ozone depletion in polar regions and to some extent in mid-latitudes (see Chapter 5 and 6). In addition, enhanced amounts of liquid sulfate aerosols impact the ozone layer (Chapter 4) and accelerates Antarctic and mid-latitude ozone depletion.[19,21]

Enhanced sulfate particles result in a decrease of the NO_x/NO_y equilibrium ratio due to increasing heterogeneous reactions:[32,33]

(1) $N_2O_5 + H_2O \rightarrow 2HNO_3$
(2) $ClONO_2 + H_2O \rightarrow HOCl + HNO_3$ $T < 200$ K as important as (1) $\rightarrow$
     increase in ClO_x and HO_x

This process decreases the abundance of reactive nitrogen in the lower stratosphere (eqn 1), but increases HOCl and therefore the abundance of

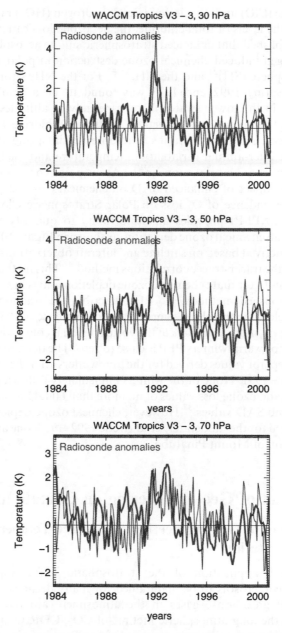

Figure 10.1 Temperature anomalies from the radiosonde observations for the period 1984–2000[55] for different pressure levels. The monthly averaged anomalies of observations are based on 14 tropical radiosonde stations between 30°N and 30°S (grey lines). In addition, temperature differences from an NCAR WACCM model simulation for the same period, once including volcanic-sized aerosols and an aerosols heating scheme in the simulations and once without volcanic aerosols, are overlayed (black lines) (Adapted from Tilmes *et al.*[22])

reactive chlorine (ClO) and, to a lesser extend, hydrogen (HO_x) radicals (eqn 2). Depending on the levels of total chlorine, the net ozone loss can be accelerated.

It is well established that enhanced stratospheric sulfate aerosol loading leads to greater halogen-induced chemical ozone destruction in polar regions of the southern hemisphere (SH)[21] and the NH.[34–37] For the NH, strongly enhanced ozone loss in spring 1992 and 1993 was found to be a result of strongly enhanced SAD in the lower stratosphere, especially at altitudes below 16 km (425 K potential temperature).[38] Further, significant chlorine activation and strongly enhanced ClO_x mixing ratios were observed in spring 1992.[39–43]

However, the inter-annual variability in temperature and the large influence of transport processes on total ozone in the Arctic polar vortex make it difficult to quantify the impact of enhanced SAD on chemical ozone depletion. Using the observed dependence of O_3 loss on Polar Stratospheric Cloud (PSC) formation potential (PFP) provides an opportunity to quantify the impact of enhanced SAD on chemical ozone depletion, as shown in Figure 10.2.[37] Chemical ozone loss was derived based on satellite and aircraft observations between 1991 and 2005 using the tracer-tracer correlations method.[41] This method was applied in various studies to estimate chemical ozone depletion and is in good agreement with other established methods.[44,45] PFP describes the fraction of the polar region that is below the threshold temperature for existence of PSCs (T_{PSC}). For cold Arctic winters, more than 10% of the vortex region is cold enough to support PSCs, whereas for warm winters PFP is close to zero. The derived linear relation is compact, except for values derived for the four winters after the eruption of Mt. Pinatubo in June 1991. None of the winters immediately following the eruption were very cold, and ozone loss values of no more than 60 DU would be expected from background SAD values.[37] However, chemical ozone depletion up to 120 DU was observed for the moderately cold Arctic 1992–1993, one and a half years after the eruption of Mount Pinatubo.

10.3   Impact of Geo-engineering on Climate and Ozone

10.3.1   Impact of Geo-engineering on Surface Temperature and Precipitation

The hypothetical application of the stratospheric sulfur approach would require enhanced aerosols to be maintained for many decades to reduce temperatures while greenhouse gases in the atmosphere are reduced by other means. Due to the long atmospheric lifetime of CO_2, CDR would be required to reduce levels significantly on timescales shorter than centuries. Model studies have been performed to investigate the impact of SRM in different ways. General Circulation Models (GCM) were employed to calculate the response of reduced incoming solar radiation on the climate system and therefore the impact on surface temperatures using medium complexity General Circulation Models (GCMs).[46–48] The enhanced planetary albedo leads to a reduction of global temperature and, therefore, counteracts global warming; however, local

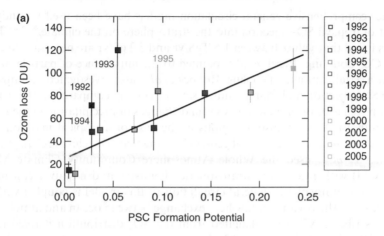

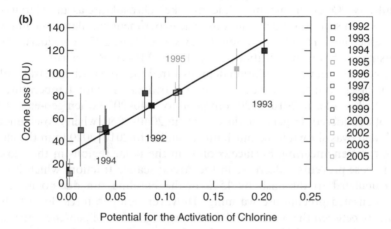

Figure 10.2 Panel A: relationship between chemical loss of Arctic ozone in DU (Dobson Unit, which equals 2.687×10^{16} molecules/cm^2) and PSC formation potential (PFP), averaged between 380–550 K potential temperature and mid-December and March, for several winters between 1992 to 2005. A linear fit (black line) was derived, excluding the years 1992 and 1993 after the Mt Pinatubo eruption. Panel B: as in A, but using the potential for the activation of chlorine (PACl) (Tilmes *et al.*, 2007) instead of PFP as the abscissa. PACl includes sulfate aerosol density (SAD) in its formulation. Data for winters 1992 to 1995 (denoted), which had high SAD due to the eruption of Mt. Pinatubo, now fall along the compact, near linear relation once the effect of SAD on chlorine activation is considered. Adapted from Tilmes *et al.*[44]

temperature changes, can differ from the global-mean behavior for example, increasing temperatures in high latitudes, were calculated. Bala *et al.*[17] investigated the impact of solar dimming on precipitation and found a decrease in global mean precipitation.

Other comprehensive general circulation models have been used to study the impact of tropical SO_2 injection into the stratosphere on the climate.[8,16,23] These studies found that values between 1.5 TgS/yr and 5 TgS/yr are necessary to counteract CO_2 doubling. Rasch et al.[23] pointed to the importance of particle sizes to achieve a certain amount of cooling. Robock et al.[16] also investigated the impact of Arctic SO_2 injection and found a much smaller cooling compared to the tropical injection scenario. Both these studies found changes of precipitation in the Indian and Southeast Asian monsoon regions in the case of a tropical injection, with perturbations smaller in the case of geo-engineering than with climate change alone.

Tilmes et al.[22] used the Whole Atmosphere Community Climate Model (WACCM) with interactive stratospheric chemistry and a slab ocean model (SOM). The atmospheric general circulation in this model is coupled with the upper layer of the ocean to allow heat exchange between ocean and atmosphere. The prescribed SAD was calculated from the SO_4 distribution derived in the study of Rasch et al.[23] assuming 2 Tg S/a of volcanic aerosols to be injected into a present-day CO_2 environment. "Volcanic-like" aerosols are assumed to have an effective radius of about 0.43 microns. The experiment simulated the impact of geo-engineering on future conditions (between 2020 and 2050) considering future chemistry and halogen conditions of the IPCC AR4 A1B scenario.

The geo-engineering simulation demonstrated that the artificial stratospheric aerosol layer counteracts global surface warming. Surface temperatures are about 0.5 K cooler in 2010–2020 compared to 2040–2050 when geo-engineering was not considered. In general, the climate in 2040–2050, when geo-engineering is included, was shown to be much more similar to 2010–2020 than the climate without geo-engineering. Surface cooling in the polar regions in the geo-engineering case prevents a decrease in the Arctic sea-ice fraction, which is otherwise calculated to diminish by 15% in the baseline run due to increasing, uncompensated greenhouse warming. However, temperatures do not change uniformly between the present and the future in the case of geo-engineering, but local changes up to 3 K occur, as also simulated by Robock et al.[23] A comparison of model results between the Community Atmospheric Model (CAM), which does not include any changes of stratospheric ozone due to geo-engineering, and WACCM shows strong differences in the tropospheric temperature response at high latitudes, particularly in winter (Chapter 7). This illustrates the importance of stratospheric processes in simulating the impact on the climate of this specific geo-engineering approach.

The studies discussed are based on different model setups, which can account at least in part for differences in predicted temperature changes and precipitation patterns. A Geoengineering Model Intercomparison Project has been proposed (GeoPIP) that prescribes uniform experiments and should lead to a less ambiguous comparison of results.[49]

10.3.2 Impact on Atmospheric Temperatures and Dynamics

The injection of geo-engineered aerosols would result in a cooling in the troposphere and a warming in the stratosphere in the layer of enhanced SAD.

The constant enhanced aerosol loading in the stratosphere, starting in 2020, as simulated by Tilmes *et al.*,[22] reduced the tropospheric temperatures in 2040–2050 to values slightly below 2010–2020 conditions (see Figure 10.3). On the other hand, the heating of the stratosphere, especially at about 20–30 km in middle and low latitudes, counteracts the cooling of the stratosphere due to increasing greenhouse gases. Depending on altitude, the baseline run shows monotonic temperature evolution with increasing tropospheric and decreasing stratospheric temperatures, depending on altitude. With the start of geo-engineering in 2020, tropospheric temperatures decrease for five years following the abrupt change of incoming solar radiation at the Earth's surface, until they stabilize with anomalies of about −0.8 K near the surface and −2 K near 10 km, compared to the baseline run. The stabilization of tropospheric temperatures after five years reflects the time-scale of the temperature response of the surface layers of the ocean to the aerosol-induced change in incoming radiation. The achieved cooling of tropospheric temperatures, as a result of a constant amount of sulfur injection, is eventually overwhelmed by 2050 due to continuously increasing greenhouse gases in this simulation. To maintain constant tropospheric temperatures, increasing injections of aerosols would be needed to counteract the continuous rise in greenhouse gases.

The study by Tilmes *et al.*,[22] as well as earlier studies, has shown that temperature changes in the stratosphere as a response to the enhancement of aerosol loading strengthen the polar vortex, as observed after volcanic eruptions. Tilmes *et al.*[22] also investigated average polar vortex temperatures and found a decrease of temperatures in the Arctic polar vortex for the geo-engineering experiment, which likely result in cooler surface temperatures as a result of geo-engineering compared to those studies that did not include stratospheric processes. In contrast, if geo-engineering was not included in the simulation, increasing vortex temperatures were found between 2010 and 2050.

10.3.3 Impact of Geo-engineering on Stratospheric Ozone

The impact of sulfur particles on ozone strongly depends on the amount of halogens in the stratosphere. Over the next half a century, the stratospheric halogen loading, commonly quantified in terms of Effective Equivalent Stratospheric Chlorine (EESC), is projected to slowly decline[50] (see also Chapter 2). The question is, therefore, how does an enhanced burden of stratospheric sulfur impact the ozone layer, especially in high latitudes, and how does it delay the expected recovery of the ozone hole brought about by the reduction of the atmospheric halogen burden, as prescribed by the Montreal Protocol (see Chapter 2).

Further, stratospheric chemistry and dynamics strongly depend on the amount and location of aerosols injected into the stratosphere and the resulting particle size distribution of the aerosols, which defines the SAD of sulfuric aerosols. Detailed studies using a 2D and 3D microphysical model investigated different injection scenarios of SO_2.[8,51] Those studies found that the tropical

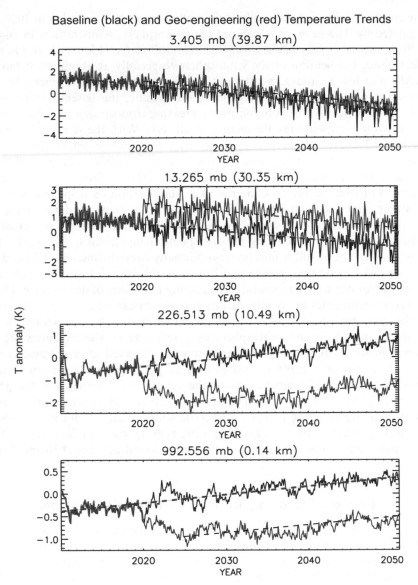

Figure 10.3   Temperature anomalies (between 2010 and 2050) for the geo-engineering
run (red) starting in 2020, and the baseline run (black) in the tropics
(between 22°N and 22°S) for different altitudes. The reference state is the
mean temperature of the baseline model simulation between 2010 and
2050. Adapted from Tilmes et al.[22]

injection of SO_2 would likely result in aerosol size similar to or larger than those
observed after volcanic eruptions. The coagulation processes might limit the
amount of cooling possible to counteract global warming, because larger
particles scatter incoming short-wave radiation less efficiently. The size of the

potentially injected particles and, therefore, the amount of reduced solar forcing depends on injection scenarios like timing and location in the horizontal and vertical. Pierce *at al.*[14] and Niemeyer *et al.*[51] found that the injection of H_2SO_4 into the stratosphere instead of SO_2 would result in smaller particles that would produce more cooling. Also, an injection in a wider area would help to reduce the particle size.[13]

The impact of enhanced SAD on stratospheric chemistry and dynamics, assuming a constant size distribution of liquid sulfate aerosol particles, was studied by Tilmes *et al.*,[22] as described above. This study is a simple approximation to reality, considering the large standard deviation of the size distribution of volcanic-sized aerosols. Heckendorn *et al.*[13] considered a different aerosol distribution, as derived from their microphysical model; however, climate and temperature response are unrealistic due to fixed ocean temperatures. The results of these studies are summarized in the following.

Global changes in stratospheric ozone as a result of geo-engineering were shown to be a combination of changes in chemistry and dynamics.[13,22] In the tropics and mid-latitudes, chemical changes are dominated the NO_x cycle. In the geo-engineering run, this cycle becomes less effective because enhanced heterogeneous reactions shift the ratio NO_x/NO_y towards NO_y (in particular $N_2O_5 + H_2O \rightarrow 2\,HNO_3$, eqn (1)), with an increasing effect towards the upper region of the enhanced aerosol layer.[32] Other chemical cycles like the ClO_x, BrO_x, and HO_x cycles become slightly more effective in the geo-engineering model run and therefore tend to reduce the total photochemical lifetime of ozone, as observed after volcanic eruptions.[20] The combined effect of the chemical cycles depends on the halogen loading in the stratosphere. For 2040–2050 halogen conditions geo-engineering results in an increase in the lifetime of ozone.[22] On the other hand, the increase of ozone due to enhanced heterogeneous reactions around 30 km, results in an enhanced absorption of UV radiation and, therefore, a reduction of ozone at altitudes 22–26 km as a result of geo-engineering. For the Tropics, ozone mixing ratios are also influenced by a weaker tropical upwelling in the geo-engineering run compared to the baseline run, because the troposphere becomes much cooler and the mechanism for the acceleration of the Brewer-Dobson circulation—namely enhanced driving of the circulation by planetary waves—proposed by Garcia and Randel[52] is less effective (see also Chapter 9). In the polar regions, significant increase in the heterogeneous chlorine activation results in accelerated halogen catalyzed ozone loss as a result of geo-engineering, centered around 16–17 km. Heckendorn *et al.*[13] simulated a significant warming of the tropical tropopause due to the presence of enhanced aerosols as a result of geo-engineering for present day conditions. Since increasing temperatures at the tropopause are correlated to the amount of water vapor in the stratosphere, geo-engineering led to a moistening of the stratosphere.

Changes in the lifetime of ozone show up in the global distribution of column ozone. In a non-geoengineered environment, Tilmes *et al.*[22] found column ozone increases in middle and high latitudes up to 25% due to increasing greenhouse gases and their impact on temperature, chemistry and transport.

In the geo-engineering case for 2040–2050 halogen conditions, the depth of the ozone column shows little change near the Equator (between 10°N and 10°S) compared to the baseline case. In the region between about 10° and 30° North and South, the impact of different processes on ozone, dominated by the reduced NO_x ozone-destroying cycle, results on average in slightly (up to 2%) larger column ozone values in the geo-engineering case compared to the baseline run. In high southern latitudes, the annually averaged column ozone for the geo-engineering run is up to 10% smaller and in high northern latitudes about 2% smaller compared to the baseline run for the period between 2040 and 2050.

In addition to the pronounced chemical ozone loss in high latitudes, Heckendorn et al.[13] found a remarkable (between 3.5 and 4.3%) decrease in column ozone in the tropics, depending on the geo-engineering scenarios, for present day halogen conditions. Enhanced water vapor in the stratosphere caused by geo-engineering further decreases the NO_y/NO_x equilibrium ratio and increases the importance of the ozone-destroying ClO_x and HO_x cycles.

10.3.4   Impact of Geo-engineering on Polar Ozone

To quantify the influence of increasing heterogeneous reactions in the geo-engineering run and changes in the stratospheric circulation and temperatures in high polar latitudes, Tilmes et al.[22] investigated the annual evolution of monthly averaged column ozone values poleward of 70° S and 70° N equivalent latitudes (Figure 10.4, b and c). For Antarctica, a significant decrease in ozone column between the geo-engineering and baseline runs occurs between October and December, reaching 15%. Over the rest of the year, 5–10% smaller column ozone values were simulated in the geo-engineering case. Geo-engineering would delay the recovery of ozone in high latitudes by about 30 years. Note that, in the model simulation, the recovery of polar ozone occurs 20 years later compared to results based on the evolution of the effective equivalent stratospheric chlorine.[50] In the Arctic, column ozone values increase significantly between 2010–2020 and 2040–2050 during winter and spring. In the geo-engineering case, column ozone values in 2040–2050 between January and March are up to 4% smaller in the baseline run. Since ozone depletion for the NH is largely underestimated in this model simulation compared to observations (as further discussed in Chapter 9), the response of geo-engineering is likely underestimated as well.

Tilmes et al.[44] have used the relationship between chemical ozone depletion and the potential for chlorine activation (PACl) for the past to estimate the future impact of geo-engineering on chemical ozone depletion in polar regions. PACl is similar to PFP (Figure 10.2 a) but uses a threshold temperature for chlorine activation, T_{ACl}, described by Drdla and Müller,[53] which is a function of temperature, ambient H_2O and SAD. For volcanically quiescent conditions and present day values of EESC, PACl is comparable to PFP. Following a strong volcanic eruption, PACl is larger than PFP because enhance SAD result

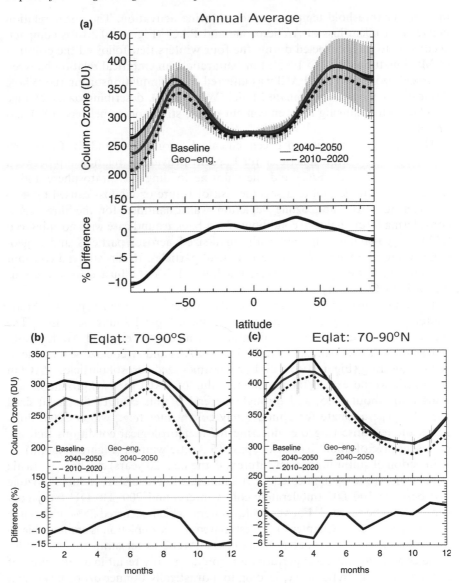

Figure 10.4 Panel (a): Decadal averages (dashed: 2010–2020; solid: 2040–2050) of column ozone (DU) as functions of latitude for the baseline (black) and geo-engineering (solid red: 2040–2050) runs, including the standard deviation (error bars), as well the difference for 2040–2050 (lower part of panel a). Panels (b) and (c): Decadal averages (dashed: 2010–202; solid: 2040–2050) of column ozone (DU) averaged between 70–90°S equivalent latitudes (panel b) and between 70–90°N equivalent latitudes (panel c) for baseline (black) and geo-engineering (solid red: 2040–2050) runs, including the standard deviation (error bars), as well as the differences for 2040–2050 (lower part of panels b and c). Adapted from Tilmes *et al.*[22]

in a higher threshold temperature for chlorine activation. The linear relation between chemical loss of Arctic ozone and PACl (Figure 10.2 b) is compact, according to data collected during the four winters that followed the eruption of Mt. Pinatubo (1992 to 1995). For Antarctica, an empirical relation between chemical ozone loss and PACl was inferred from simulations using the Whole Atmosphere Chemistry Climate Model (WACCM3), describing ozone change in relation to changing halogen content of the stratosphere between 1960 and present.

The empirical relation between chemical ozone loss and PACl (which incorporates the stratospheric sulfur burden) provides a relationship between chemical ozone loss, SAD and the chlorine loading the stratosphere and is therefore a suitable tool to assess the risk of future ozone loss caused by geo-engineering. PACl and ozone depletion was quantified for the three cases considering future halogen conditions, for a background case with no enhanced SAD, a geo-engineering case with small-sized aerosol particles and a geo-engineering case with volcanic-sized aerosol particles. In this study, a constant amount of SAD enhancement is assumed, which is not adjusted to continuous increasing CO_2 mixing ratios.

Three meteorological conditions were considered, a recent very cold Arctic winter, a moderately cold Arctic winter, and a typical Antarctic winter. The scenario using meteorological conditions for a recent, very cold Arctic winter characterizes the maximum perturbation to PACl and Arctic ozone loss due to geo-engineering (Figure 10.5a, c). For volcanic-sized aerosol particles, PACl in the Arctic would exceed the maximum value of PACl for background conditions until about 2055, and through the end of this century for small-sized particles. The estimate for a moderately cold winter (*e.g.*, winter 2003) illustrates the response to geo-engineering for meteorological conditions that are representative of about half of the past 15 Arctic winters (Figure 10.5b, d).

Injection of sulfur in the near future (*i.e.* the next 20 years) can have a drastic impact on Arctic ozone depletion. If small-sized aerosols are assumed, ozone loss between 100 DU (moderately cold winters) and 200–230 DU (very cold winters) can be reached. For very cold winters, which occurred 25% of the time in the past 15 years (Chapter 5), the estimated ozone depletion is comparable to the total amount of available ozone in the Arctic lower stratosphere. Under these conditions, the SAD perturbation could possibly result in a saturation of chemical loss of Arctic ozone, leading to a drastically thinner ozone layer than presently observed. For the SAD perturbation associated with volcanic-sized aerosols, chemical loss of Arctic ozone could exceed 150 DU during very cold winters and 70 DU for moderately cold winters.

The doubling of CO_2 with respect to pre-industrial values is expected to occur between 2050–2100. In this time frame and considering the reduced estimated halogen loading in the future, chemical ozone loss due to geo-engineering reaches 125–150 DU for very cold Arctic winters, compared to 80 DU for the background case. For either geo-engineering case, ozone depletion would exceed presently observed values because the recent spate of very cold Arctic winters occurred for low (*i.e.*, near background) values of SAD.

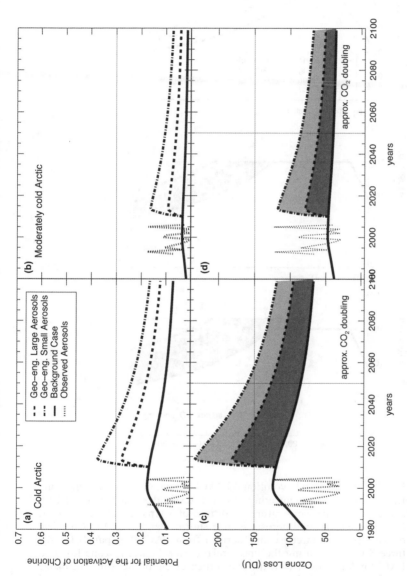

Figure 10.5 Top panels: the temporal development of PACl taking into account changing EESC for two geo-engineering cases (see text and legend), between 2010 and 2050, using observed temperatures for a very cold Arctic winter (panel a) and a moderately cold Arctic winter (panel b). The temporal evolution of PACl for background SAD (solid line), taking into account changing EESC, is also shown. Finally, values of PACl based on observed SAD, temperature, and EESC, are shown (dotted lines). Bottom panel: chemical ozone loss *versus* time, derived from PACl (top panels) for the various SAD cases, is shown for meteorological conditions corresponding to a very cold Arctic winter (panel c) and a moderately cold Arctic winter (panel d). The ozone loss estimates are based on the linear relationship between chemical loss and PACl for the Arctic (see text). Adapted from Tilmes *et al.*[44]

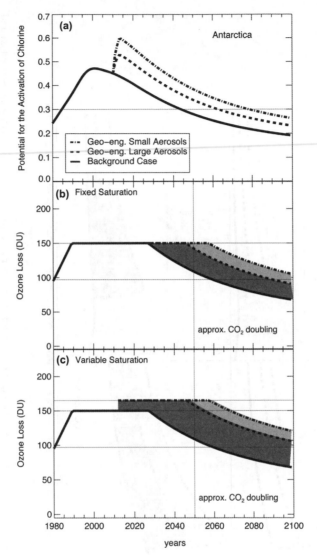

Figure 10.6   Panel a: the temporal evolution of PACl, taking into account changing
EESC, for two geo-engineering cases (see text and legend), between 2010
and 2050, for the temperature conditions of a typical Antarctic winter.
PACl for background aerosol is also shown. Panel b and c: Chemical
ozone loss of Antarctic ozone derived from PACl (top panel) for the
three SAD cases, using the linear relationship between ozone loss and
PACl for Antarctica (see text). The vertical extent of the Antarctic ozone
hole is either assumed to remain fixed at present levels, denoted as "Fixed
Saturation" (panel b) or is assumed to extend vertically downward by
~2 km (see text), denoted as "Variable Saturation" (panel c). Adapted
from Tilmes *et al.*[44]

A moderately cold winter would reach 60 to 80 DU of ozone depletion, a value observed in the past only for very cold Arctic winters. Therefore, ~75% of all winters would result in ozone loss of at least 60 to 80 DU, and possibly as high as 150 DU, if geo-engineering through stratospheric sulfur injections is used to mitigate global warming.

Results for two different assumptions for Antarctica are shown in Figure 10.6: either future chemical loss of Antarctic ozone due to geo-engineering will continue to saturate at present-day values of 150 DU[54] or else future chemical loss will saturate at higher pressure levels, due to the likely strong increase of SAD at 10–12 km of the Antarctic stratosphere, where ozone loss is presently not saturated (Figure 10.6c), as observed after the eruption of Mt Pinatubo.[56] In Figure 10.6c it is assumed that geo-engineering will lead to a downward extension of the region of ozone loss, adding another 15 DU (the amount between 10–12 km) to the total column abundance of ozone that could be lost.

The injection of aerosol particles would therefore result in a delay to the recovery of the Antarctic ozone hole.[44] The time when Antarctic ozone loss drops below saturation in late winter would be delayed by 15–45 years assuming volcanic-sized particles and by more than 30–60 years assuming small-sized particles. Taking the additional ozone loss at higher pressure levels into account (Figure 10.6b). The recovery of Antarctic ozone to conditions that prevailed in 1980 would be delayed at least until the last decade of this century by geo-engineering.

Acknowledgements

We thank Phil Rasch and Mike Milles for a careful review or this chapter and well as Rolf Müller for helpful discussions. The National Center for Atmospheric Research is operated by the University Corporation for Atmospheric Research under sponsorship of the National Science Foundation.

References

1. J. R. Fleming, *Meteorology in America, 1800–1870,* pp. 24–31, Johns Hopkins University Press, Baltimore, 1990, pp. 24–31.
2. N. Rusin and L. Flint, *Man versus climate.* (Transl. Drian Rottenberg), In Peace Publishers, Moscow, 1960.
3. *Geoengineering the Climate: science, governance and uncertainty*, September 2009, ISBN: 978-0-85403-773-5, The Royal Society, 2009, p. 82.
4. B. Franklin, *Manchester Literary and Philosophical Society Memoirs and Proc*, 1784, **2**, 122.
5. WMO (World Meteorological Organization), Scientific Assessment of Ozone Depletion: 2006, Global Ozone Research and Monitoring Project–Report 51, Geneva, Switzerland, 2010.
6. M. I. Budyko, *Tellus*, 1977, **29**, 193–204.
7. P. J. Crutzen, *Clim. Change*, 77(3–4), 211–220, DOI:10.1007/s10584-006-9101-y, 2006.

8. T. M. Wigley, *Science*, 2006, **314**, 452–454.
9. P. J. Rasch, S. Tilmes, R. P. Turco, A. Robock, L. Oman, C.-C. Chen, G. L. Stenchikov and R. R. Garcia, *Phil. Trans. Royal Soc. A.*, 1882, **366**, 4007–4037, DOI: 10.1098/rsta.2008.0131, 2008b.
10. T. M. Lenton and N. E. Vaughan, *Atmos. Chem. Phys.*, 2009, **9**(15), 5539–5561, DOI:10.5194/acp-9-5539-2009.
11. A. Robock, A. B. Marquardt, B. Kravitz and G. Stenchikov, *Geophys. Res. Lett.*, 2009, **36**, L19703, DOI: 10.1029/2009GL039209.
12. D. W. Keith, E. Parson and M. G. Morgan, *Nature*, **463**, 426–427, DOI: 10.1038/463426a, 2010.
13. P. Heckendorn, D. Weisenstein, S. Fueglistaler, B. P. Luo, E. Rozanov, M. Schraner, L. W. Thomason and T. Peter, *Env. Res. Lett.*, 2009, **4**, 045108.
14. J. R. Pierce, D. K. Weisenstein, P. Heckendorn, T. Peter and D. W. Keith, *Geophys. Res. Lett.*, 2010, **37**, L18805, DOI:10.1029/2010GL043975.
15. K. E. Trenberth and A. Dai, *Geophys. Res. Lett.*, **34**, L15702. (DOI:10.1029/2007GL030524).
16. A. Robock, L. Oman and G. Stenchikov, *J. Geophys. Res.*, 2008, **113**, (D16101).
17. G. Bala, P. B. Duffy and K. E. Taylor, *Proc. Natl. Acad. Sci., U. S. A.*, 2008, **105**(22) 7664–7669, DOI:10.1073/pnas.0711648105.
18. G. C. Hegerl and S. Solomon, *Science*, 2009, **325**(5943), 955–956, DOI: 10.1126/science.1178530.
19. S. Solomon, R. Portmann, R. R. Garcia, L. Thomason, L. R. Poole and M. P. McCormick, *J. Geophys. Res.*, **101**, 6713–6727.
20. S. Solomon, *Rev. Geophys.*, 1999, **37**, 275–316.
21. R. W. Portmann, S. Solomon, R. R. Garcia, L. W. Thomason, L. R. Poole and M. P. McCormick, *J. Geophys. Res.*, 1996, **101**(D17), 22991–23006.
22. S. Tilmes, R. R. Garcia, D. E. Kinnison, A. Gettelman and P. J. Rasch, *J. Geophys. Res.*, 2009, **114**, D12305, DOI: 10.1029/2008JD011420.
23. P. J. Rasch, P. J. Crutzen and D. B. Coleman, *Geophys. Res. Lett.*, 2008a, **35**, L02809, DOI: 10.1029/2007GL032179.
24. A. Robock, *Rev. Geophys.*, **38**, 191–219. DOI:10.1029/1998RG000054, 2000.
25. J. J. Bauman, P. B. Russell, M. A. Geller and P. Hamill, *J. Geophys. Res.*, 2003, **108**, 4383. (DOI:10.1029/2002JD002993).
26. G. L. Stenchikov, I. Kirchner, A. Robock, H. F. Graf, J. C. Antuna, R. G. Grainger, A. Lambert and L. Thomason, *J. Geophys. Res.*, 1998, **103**, 13 837–13 857. (DOI:10.1029/98JD00693).
27. W. D. Collins, P. J. Rasch, B. A. Boville, J. R. McCaa, D. L. Williamson, J. T. Kiehl, B. Briegleb, C. Bitz, S.-J. Lin, M. Zhang and Y. Dai, Description of the NCAR community atmosphere model: CAM3.0. Technical report NCAR/TN-464CSTR, National Center for Atmospheric Research, Boulder, Colorado, USA, pp. 226. See http://www.ccsm.ucar.edu/models/atm-cam, 2004.

28. H. M. Steele and R. P. Turco, *J. Geophys. Res.*, 1997, **102**, 19 665–19 681. (DOI:10.1029/97JD01263).

29. M. H. Hitchman, M. McKay and C. R. Trepte, *J. Geophys. Res.*, 1994, **99**, 20 689–20 700.

30. B. J. Soden, R. T. Wetherald, G. L. Stenchikov and A. Robock, *Science*, 2002, **296**(5568), 727–730, DOI: 10.1126/science.296.5568.727.

31. G. L. Stenchikov, A. Robock, V. Ramaswamy, M. D. Schwarzkopf, K. Hamilton and S. Ramachandran, *J. Geophys. Res.* **107**, 0.1029/2002JD002090, 2002.

32. D. W. Fahey, S. R. Kawa, E. L. Woodbridge, P. Tin, J. C. Wilson, H. H. Jonsson, J. E. Dye, D. Baumgardner, S. Borrmann and D. W. Toohey, *Nature*, 1993, **363**, 509–514.

33. M. J. Mills, A. O. Langford, T. J. O'Leary, K. Arpag, H. L. Miller, M. H. Proffitt, R. W. Sanders and S. Solomon, *Geophys. Res. Lett.*, 1993, **20**, 1187–1190.

34. R. Cox, A. MacKenzie, R. Muller, T. Peter and P. Crutzen, *Geophys. Res. Lett.*, 1994, **21**(13), 1439–1442.

35. A. Tabazadeh, K. Drdla, M. R. Schoeberl, P. Hamill and O. B. Toon, *Proc. Natl. Acad. Sci., U. S. A.*, 2002, **99**(5), 2609–2612.

36. M. Rex, R. J. Salawitch, H. Deckelmann, P. von der Gathen, N. R. P. Harris, M. P. Chipperfield, B. Naujokat, E. Reimer, M. Allaart, S. B. Andersen, R. Bevilacqua, G. O. Braathen, H. Claude, J. Davies, H. De~Backer, H. Dier, V. Dorokov, H. Fast, M. Gerding, S. Godin-Beekmann, K. Hoppel, B. Johnson, E. Kyrö, Z. Litynska, D. Moore, H. Nakane, M. C. Parrondo, A. D. Risley Jr., P. Skrivankova, R. Stübi, P. Viatte, V. Yushkov and C. Zerefos, *Geophys. Res. Lett.*, 2006, **33**.

37. S. Tilmes, R. Müller, R. J. Salawitch, U. Schmidt, C. R. Webster, H. Oelhaf, C. C. Camy-Peyret and J. M. Russell III, *Atmos. Chem. Phys.*, 2008, **8**, 1897–1910, DOI:10.5194/acp-8-1897-2008.

38. E. V. Browell, C. F. Butler, M. A. Fenn, W. B. Grant, S. Ismail, M. Schoeberl, O. B. Toon, M. Loewenstein and J. R. Podolske, *Science*, 1993, **261**, 1155–1158.

39. J. W. Waters, L. Froidevaux, W. G. Read, G. L. Manney, L. S. Elson, D. A. Flower, R. F. Jarnot and R. S. Harwood, *Nature*, 1993, **362**, 597–602.

40. D. W. Toohey, L. M. Avallone, L. R. Lait, P. A. Newman, M. R. Schoeberl, D. W. Fahey, E. L. Woodbrige and J. G. Anderson, *Science*, 1993, **261**, 1134–1136.

41. M. H. Proffitt, K. Aikin, J. J. Margitan, M. Loewenstein, J. R. Podolske, A. Weaver, K. R. Chan, H. Fast and J. W. Elkins, *Science*, 1993, **261**, 1150–1154.

42. R. J. Salawitch, S. C. Wofsy, E. W. Gottlieb, L. R. Lait, P. A. Newman, M. R. Schoeberl, M. Loewenstein, J. R. Podolske, S. E. Strahan, M. H. Proffitt, C. R. Webster, R. D. May, D. W. Fahey, D. Baumgardner, J. Dye, J. C. Wilson, K. K. Kelly, J. W. Elkins, K. R. Chan and J. G. Anderson, *Science*, 1993, **261**, 1146–1154.

43. R. Brandtjen, T. Klüpfel, D. Perner and B. Knudsen, *Geophys. Res. Lett.*, 1994, **21**(13), 1363–1366.
44. S. Tilmes, R. Müller, J.-U. Grooß and J. M. Russell, *Atmos. Chem. Phys.*, 2004, **4**(8), 2181–2213.
45. R. Müller, S. Tilmes, P. Konopka, J.-U. Grooß and H.-J. Jost, *Atmos. Chem. Phys.*, 2005, **5**, 3139–3151.
46. B. Govindasamy and K. Caldeira, *Geoph. Res. Lett.*, 2000, **27**, 14.
47. H. D. Matthews and K. Caldeira, *Proc. Natl. Acad. Sci., U. S. A.*, 2007, **104**(24), 9949–9954.
48. K Caldeira, L. Wood, *Philos. Trans. R. Soc. London, Ser. A*, 2008, **366**(1882), 4039–4056, DOI: 10.1098/rsta.2008.0132.
49. B. Kravitz and A. Robock, *J. Geophys. Res.*, 2011, **116**, D01105, DOI:10.1029/2010JD014448.
50. P. A. Newman, J. S. Daniel, D. W. Waugh and E. R. Nash, *Atmos. Chem. Phys.*, 2007, **7**, 4537–4552.
51. U. Niemeier, H. Schmidt and C. Timmreck, *Atmos. Sci. Lett.*, 2010, DOI: 10.1002/asl.304.
52. R. R. Garcia and W. J. Randel, *J. Atmos. Sci.*, 2008, **65**, 2731–2739.
53. K. Dradla and R. Müeller, *Atmos. Chem. Phys. Discuss.*, 2010, **10**, 28687–28720.
54. S. Tilmes, R. Müller, A. Engel, M. Rex and J. M. Russell III, *Geophys. Res. Lett.*, 2006, **33**, L20812, DOI:10.1029/2006GL026925.
55. W. J. Randel, M. Park and F. Wu, *J. Atmos. Sci.*, 2008, **64**, 4479–4488.
56. D. Hofmann and S. Oltmans, *J. Geophys. Res.*, 1993, **98** (D10), DOI: 10.1029/93JD02092.

Subject Index

AAOE (Airborne Antarctic Ozone Experiment), 18

Abbatt, J. P. D., 135–137

Absolute absorption cross, 94, 95

Absolute absorption cross sections, 94

Absorption cross sections
chlorine peroxide, 93–95
defined, 43

Abundances, 83–85

ACC (Antarctic Circumpolar Current), 199–200

ACE-FTS. *See* Atmosphere Chemistry Experiment-Fourier Transform Spectrometer

Actinic flux, 93

Active chlorine (ClO$_x$), 4
in catalytic cycles, 90
conversion into, 98–99
deactivation of, 146
in denitrified air masses, 129
from ozone destruction cycles, 90
partitioning of, 92
and polar ozone loss, 146
in polar regions, 147

Active hydrogen (HO$_x$), 4
in catalytic chemical loss cycles, 169
from HOBr photolysis, 89
from ozone destruction cycles, 90
source of, 183

Active nitrogen (NO$_x$), 4
in catalytic cycles, 90, 169
from nitrous oxide, 63

NO$_x$/NO$_y$ equilibrium ratio, 282, 284, 289, 290
from photochemical destruction of nitrous oxide, 66
in polar ozone loss cycle, 132, 133
and sedimentation of PSC particles, 18
and tropospheric ozone production, 63

Adiabatic heating, 223

Adriani, A., 124

Advanced Global Atmospheric Gas Experiment (AGAGE), 44, 46, 47, 53, 61

Advection processes, 7

Aerosol layer, 109–110
artificial, 280, 286
Junge, 6, 67, 110, 113

Aerosol particles, 62–63
in chemical ozone destruction, 108
as cloud condensation nuclei, 109
enhancing albedo with, 280
heterogeneous reactions on, 6, 19–20
homogeneous nucleation rates of, 121
and inorganic brominated chemicals, 67
and inorganic chlorine, 66–67
sulfur-containing gases in, 37
transformation of chlorine by, 108

Aerosols. *See also* sulfate aerosols
　　and chemistry of lower
　　　　stratosphere, 256
　　and chlorine activation, 112
　　in lower stratosphere, 177, 178
　　"volcanic-like," 286
AFEAS (Alternative Fluorocarbon
　　Environmental Acceptability
　　Study), 40–41, 44
Age of air, 87, 230. *See also* mean age
　　of air
Agriculture-related emissions, 63, 66
Airborne Antarctic Ozone
　　Experiment (AAOE), 18
AIRS (Atmospheric Infrared
　　Sounder), 123
Albedo, enhancing, 280–281,
　　284–285
ALE (Atmospheric Lifetime
　　Experiment), 46
Alpha, 59
Alternative Fluorocarbon
　　Environmental Acceptability Study
　　(AFEAS), 40–41, 44
Altitude
　　and abundance of atomic
　　　　oxygen, 89
　　and cooling effect of greenhouse
　　　　gases, 216
　　and cycles for ozone loss, 4, 6
　　and ozone mixing ratio, 8, 160
　　and stratospheric ozone loss, 174
　　and variation of ozone loss, 14
Altitude profile (stratospheric
　　ozone), 1, 3
Annular modes, 234
　　Northern, 203, 208, 234–236
　　Southern, 191, 197–199, 200,
　　　　234–236
Antarctic
　　aerosol extinction in, 110
　　climate change in, 236
　　denitrification in, 122, 128–129
　　EESC in, 60, 62
　　expected stratospheric temperature
　　　　changes in, 222

firn air samples from, 41, 45–46
geo-engineering and ozone in, 290,
　　295
lower stratospheric temperatures
　　in, 19
ozone recovery in, 207
PSC diversity in, 115–117
PSC formation in, 122–123
springtime ozone losses in, 108
and stratospheric
　　circulation, 145–146
total ozone and ESC over, 22, 23
total ozone over, 146, 147
trends in PSC occurrence
　　frequencies, 112
21st century springtime ozone
　　in, 269–271
Antarctic Circumpolar Current
　　(ACC), 199–200
Antarctic ozone depletion
　　and geo-engineering, 295
　　impact on the
　　　　troposphere, 191–207
　　potential for chlorine activation
　　　　and, 292
　　stratospheric, 192, 193
　　　　future changes in, 206–207
　　　　impact on the
　　　　　　troposphere, 192–200
　　　　mechanisms of tropospheric
　　　　　　response to, 201–206
　　　　and ocean, 199–200
　　　　and surface climate, 197–199
　　　　and tropospheric
　　　　　　circulation, 192–197
Antarctic ozone hole, 15–16,
　　149–154, 162, 163
　　area of, 269, 270
　　chemical loss in Antarctic
　　　　vortex, 150–154
　　discovery of, 34, 90–91, 110, 111
　　geo-engineering and recovery
　　　　of, 295
　　main features of, 149
　　projections of recovery of, 271
　　in "world avoided" simulation, 24

Antarctic vortex, 15, 16, 91
　chemical ozone loss in, 150–154
　cooling of, 192
　main chemical species in, 146, 148
　mix with mid-latitude air, 178
　and ozone column density, 146,
　　147
　and ozone loss, 145–147
　total ozone value in, 149–151
Anthropocene, vii
Anthropogenic emissions
　of brominated source gases, 56
　of chlorinated VSLS, 55–56
　of halogenated chemicals, 38
　of long-lived, synthetic chlorinated
　　chemicals, 54, 55
　methane from, 63
　of methyl chloride, 51
　of nitrous oxides, 66
　and recovery from effects of
　　halogen source gases, 21–24
Anthropogenic influence, 9–15
　and Antarctic ozone hole, 15
　first concern about, 9–10
　increase in halogen source
　　gases, 10–13
　on stratosphere-troposphere
　　coupling, 235–236
　upper stratospheric ozone
　　depletion, 13–15
AOGCMs. *See* Atmosphere-Ocean
　General Circulation Models
A1B GHG scenario (SRES), 257,
　258, 264, 286
Arctic
　aerosol extinction in, 110
　denitrification in, 122, 128, 129
　expected stratospheric temperature
　　changes in, 222
　geo-engineering and ozone in, 290
　lower stratospheric temperatures
　　in, 19
　methane emissions in, 65
　NAT nucleation without ice
　　in, 125
　NAX particles in, 121

ozone loss rates in, 147, 148
　PSCs in, 111
　renitrification in, 129
　stratospheric temperature in, 220
　STS and NAT measurements
　　in, 117–119
　total ozone and ESC over, 22, 23
　total ozone over, 146, 147
　trends in PSC occurrence
　　frequencies, 112
　21st century springtime ozone
　　in, 271
Arctic ozone depletion, 16–18,
　154–164
　as Arctic ozone hole, 162–163
　chemical loss in Arctic
　　vortex, 155–160
　and denitrification, 129
　impact on the troposphere, 191,
　　207–208
　and injection of sulfur, 292, 295
　interannual variability in, 160–164
　magnitude of, 207–208
　and natural variability in
　　stratospheric ozone, 154–155
　and PSC formation, 112
　in winter, 147
Arctic ozone hole, 162–163, 271
Arctic vortex, 16
　characteristics of, 154
　chemical ozone loss in, 155–160
　main chemical species in, 146, 148
　mix with mid-latitude air,
　　178–180
　ozone column density in, 146, 147
　and ozone loss, 146
　temperature decrease in, 287
Area of the ozone hole, 269, 270
Arnold, F., 18, 128
Aschmann, J., 97
Assessment models, 183–186. *See
　also specific models*
Astatine, 78
Atmosphere Chemistry Experiment-
　Fourier Transform Spectrometer
　(ACE-FTS), 82, 87, 135, 148, 154

Atmosphere-Ocean General
    Circulation Models
    (AOGCMs), 231–232, 237, 240
Atmospheric dynamics. *See*
    dynamics/dynamical processes
Atmospheric general circulation,
    modification of, 199
Atmospheric Infrared Sounder
    (AIRS), 123
Atmospheric lifetime (trace gases),
    42. *See also* lifetimes
Atmospheric Lifetime Experiment
    (ALE), 46
Atmospheric temperatures,
    geo-engineering and, 286–287
Atmospheric waves, 223–225, 231
Atomic oxygen (O)
    in odd oxygen, 3–4
    and ozone destroying catalytic
        cycle, 89
    in ozone formation, 3
A2 GHG scenario (SRES), 264
Aura satellite, 82, 153–154
Austin, J., 231, 259, 271
Aydin, M., 45

Backscatter Ultraviolet (BUV)
    instrument, 172
Bala, G., 285
Balard, Antoine-Jerome, 79–80
Baldwin, M., 235
Baldwin, M. P., 235
Balloon-borne measurements, 3, 109,
    110
Banks, 10, 12
    emissions from, 36, 44
    growth of, 50
Baroclinic instability, 236–237
Baroclinic waves, 203
Bates, D. R., 4
BD circulation. *See* Brewer-Dobson
    circulation
Bedjanian, Y., 83
Beijing amendments (Montreal
    Protocol, 1999), 11
Berzelius, J. J., 79

Biele, J., 113–115
Biermann, U. M., 121, 125, 128
Biological processing, ODSs removed
    by, 38
Biomass burning, 51
Black, R. X., 236
Bodenstein, M., 80
Böning, C. W., 200
Boundary layer
    bromine monoxide in, 98
    chemical reactions in, 79
Branching ratios (ClO/BrO cycle), 93
Brewer, A. W., 7, 225
Brewer-Dobson (BD)
    circulation, 7–9, 145, 225–231
    and Arctic ozone projections, 271
    and coupling of stratosphere and
        troposphere, 233
    defined, 224
    and direct radiative forcing of solar
        cycle, 219
    enhanced driving of, 289
    interannual/long-term changes
        in, 181
    and northern hemisphere increases
        in greenhouse gases, 207
    trends in strength of, 226–228, 266
    and tropical tropopause layer
        temperatures, 238, 239
    variability and long-term changes
        of, 228–231
Bromine (Br), 78
    and Antarctic ozone depletion, 34
    and catalytic ozone loss
        cycles, 89–93, 177–178
    from ClO/BrO cycle, 93
    first production of, 79–80
    from human activities, 38
    inorganic bromine budget, 96–98
    from long-lived ODSs, 40–46,
        52–54
    measurement of, 83
    from natural processes, 37, 50–52
    ozone chemistry role of, 78
    and polar ozone depletion, 20–21
    reactivity of, 86

relative to chlorine, 59–60
short-lived species of, 178, 179
sources of, 33–34
stratospheric, 260
surface total, 260
total atmospheric levels and
trends, 54
transport from surface to
stratosphere, 40
Bromine chemistry, chlorine
chemistry *vs.*, 87–89. *See also*
stratospheric halogen chemistry
Bromine dioxide (OBrO), 88
Bromine monochloride (BrCl)
from ClO/BrO cycle, 93
formation of, 88–89
measurement accuracy for, 81, 82
production and photolysis of, 90
Bromine monoxide (BrO), 54, 88
and ClO/BrO cycle, 93
gas phase reactions of, 88
measurement accuracy for, 81–83,
97–98
in ozone loss studies, 153
partitioning of, 87, 89
as reactive intermediate, 80
and total stratospheric
bromine, 56, 57
Bromine nitrate (BrONO$_2$)
hydrolysis of, 88–90
measurement accuracy for, 82, 83
photolysis of, 90
Bromoform (CHBr$_3$), 58
Brönnimann, S., 181, 233
BrO$_x$ cycle, 289
Br$_x$, 169
Budyko, M. I., 280
Butchart, N., 240–241
BUV (Backscatter Ultraviolet)
instrument, 172

Calibration, 43, 48
CALIOP instrument, 115
CALIPSO (Cloud-Aerosol LiDAR
and Infrared Pathfinder Satellite
Observations), 115–117, 125, 126

Callaghan, P., 228
Callaghan, P. F., 181
CAM (Community Atmospheric
Model), 286
Carbon dioxide (CO$_2$)
CDR approach, 280, 284
and chemistry of upper
stratosphere, 176
and CO$_2$-equivalent
emissions, 68–69
and direct radiative forcing, 68
and Ekman-driven circulation of
Southern Ocean, 200
expected increase of, 215, 240, 292
and stratosphere-troposphere
coupling, 236
in the 21st century, 257
and upper stratospheric ozone, 14
from volcanic eruptions, 281
Carbon dioxide removal (CDR), 280,
284
Carbon tetrachloride (CCl$_4$), 33
concentrations and abundance
of, 48
global surface mixing ratio of, 47
natural sources of, 41, 45–46
oceanic losses of, 38
properties and atmospheric
abundance of, 39
removed by photolysis, 38
sources of atmospheric increases
in, 40–41
Carbonyl sulfide (COS), 67
Carslaw, K. S., 117
Catalytic ozone loss cycles, 4–6
active nitrogen in, 66
and anthropogenic CFC
accumulation, 13
bromine in, 34
chlorine in, 34
ClO in, 80
and denitrification, 122
and distribution of stratospheric
ozone, 7
halogen catalyzed, 89–93
ClO/BrO cycle, 93

Catalytic ozone loss cycles (*continued*)
    ClO dimer cycle, 92
    ClOOCl photolysis, 93–95
    and heterogeneous chlorine
        activation, 132, 177
    mixed, 89–90
    and non-halogenated gases, 37
    in polar regions, 20–21
    rate-limiting reactions for, 221
    simplest, 89
    and temperature changes, 221
CCMs. *See* Chemistry-Climate
    Models
CCMVal. *See* Chemistry-Climate
    Model Validation Activity
CFC-11 (CFCl$_3$)
    emissions history for, 41
    global surface mixing ratio for, 47
    lifetime of, 39, 43
    mean hemispheric difference in,
        50
    Ozone Depletion Potential of, 40
    total global emissions of, 44
CFC-12 (CF$_2$Cl$_2$)
    global surface mixing ratio for, 47
    lifetime of, 39
    mean hemispheric difference, 50
    natural sources of, 46
    peak level of, 13
    total global emissions of, 44
CFC-13 (CF$_3$Cl), 39
CFC-113 (CCl$_2$FCClF$_2$)
    emissions *vs.* atmospheric
        abundance of, 49
    global surface mixing ratio for, 47
    lifetime of, 49
    N–S mixing ratio for, 50
    total global emissions of, 44
CFCs. *See* chlorofluorocarbons
Chabot, H., 80
Chapman, S., 4
Chapman reactions, 4, 5
Chappuis, J., 80
CH$_2$ClCH$_2$Cl, 55–56
Chemical ionization mass
    spectrometry (CIMS), 98

Chemical Lagrangian Model of the
    Stratosphere (CLaMS), 130, 131
Chemical mechanisms of ozone
    depletion, 18–21
Chemical-transport models (CTMs)
    and accelerated BD
        circulation, 229, 230
    REPROBUS, 156
    SLIMCAT, 154, 156
Chemistry
    bromine, 87–89
    chlorine
        bromine chemistry *vs.*, 87–89
        coupled with dynamics, 256
        impact of enhanced greenhouse
            gas concentrations
            on, 216–222
        simulation of, 135
        in tropopause region, 98–99
    stratospheric, 287–288
    of stratospheric ozone, 3–7
Chemistry-Climate Models
    (CCMs), 21–24, 254. *See also*
    Climate-Chemistry Models
    3D, 183–185
    of 21st century stratospheric
        ozone, 254–256
    of Antarctic vortex, 154
    of Arctic vortex, 160
    and cooler winter
        temperatures, 161
    evaluation of, 254, 259–260
    of global mean temperature, 218
    schematic of, 255
    simulations with, 256–260
    of southern hemisphere circulation
        trends, 205
    of stratospheric ozone in the 21st
        century, 254–256
Chemistry-Climate Model Validation
    (CCMVal) Activity, 227, 231, 254,
    259–260
    phase 1 (CCMVal-1), 257,
        259–260, 271
    phase 2 (CCMVal-2), 194, 196,
        197, 202, 203, 208, 257–261, 271

Chen, G., 203
Chen, P., 232
Chipperfield, M. P., 180
Chlorine (Cl), 78
  and Antarctic ozone depletion, 34
  and catalytic ozone loss
    cycles, 89–93, 177–178
  from ClO/BrO cycle, 93
  discovery of, 79
  EESC, 11–13
  ESC, 22
  history of, 79–80
  from human activities, 38
  from long-lived ODSs, 40–46,
    52–54
  measurement of, 83
  methane as sink for, 63
  from natural processes, 37, 50–51
  oxides of, 80
  ozone chemistry role of, 78
  in photochemical breakdown of
    halogen source gases, 10, 11
  and polar ozone depletion, 18–21
  reactivity of, 86
  relative to bromine, 59–60
  in reservoir species, 4
  sources of, 33–34
  stratospheric, 260
  and stratospheric aerosol
    particles, 66–67
  surface total, 260
  transformation from PSCs and
    cold stratospheric aerosol
    particles, 108
  transport from surface to
    stratosphere, 40
  in upper stratosphere, 54
  and upper stratospheric ozone
    depletion, 13–15
Chlorine activation
  defined, 131, 132
  by heterogeneous reactions, 91,
    131, 132, 135–136, 177
  in lower stratosphere, 177
  open issues concerning, 138–139
  potential for, 285, 290, 292–294

  in presence of supercooled
    liquids, 111–112
  by solid NAT *vs.* liquid
    STS, 135–138
  on stratospheric sulfate
    aerosol, 19–20
Chlorine chemistry. *See also*
  stratospheric halogen chemistry
  bromine chemistry *vs.*, 87–89
  simulation of, 135
  in tropopause region, 98–99
Chlorine dioxide (OClO), 80
  from ClO/BrO cycle, 93
  generation of, 89
  measurement accuracy for, 81, 82
Chlorine oxide (ClO), 80
  in ClO dimer cycle, 92
  deactivation of, 164
  gas phase reactions of, 88
  in tropical UTLS, 99
Chlorine peroxide (ClOOCl)
  in ClO dimer cycle, 92
  photolysis of, 93–95
Chlorofluorocarbons (CFCs), 10
  anthropogenic
    and polar ozone depletion, 18
    and upper stratospheric ozone
     loss, 13–15
  concentrations and abundance, 48
  defined, 33
  destruction of, 38–39
  early recognition of threat posed
    by, 80
  in firn air samples, 45
  human-produced, 34
  lifetimes for, 39
  and mean age of air, 230–231
  natural sources of, 41, 45–46
  non-propellant uses of, 34
  under non-steady-state
    conditions, 43
  past and projected abundances
    of, 12
  removed by photolysis, 38
  sources of atmospheric increases
    in, 40–41

Chlorofluorocarbons (CFCs)
(*continued*)
  and tropospheric chlorine
    decline, 53
  in the 21st century, 255
Chloroform (CHCl₃), 55–56
Cicerone, R. J., 4, 34, 176
CIMS (chemical ionization mass
  spectrometry), 98
Circulation. *See also* transport
  Brewer–Dobson (*See* Brewer-
    Dobson circulation)
  due to increasing greenhouse
    gases, 216
  in southern hemisphere, 199
  stratospheric, 7–9, 215
    expected changes in, 262–264
    lower stratosphere, 7–9
    and planetary-scale Rossby
      waves, 191
    and planetary waves, 8, 9
    upper stratosphere, 7
    and weather events, 235
  tropospheric, 192–197, 203–205, 215
Cl. *See* Chlorine
Cl₂, 94
  formation of, 91
  measurement accuracy for, 81, 82
  and polar ozone loss, 146
ClO. *See* Chlorine oxide
ClOOCl. *See* Chlorine peroxide
Cl₂O₂
  in ozone loss studies, 153
  photolysis rate of, 164
CLaMS (Chemical Lagrangian Model
  of the Stratosphere), 130, 131
Climate
  Antarctic surface, 197–199
  and future ozone
    concentrations, 222
  impact of volcanic eruptions
    on, 282, 283
  and ozone in tropopause
    region, 95, 96
  polar, 191
  tropospheric, 190, 191

Climate change
  and Antarctic ozone depletion, 191
  Antarctic surface climate, 197–199
  and Arctic denitrification and
    ozone loss, 112
  and Brewer-Dobson
    circulation, 226
  expected tropopause
    changes, 240–241
  geo-engineering for, 284–295
    atmospheric temperatures and
      dynamics, 286–287
    motivation for proposed
      approaches, 279–281
    surface temperature and
      precipitation, 284–286
  from greenhouse gases, 38
  and HCFC-22 production, 49
  impact on stratospheric ozone
    layer, 214–243
    coupling of stratosphere and
      troposphere, 203, 233–241
    radiation and
      chemistry, 216–222
    stratospheric dynamics, 222–233
  and methane emissions, 65
  and naturally-emitted gases, 37
  from ozone-depleting gases, 67–69
  and ozone recovery, 21–23, 253
  in southern hemisphere, 193–195
  stratosphere-troposphere
    coupling, 203, 233–241
  tropical and extra-tropical
    tropopause layer, 238–240
  and upper stratosphere
    cooling, 177
Climate-Chemistry Models (CCMs).
  See also Chemistry-Climate Models
  and acceleration of BD
    circulation, 226, 227
  BD circulation and mean age of
    air, 230
  increased GHG concentrations
    and ozone recovery, 241
  of projected ozone layer
    evolution, 215

sea surface temperatures in, 231,
232
and tropical tropopause layer
changes, 239
tropopause changes, 241
Climate engineering, 279. *See also*
geo-engineering
Climate forcing. *See* radiative forcing
Climate models, 193–195. *See also*
specific models
Climatological jets, 194, 196
ClO/BrO cycle, 93
ClO dimer cycle, 92, 93, 132, 133
$ClONO_2$
at altitudes above 30 km, 87
in Arctic, 164
and chlorine activation, 132–135
heterogeneous activation of
chlorine from, 91
heterogeneous reaction on sulfate
aerosol, 88
measurement accuracy for, 81–83
photolysis of, 90
and polar ozone depletion,
18, 146
reactivities on HCl-doped
surfaces, 111
total column abundance, 55
in upper stratosphere, 54
Cloud particles, heterogeneous
reactions on, 6
ClO_x cycle, 289
Cl_x, 169
CMIP3 (Coupled Model
Intercomparison Project phase 3)
models, 193–196, 198
CO_2-equivalent (eq) emissions, 68–69
Cold liquid aerosols, 112
Cold point temperature, 265
Cold point tropopause
(CPT), 238–239
Cold pole problem, 259
Cold sulfate aerosols
and chlorine activation, 177
heterogeneous reactions
on, 131–135

Column ozone, 169
accumulated chemical loss in, 157,
158
in Antarctic ozone hole, 15
in Arctic, 17, 161
decadal variation in, 182–183
in extra-tropics, 8
as function of latitude and
month, 8, 9
geo-engineering influence
on, 289–291
interhemispheric differences in
evolution of, 269
measuring, 2–3
mid-latitude, 268, 269
observed and modeled trends
in, 184–186
past changes in, 172–175
total thickness, 2–3
tropical, 267, 268
and tropopause height, 181, 182
variations in, 170, 171
Community Atmospheric Model
(CAM), 286
Concentration(s)
converting molar mixing ratio
to, 2
defined, 2
in DU per km, 3
expression of, 2
of greenhouse gases, 253
and expected temperature
changes, 220–221
impact on radiation and
chemistry, 216–222
impact on stratospheric
dynamics, 222–233
and measurement of global
emissions, 42, 43
of ozone, determinants of, 152
Cooling. *See also* climate change;
temperature
in Antarctic lower
stratosphere, 205–206
of Antarctic surface, 197–199
in Arctic stratosphere, 208

Cooling (*continued*)
  from greenhouse gases, 215,
    216
  in stratosphere, 192, 207
  in troposphere, 202
  in upper stratosphere, 177
  in upper troposphere/lowermost
    stratosphere, 190
Copenhagen amendments (Montreal
  Protocol, 1992), 10, 11
Coriolis force, 15, 16
Coupled Model Intercomparison
  Project phase 3 (CMIP3)
  models, 193–196, 198
Courtois, Bernard, 79
CPT (cold point
  tropopause), 238–239
Crutzen, P. J., vii, 4, 13, 14, 18, 128,
  280
CTMs. *See* chemical-transport
  models

Daerden, F., 131
Daily minimum ozone, 269, 270
Dall'Amico, M., 220
Dameris, M., 230, 232, 239
Davies, S., 129, 130
Davy, Humphrey, 79
Decadal ozone changes, causes
  of, 181, 182
Deckert, R., 230, 232, 239
Deep convection, 232, 239, 240
Denitrification, 122
  in Antarctic, 122
  in Antarctic vortex, 146
  in Arctic, 112, 122, 154
  defined, 122, 128
  of polar stratosphere, 18
  in polar winters, 146
  requirements for, 112
  simulation of, 128–131
    Lagrangian modeling,
      130–131
    simple fixed-grid, 129–130
Developed countries, phase-out of
  ODSs in, 36

Developing countries, phase-out of
  ODSs in, 36
Diabatic radiative heating, 223
Dibromomethane (CH_2Br_2), 58
Dichlorine oxide (Cl_2O), 80
Dichlorine peroxide (ClOOl)
  formation of, 88
  photolysis of, 93–95
Dichloromethane (CH_2Cl_2), 55–56
Dickinson, R. E., 202
Differential Optical Absorption
  Spectroscopy (DOAS), 97, 98
Dilution effect, 178
Diode array spectroscopy, 94
Direct climate forcing, 68
DLAPSE model, 130, 132
DOAS (Differential Optical
  Absorption Spectroscopy), 97, 98
Dobson, G. M. B., 1, 3, 7, 15, 225
Dobson units (DU), 1–3
Downward control, 203, 232, 236
Downwelling long-wave (LW)
  radiation, 190, 202
Downwelling short-wave (SW)
  radiation, 190, 192, 202
Drdla, K., 137, 290
DU (Dobson units), 1–3
"dust layer," 109
Dye, J. E., 117, 127
Dynamics/dynamical processes.
  See also stratospheric dynamics
  and Arctic ozone projections, 271
  and climate change, 215
  coupled with chemistry, 256
  and effect on troposphere of
    stratospheric ozone loss, 191
  impact of geo-engineering
    on, 286–287
  and mid-latitude ozone
    depletion, 180–183
  zonal-mean, 202–205

Easterly winds, 224
ECI (Equivalent Chlorine), 37,
  59–60
Eckermann, S. D., 123

ECMWF. *See* European Centre for
Medium-Range Weather Forecasts
Eddy dissipation, SST-related, 232
EECI (Equivalent Effective
Chlorine), 37, 60
Effective Equivalent Stratospheric
Chlorine (EESC). *See also*
Equivalent Effective Stratospheric
Chlorine
and Antarctic ozone hole, 149
expected change in, 260–262
and global mean column
ozone, 170, 171
projected decline in, 287
Eigen, Manfred, 80
Ekman flow, 199
Ekman transport, 200
11-year solar cycle, 169–171, 183,
218–220, 235
El Niño/Southern Oscillation
(ENSO), 233
Emissions histories, 41, 49
Engel, A., 231, 263
ENSO (El Niño/Southern
Oscillation), 233
ENVISAT satellite, 83, 119,
122–123
Equivalent Chlorine (ECl), 37,
59–60
Equivalent Effective Chlorine
(EECl), 37, 60
Equivalent Effective Stratospheric
Chlorine (EESC), 11–13, 37.
See also Effective Equivalent
Stratospheric Chlorine (EESC)
in Antarctic springtime, 271
calculating, 11, 60
and Montreal Protocol, 11–13,
61–62
past and projected abundances
of, 12
in polar and mid-latitude
stratospheres, 61
recovery, 62
Equivalent Stratospheric Chlorine
(ESC), 22–24, 37, 270

Erle, F., 89
Erythemal UV, over
Antarctica, 201–202
ESC. *See* Equivalent Stratospheric
Chlorine
Espy, J. P., 279
Euchlorine, 80
EUPLEX campaign, 124–125
European Centre for Medium-Range
Weather Forecasts
(ECMWF), 115, 116, 123, 125,
162, 181, 229
ExTL (extra-tropical tropopause
layer), 238–240
Extra-polar regions, expected cooling
in, 262
Extra-polar stratosphere, expected
ozone loss in, 264–265
Extra-tropical regions
generation of planetary
waves, 230, 233
ozone transport in, 8, 9
sea surface temperature anomalies
in, 232, 233
and stronger BD circulation, 227
transport of chemical species and
particles, 234
Extra-tropical stratosphere
climatology of, 224
planetary wave forcing in, 225
Extra-tropical tropopause layer
(ExTL), 238–240
Extreme weather events, 235, 237
Eyring, V., 183, 271

Fahey, D. W., 121, 124, 128, 129
Farman, J. C., 15, 18, 34, 110, 149
Feng, W., 181
Fingerprint (in models), 183
Firn air samples, 41, 45–46, 48,
63, 66
Fixed-grid simulation (NAT-rock
formation and
denitrification), 129–130
Flash photolysis technique, 80
Flit, L., 279

Fluorine, 78
  first isolation of, 80
  from human activities, 38
  and ozone budget, 78
  and ozone destruction, 89
  reactivity of, 86
  and stratospheric ozone
    destruction, 83, 84
Fomichev, V., 232
Forster, P., 233, 236
Fourier transform absorption
  spectroscopy, 48
Fractional release rates (ODSs), 260
Free chlorine, 79
Fueglistaler, S., 122, 238
Fusco, A. C., 181
Future reference simulation, 257
Fyfe, J., 237

G (global atmospheric
  abundance), 42–43
γ (reactive uptake
  coefficient), 133–135
Garcia, R. R., 289
Gardiner, B. G., 15
Gas chromatography, 83, 97
Gas-phase reactions
  expected decrease in, 262
  projected, 215
Gas-to-particle conversion, 109
GCM-Reality Intercomparison
  Project (GRIPS), 254
GCMs. *See* General Circulation
  Models
General Circulation Models
  (GCMs), 226, 230
  Atmosphere-Ocean, 231–232, 237,
    240
  Chemistry-Climate Models based
    on, 255, 259
  effects of increasing GHG
    concentrations, 237
  international projects
    evaluating, 254
  sea surface temperatures in, 231,
    232

of solar radiation
  management, 284–286
of stratosphere-troposphere
  coupling, 235
of tropospheric SH circumpolar
  circulation, 236
Geo-engineering, 279–281, 284–295
  and atmospheric temperature and
    dynamics, 286–287
  defined, 279
  motivation for proposed
    approaches, 279–281
  and polar ozone, 290–295
  and stratospheric ozone, 287–290
  and surface temperature and
    precipitation, 284–286
Geoengineering Model
  Intercomparison Project
  (GeoPIP), 286
GEOS. *See* Goddard Earth Observing
  System Model
Geostrophic adjustment of
  tropospheric flow, 236–237
Gettelman, A., 97, 259
GHGs. *See* greenhouse gases
Gillett, N., 237
Glauber, Johann Rudolph, 79
Global atmospheric abundance
  (G), 42–43
Global emissions. *See also*
  anthropogenic emissions
  declines in, 49–50
  deriving, 42–44
  and global mean mixing ratio,
    47, 49
Global Ozone Monitoring
  Experiment (GOME), 172
Global total ozone, 22, 23, 172–173
Global Warming Potential
  (GWP), 68–69
Gnanadesikan, A., 200
Goddard Earth Observing System
  Model (GEOS), 125, 126, 206, 264
Godin, S., 179, 180
GOME (Global Ozone Monitoring
  Experiment), 172

Götz, F. W. P, 1
Gravity waves, 225, 231
Greenhouse effect, 215
Greenhouse gases (GHGs), 38.
See also specific gases
atmospheric concentrations
of, 253
circulation changes due to, 216
and climate forcing, 68
CO_2-equivalent emissions
of, 68–69
and decadal ozone changes, 181,
183
and expected temperature
changes, 220–221
impact on radiation and
chemistry, 216–222
expected temperature
changes, 220–221
past temperature
changes, 217–220
temperature ozone
feedback, 221–222
and increased upwelling, 263–264
lifetimes of, 226
northern hemisphere increases
in, 207
projected increase of, 215
response of stratospheric climate
and circulation to, 241
simulations for 21st
century, 256–258
and stratosphere-troposphere
coupling, 235–237
and stratospheric
dynamics, 222–233
Brewer-Dobson circulation and
mean age of air, 225–231
importance of atmospheric
waves, 223–225
role of sea surface
temperatures, 231–233
and tropospheric circulation, 191
Greenland, firn air samples from, 41,
45–46
Grenfell, J. L., 90

GRIPS (GCM-Reality
Intercomparison Project), 254
Grise, K. M., 202, 208
Grooß, J.-U., 129, 130
GWP (Global Warming
Potential), 68–69

Hadjinicolaou, P., 181
Hadley cell, 196, 198, 199
Hallberg, R., 200
HALOE. See Halogen Occultation
Experiment
Halogen catalyzed ozone loss
cycles, 89–93
ClO/BrO cycle, 93
ClO dimer cycle, 92
ClOOCl photolysis, 93–95
Halogen chemistry, 79–80. See also
stratospheric halogen chemistry
Halogen emissions
anthropogenic, 10–13
from human activities, 38
and "world avoided"
simulation, 24
Halogen monoxides (XO), 86
Halogen nitrate reactions, 88–89
Halogen nitrates ($XONO_2$), 86
formation of, 88
Halogen Occultation Experiment
(HALOE), 157, 158, 220,
228, 229
Halogen oxides, first synthesis and
description of, 80
Halogens, 78
decrease in, 253
heterogeneous activation of, 281
and impact of sulfur particles on
ozone, 287
industrial activity as source
of, 40–41, 44
lifetimes of, 178
production of, 90
total atmospheric halogen loading
changes, 59–62
transport of, 56, 178
21st century changes in, 260–261

Halogen source gases
  "banks" of, 10
  increase in, 10–13
  legally binding controls on, 13
  longer-lived, 33–34, 38–54
    human *vs.* natural sources
      of, 40–46
    measuring/interpreting changes
      in abundance of, 46–50
    from natural processes, 50–52
    systematic changes in total
      tropospheric chlorine and
      bromine from, 52–54
    timescales and processes that
      remove, 38–40
  as ozone-depleting substances, 10
  past and projected abundance
    of, 12–13
  photochemical breakdown of,
    10, 11
  recovery from effect of, 21–24
Halon-1211 (CF$_2$ClBr), 34
  in firn air samples, 45
  global surface mixing ratio for, 47
  properties and atmospheric
    abundance of, 39
  total global emissions, 44
Halon-1301 (CF$_3$ClBr), 34
  global surface mixing ratio for, 47
  increasing concentrations of, 48
  properties and atmospheric
    abundance of, 39
Halon-2402 (CF$_2$BrCF$_2$Br), 34
  global surface mixing ratio for,
    47
  properties and atmospheric
    abundance of, 39
Halon fire-extinguishing
  agents, 33–34
Halons
  natural sources of, 41, 45–46
  Ozone Depletion Potential of, 40
  properties and atmospheric
    abundance of, 39
  sources of atmospheric increases
    in, 40–41

and total stratospheric bromine,
  56
and tropospheric bromine
  decline, 53, 54
Hampson, J., 4
Hanson, D. R., 111, 134–137
Harnik, N., 236
Harris, N. R. P., 181
Harrison, D. N., 3
Hartley, D., 236
Hautefeuille, P., 80
Haynes, P. H., 236
HCFC-22 (CHClF$_2$)
  feedstock production of, 49
  global surface mixing ratio for, 47
  properties and atmospheric
    abundance, 39
  total global emissions, 44
HCFC-141b (CH$_3$CFCl$_2$)
  global surface mixing ratio for, 47
  lifetime of, 39
  properties and atmospheric
    abundance, 39
HCFC-142b (CH$_3$CF$_2$Cl)
  global surface mixing ratio for, 47
  lifetime of, 39
  properties and atmospheric
    abundance, 39
HCFCs. *See*
  hydrochlorofluorocarbons
Heckendorn, P., 289, 290
Held, I. M., 203
Henry's law, 134
Heterogeneous nucleation (NAT)
  after preactivating SAT, 127, 128
  on SAT, 126, 127
  on solid inclusions, 127, 128
Heterogeneous reactions, 6–7, 177
  on aerosol particles, 6, 19–20
  and Antarctic ozone hole, 111
  and chlorine activation, 131, 132,
    135–136
  on cold sulfate aerosols, 131–135
  expected increase in, 262
  geo-engineering enhancement
    of, 289, 290

on liquid particles, 112
in nitric acid fluxes, 129–130
for polar heterogeneous chlorine
activation, 19–20, 91
in polar stratosphere, 133
on polar stratospheric clouds, 111,
131–135
on stratospheric sulfate, 19–20
on sulfate aerosols, 19–20, 88,
131–135, 177
HFC-23, 45
HFC-125, 45
HFC-134a, 45
Hobe, M. von, 92, 95, 96
HOCl cycle, 90, 91
Holton, J., 224
Homogeneous nucleation
of aerosol particles, 121
nitric acid trihydrate
of glassy aerosols, 127, 128
interface-induced, 127, 128
of non-equilibrium
aerosol, 127, 128
of STS, 126, 127
Hood, L. L., 181, 183
Höpfer, M., 122, 124
Hossaini, R., 97
HO$_x$ cycle, 289
Hu, Y., 228
Hydrochlorofluorocarbons (HCFCs)
CO$_2$-equivalent emissions of, 69
concentrations and abundance,
48
and direct radiative forcing, 68
in firn air samples, 45
lifetime of, 39
under Montreal Protocol, 35
natural sources of, 41, 45–46
as ozone-depleting substances, 34
Ozone Depletion Potential of, 40
past and projected abundances
of, 12
phase-out of, 257
sources of atmospheric increases
in, 40–41
as substitutes for main ODSs, 34

Hydrofluorocarbons (HFCs)
CO$_2$-equivalent emissions of, 69
and direct radiative forcing, 68
in firn air samples, 45
Hydrogen chloride (HCl)
and chlorine activation, 111–112,
132–135
decline in, 146, 147
formation of, 86
heterogeneous activation of
chlorine from, 91
heterogeneous reaction on sulfate
aerosol, 88
measurement accuracy for, 81–83
and polar ozone depletion, 18,
146
total column abundance, 55
in upper stratosphere, 54, 174
Hydrogen fluoride (HF)
formation of, 86
measurement accuracy for, 83
stability of, 89
Hydrogen halides (HX), 86
Hydrogen oxide (HO$_x$), 264
Hydrogen sulfide (H$_2$S), 281
Hydrological cycle, volcanic
eruptions and, 282
Hydrology, 199
Hydrolysis, ODSs removed by, 38
Hydrostatic adjustment of
tropospheric flow, 236–237
Hydroxide (OH), 43, 65
Hydroxyl radical
global concentrations of, 43
in methane removal, 63, 64
Hypobromous acid (HOBr), 88–90
Hypochlorous acid (HOCl), 111, 146

Ice-assisted NAT
nucleation, 121–123, 130–131
Ice bubble samples, 41, 45–46, 51, 63
Ice nucleation, 109
Ice PSCs. *See* Type-II (ice) PSCs
Ideal gas, molar and volume mixing
ratios for, 2
Ideal gas law, 2

Industrial activity
  as source of methyl bromide, 52
  as source of ozone-depleting
    halogens, 40–41, 44
Infrared (IR) radiation, 216, 282
Inorganic bromine (Br$_y$), 83, 85, 88
  budget of, 96–98
  calculating from BrO, 98
  at Earth's surface, 57
  expected changes in, 260, 261,
    267
  heterogeneous reactions, 88
  measurement accuracy for, 81–83
  partitioning of, 87
  springtime changes in, 270, 271
  total stratospheric, 57
Inorganic chlorine (Cl$_y$), 4, 5, 83, 84
  altitude and increase in, 164
  in CCM simulations, 259
  expected changes in, 260, 261,
    267
  heterogeneous reactions, 88
  partitioning of, 86
  springtime changes in, 270, 271
  total column abundance, 55
  in the UTLS region, 98–99
In situ (IS) measurement, 81–83
  of Arctic NAX particles, 121
  of HNO$_3$-containing particles,
    129
  of inorganic bromine, 98
  NAT-PSC nucleation, 121
  of PSCs, 119
  of stratospheric aerosol
    particles, 110
Intergovernmental Panel on Climate
  Change (IPCC), 237
Intergovernmental Panel on Climate
  Change Special Report on
  Emission Scenarios (IPCC
  SRES), 257, 258, 264
Inter-tropical convergence zone, air
  mixing in, 50
Iodine, 78
  discovery of, 79–80
  from human activities, 38

and ozone budget, 78
and ozone destruction, 58, 83,
  84, 89
reactivity of, 86
IPCC, 237
IPCC SRES. *See* Intergovernmental
  Panel on Climate Change Special
  Report on Emission Scenarios
IR (infrared) radiation, 216, 282
Isopychals, 200

Junge, Christian, 109, 110
Junge Layer, 6, 67, 110, 113

Kang, S. M., 199
Kawa, S. R., 93
Keeley, S. P. E., 202
Kinetic theory, 80
Knopf, D. A., 128
Knudsen, B. M., 129
Koch, G., 181
Kodera, K., 235
Koop, T., 113, 118, 120, 121, 127,
  128, 130
Kyoto Protocol, 54, 68, 69

Lagrangian modeling
  (PSCs), 130–131
Lal, M., 202
Larsen, N., 124
Lary, D. J., 83
Le Quéré, C., 200
Lewis, B., 80
Li, F., 228, 231
Li, S., 230
LiDAR. *See* Light Detection And
  Ranging
Lien, C. Y., 94, 95
Lifetimes
  CFC-113, 49
  chlorofluorocarbons, 39
  and climate forcing, 68
  of greenhouse gases, 226
  of halogens, 178
  hydrochlorofluorocarbons, 39
  and hydroxyl loss, 43

of long-lived chemicals, 223
of lower stratospheric ozone, 177
methyl chloride, 51
methyl chloroform, 49
nitrous oxide, 66
of ozone, 289–290
of ozone anthropogenic
ODSs, 226
ozone-depleting substances, 36,
39, 42–43
photochemical, 256
and steady-state *vs.* non-steady-
state conditions, 43
and stratospheric photolysis, 43
of trace gases, 42
uncertainties in, 50
Light Detection And Ranging
(LiDAR), 111, 172
Arctic STS and NAT
measurements, 118–119, 121
CALIPSO, 115–117, 125, 126
NAT nucleation without ice, 124,
125
PSC classification based
on, 113–115
Limpasuvan, V., 181
Liquid PSCs, 111–114
Liquid sulfate aerosols, 281–282
London amendments (Montreal
Protocol, 1990), 10, 11
Longer-lived halogenated source
gases, 33–34, 38–54
human *vs.* natural sources
of, 40–46
measuring/interpreting changes in
abundance of, 46–50
from natural processes, 50–52
systematic changes in total
tropospheric chlorine and
bromine from, 52–54
timescales and processes that
remove, 38–40
Long-lived chemicals
air mass exchange rates and
lifetimes of, 223
and climate change, 67–69

Montreal Protocol control of,
36
ozone-depleting substances
changes in threat to ozone
from, 37
measurement of, 42
Long-wave (LW) radiation
and concentrations of radiatively
active gases, 216
downwelling, 190, 202
forcing of, 281
upwelling, 192
Lower mesosphere, ozone
concentrations in, 221
Lower stratosphere. *See also* upper
troposphere/lowermost
stratosphere (UTLS)
air motions in, 155
Arctic, 220
and Brewer–Dobson
circulation, 7–9
chemical processes in, 177–178
cooling of, 192
expected temperature trends
in, 262
ozone distribution in, 6–7
ozone loss in, 174
polar
and expected temperature
changes, 221–222
springtime ozone depletion
in, 269–271
and sea surface temperatures, 232,
233
simulated total ozone in, 25
temperature changes in, 220
temperatures in, 215
tropical, 265–267
Löwig, Carl, 79–80
LW radiation. *See* long-wave (LW)
radiation

MacKenzie, A. R., 121
Mangold, A., 181
Mann, G. W., 130
Marchand, M., 179

Marine environments
  loss of long-lived halogenated trace
    gases in, 38
  nitrous oxide emissions in, 65–66
Mass, C., 224
Mass mixing ratios, 2
Match technique, 152–153, 156,
    158–159
Matthes, K., 235
Mauersberger, K., 111
McElroy, M. B., 91
Mean age of air, 56
  and Brewer-Dobson
    circulation, 230–231
  defined, 230
  long-term change of, 231
  past decrease in, 263
Mean stratospheric age, 56
Measurement
  of bromine monoxide, 97–98
  calibration uncertainty in, 43
  of changes in ODS atmospheric
    abundance, 46–50
  estimates of ozone loss based
    on, 155
  of inorganic bromine
    compounds, 57, 58
  of methane, 63–65
  of mixing ratios, 15
  of source gases, 42
  of stratospheric aerosol
    particles, 109, 110
  of stratospheric halogen
    species, 81–83
  of STS and NAT
    Antarctic, 115–117
    Arctic, 117–119
  of total ozone, 1
  of trace gases, 43
  of VSLS concentrations, 97
M-83 ozonometer, 1
Meteorological conditions
  and PACl and polar ozone
    loss, 292
  variation of ozone with, 1
Methane (CH_4), 63–65

CO_2-equivalent emissions of, 69
  and direct radiative forcing, 68
  in firn air samples, 45
  human-influenced emissions
    of, 37
  and increase in stratospheric water
    vapor, 265
  mixing ratio, 63, 64
  projected increase of, 215
  reactions of fluorine atoms
    with, 89
  in the 21st century, 255, 257
  in upper stratosphere, 176
Methyl bromide (CH_3Br), 34
  anthropogenic sources of, 51–52
  global surface mixing ratio for, 47
  as natural contributor to
    stratospheric bromine, 37
  natural sources of, 41, 45–46,
    51–52
  past and projected abundance
    of, 12
  pre-industrial levels of, 46
  and total stratospheric
    bromine, 56
  and tropospheric bromine
    decline, 53–54
Methyl chloride (CH_3Cl), 10
  in firn air samples, 45
  from human activities, 51
  as natural contributor to
    stratospheric chlorine, 37
  from natural processes, 51
  past and projected abundance
    of, 12
  pre-industrial levels of, 46
Methyl chloroform (CH_3CCl_3), 33
  emissions *vs.* atmospheric
    abundance of, 49
  global surface mixing ratio for, 47
  lifetime of, 49
  natural sources of, 41, 45–46
  N–S mixing ratio for, 50
  oceanic losses of, 38
  properties and atmospheric
    abundance of, 39

sources of atmospheric increases in, 40–41
total global emissions, 44
and total tropospheric chlorine, 52–53
Methyl halides, oceanic losses of, 38
Methyl iodide (CH$_3$I), 58–59
Michelson Interferometer for Passive Atmospheric Sounding (MIPAS), 83, 119, 122–123
Microwave Limb Sounder (MLS-Aura), 82, 153–154, 229
Microwave Sounding Unit (MSU), 218, 219
Middle stratosphere
  Arctic, 220
  expected cooling in, 262
  mid-latitude, 267, 269
  temperature changes in, 220
Mid-latitude ozone
  causes of changes in, 170
  and chlorine catalysed reaction cycles, 14
  reactions on aerosol particles in, 6–7
  total ozone and ESC, 1960 to 2100, 22–24
  in the 21st century, 267–269
Mid-latitude ozone depletion, 89, 169–186
  assessment model calculations, 183–186
  from BrONO$_2$ hydrolysis cycle, 90
  and changes in source gases, 183
  chemical processes, 174, 176–180
    effect of polar vortex loss, 178–180
    lower stratosphere, 177–178
    upper stratosphere, 174, 176–177
  dynamical contributions, 180–183
  past change observations, 172–174
    column ozone, 172–175
    ozone profile, 174, 176
  and solar cycle, 183

Mid-latitudes
  chlorine inventory for, 87
  defined, 170
  EESC in, 60, 62, 260
  mean stratospheric age in, 56
  stratosphere aerosol loading at, 110
Millard, G. A., 179
Miller, R., 237
Millon, E., 80
MIPAS. *See* Michelson Interferometer for Passive Atmospheric Sounding
Mixed-phase PSCs, 113–115, 118, 121, 125, 126
Mixing processes, 7, 256
  and mid-latitude ozone changes, 170
  for polar and tropical latitudes, 8
Mixing ratios
  defined, 2
  global mean, 46–50
  ideal gases, 2
  measurements of, 15
  methane, 63, 64
  methyl bromide, 52
  methyl chloride, 51
  nitrous oxide, 64, 66
  non-ideal gases, 2
  ozone, 160, 289
  in polar region winters, 18
  in tropics, 8, 289
MLS-Aura. *See* Microwave Limb Sounder
Moissan, H., 80
Molar mixing ratios (μx)
  converting to concentration, 2
  for non-ideal gases, 2
Molecular oxygen (O$_2$)
  photolysis of, 3
  UV-photolysis of, 89
Mole fractions, 2
Molina, L. T., 20, 91
Molina, M. J., 10, 13, 20, 34, 80, 91, 112, 135–137, 176
Monge-Sanz, B., 181
Monte Carlo simulations, 153

Month, total column ozone and, 8–9
Montreal amendments (Montreal
     Protocol, 2007), 11, 13
Montreal Protocol on Substances that
     Deplete the Ozone Layer
     (1987), 24–26
  amendments and adjustments
     to, 10, 11, 34–36
  climate benefit from, 38
  and decline in chlorine and
     bromine concentrations, 52–54
  and EESC change, 11–13, 61, 62
  and greenhouse emissions, 69
  and halogen burden, 10
  hydrochlorofluorocarbons
     under, 48, 257
  and natural processes, 50
  ODSs controlled by, 48
  ODSs not controlled by, 49
  original restrictions under, 34
  and production of ozone-depleting
     substances, 10
  success of, 35–36
Morgenstern, O., 24
"Mother Cloud/NAT Rock
     mechanism," 122, 124, 130–132
Mother-of-pearl clouds, 109
Mountain wave-induced ice
     clouds, 122–125
MSU (Microwave Sounding
     Unit), 218, 219
Müller, R., 137, 290

Nacreous clouds, 109
NAM. *See* Northern Annular Mode
NAO (North Atlantic
     Oscillation), 234–235, 237
Nash, E. R., 115
National Oceanic and Atmospheric
     Administration (NOAA), 44, 46,
     47, 53, 61, 64
National Ozone Expedition
     (NOZE), 18
NAT PSCs, 18–19, 113
  nucleation mechanisms
     of, 119–128

  ice-assisted, 121–123, 130–131
  kinetic suppression of, 112
  without ice, 124–125
NAT rocks
  "Mother Cloud/NAT Rock
     mechanism," 122, 130–132
  simulated formation of, 128–131
NAT temperature, 112, 115, 135, 136
Natural gas, 63
Natural processes
  chlorine and bromine contributed
     by, 37
  decomposition of halogenated
     source gases by, 38–40
  longer-lived halogenated source
     gases from, 50–52
  methane from, 63
  ODSs destruction by, 36
Natural sources
  of ODSs, 41–46
  of VSLS, 55, 56
Nature, 109
Nedoluha, G. E., 228
Newman, P. A., 24
NH. *See* northern hemisphere
Nicolet, M., 4
Niemeyer, U., 289
Nitric acid (HNO_3)
  and Arctic STS and
     NAT, 117–119
  conversion of NO_y species to, 177
  and de-/renitrification, 129–130
  "dew point" of, 113
  photolysis rate of, 164
  in PSC particles, 112–113
  in PSCs, 111
  vertical transport of, 122
Nitric acid hydrate (NAX), 111, 120,
     121
  and de-/renitrification, 129
  heterogeneous nucleation rate on
     ice, 121
  nucleation, 124, 125
Nitric acid trihydrate (NAT)
  Arctic measurements of, 117–119
  in chlorine activation, 135–138

coexistence of STS and, 113, 115
early research on, 111
NAT rocks, 18, 128–131
NAT temperature, 112, 115, 135, 136
nucleation, 117–128
ice-assisted, 121–123, 130–131
kinetic suppression of, 112
without ice, 124–125, 138
PSCs as, 18–19
and PSC types, 113
Nitrogen dioxide (NO$_2$), 164
Nitrogen gas (N$_2$), 4, 65, 281
Nitrogen trifluoride (NF$_3$), 68
Nitrous oxide (N$_2$O), 65–66
CO$_2$-equivalent emissions of, 69
and direct radiative forcing, 68
expected increases in, 264
in firn air samples, 45
in halogen nitrates formation, 86
human-influenced emissions of, 37
and mean age of air, 230–231
mixing ratio, 64
and ozone destruction, 63
projected increase of, 215
stratospheric reactive nitrogen from, 183
in the 21st century, 255, 257
in upper stratosphere, 176
NOAA. *See* National Oceanic and Atmospheric Administration
Non-halogenated gases, human-influenced emissions of, 37
Non-halogenated source gases, 62–67
methane, 63–65
nitrous oxide, 65–66
sulfur compounds, 66–67
Non-ideal gases, mixing ratios for, 2
Norrish, Ronald G. W., 80
North Atlantic Oscillation (NAO), 234–235, 237
Northern Annular Mode (NAM), 203, 208, 234–236

Northern hemisphere (NH). *See also* Arctic
air mix between southern hemisphere and, 50–52
Arctic ozone depletion, 16–18
Brewer-Dobson circulation in, 225, 226
mid-latitude column ozone in, 269
ozone concentration in winter, 7
planetary waves in, 8, 9
polar night jet and NAO in, 234, 235
sulfate aerosol loading and ozone destruction in, 284
volcanic eruptions in, 281
winter stratosphere in, 146
Northern mid-latitudes
erythermal radiation in, 25
ozone depletion in, 173–174
total ozone and ESC in, 22–24
NOZE (National Ozone Expedition), 18

OBrO. *See* Bromine dioxide
Ocean, loss of long-lived halogenated trace gases in, 38. *See also* sea surface temperatures (SSTs)
OClO. *See* Chlorine dioxide
Odd oxygen (odd oxygen family) (O$_x$), 3–4, 169
ODP. *See* Ozone Depletion Potential
ODSs. *See* ozone-depleting substances
OMI column observations, 98
Optical Particle Counter (OPC), 110
Orsolini, Y. J., 181
OSIRIS, 227
Oxidation, ODSs removed by, 38, 43
Ozone (O$_3$). *See also* column ozone
dissociative absorption of short-wave solar radiation by, 3
formation of, 3
mid-latitude (*See* mid-latitude ozone)
in odd oxygen, 3–4
polar, 162, 290–295
stratospheric (*See* stratospheric ozone)

Ozone (O_3) (*continued*)
   total, 1, 22–25, 163
   tropical, 265–268
   UV-photolysis of, 89
Ozone-depleting substances (ODSs).
   See also specific substances
   and Antarctic ozone hole, 150,
      152
   CO_2-equivalent emissions of, 69
   decline in, 253
   delayed response to production
      decreases, 36
   destruction by natural processes in
      atmosphere, 36
   and direct radiative forcing, 68
   emissions histories for, 41
   fractional release rates for, 260
   global emissions of, 42–44
   halogen source gases as, 10, 40–46
   1960 levels of, 215
   lifetimes of, 36, 226
   measuring/interpreting changes in
      abundance of, 46–50
   and Montreal Protocol, 10
   for non-dispersive uses, 36
   peak total abundance of, 253
   production of substitutes for, 36
   properties and atmospheric
      abundances of, 39
   quantifying depletion attributable
      to, 182
   recovery from effect of, 21–24
   simulations for 21st
      century, 256–258
   success in control of, 35–36
   tracking changes in abundance
      of, 36
Ozone depletion
   and accelerated BD
      circulation, 229
   defined, 170
   and stratosphere-troposphere
      coupling, 235–236
Ozone Depletion Potential (ODP), 40
   of nitrous oxide, 66
   for VSLS emissions, 58–59

"ozone hole," 15. *See also* Antarctic
   ozone hole; Arctic ozone hole
Ozone loss, defined, 170
Ozone mass, 269
Ozone profiles, 156
   observation of, 172
   observed and modeled trends at
      northern mid-latitudes, 184–185
   in ozone/tracer correlation, 157,
      158
   past changes in, 174, 176
Ozone recovery
   accelerated by climate
      change, 215–216
   in Antarctic, 207, 295
   in Arctic, 207
   CCM-simulated, 259
   and climate change, 21–23, 253
   and EESC, 62
   and ESC, 22–24
   projections of, 21–24
   response of stratospheric climate
      and circulation to, 241
   "super-recovery," 216
Ozone/tracer correlation
   method, 156–157

Pacific Ocean, 233
PACl. *See* potential for chlorine
   activation
Pagan, K. L., 124
Pannetier, G., 80
Papanastasiou, D. K., 94
Partitioning, 85–87, 92
Parts per billion (ppb), 2
Parts per billion by volume (ppbv), 2
Parts per million (ppm), 2
Parts per million by volume
   (ppmv), 2
Parts per trillion (ppt), 2
Parts per trillion by volume (pptv), 2
Past reference simulation, 257
Perfluorocarbons (PFCs)
   CO_2-equivalent emissions of, 69
   and direct radiative forcing, 68
Perlwitz, J., 236

PFP. *See* polar stratospheric cloud
  formation potential
PGI. *See* product gas injection
Photochemical destruction
  in Arctic vortex, 154–155
  by catalytic cycles, 7
  of nitrous oxide, 66
  of source gases, 86
Photochemical loss, 170, 174
Photochemical processes, 146
Photochemical production, 7
Photochemical reactions, 80
Photochemical timescales, 256
Photochemistry, of polar ozone
  loss, 153–154
Photolysis
  of CFCs, 38–39
  in Chemistry-Climate
    Models, 255–256
  of chlorine peroxide, 93–95
  of Cl$_2$, 91
  of ClONO$_2$ and BrONO$_2$, 90
  of dichlorine peroxide, 93–95
  flash photolysis technique, 80
  of HOBr, 89
  lifetimes associated with, 43
  of long-lived ODSs, 38–40
  of molecular oxygen and ozone, 89
  ozone production by, 3, 5, 6
  and speed of ozone loss, 164
Photolysis cross section
  ClOOCl, 95
  in photolysis rate, 93
Pierce, J. R., 289
Pitts, M. C., 114, 115, 126
Planetary scale wave-mean flow
  interaction, 236
Planetary (Rossby) waves, 191, 203,
  207, 223
  and Brewer-Dobson circulation, 8,
    9, 225, 228, 229
  and change in stratospheric
    circulation, 191
  and climatology of
    stratosphere, 225
  extra-tropical generation of, 230

and meridional temperature
  gradient, 218, 219
and polar vortices, 17
propagation of, 194
and seasonal radiative forcing, 224
and sea surface
  temperatures, 231–233
and stratosphere-troposphere
  coupling, 236
POAM satellite, 156, 157
Polar cap average, 269, 270
Polar night jet, 219, 223, 234, 235
Polar ozone, 162
  chemical destruction of, 162
  impact of geo-engineering
    on, 290–295
Polar ozone loss, 15–21, 145–165
  Antarctic, 15–16, 149–154,
    191–207
    chemical loss in Antarctic
      vortex, 150–154
    future changes in, 206–207
    impact on the
      troposphere, 192–200
    mechanisms of tropospheric
      response to, 201–206
    and ocean, 199–200
    in springtime, 269–271
    stratospheric, 192–207
    and surface climate, 197–199
    and tropospheric
      circulation, 192–197
  Arctic, 16–18, 154–164
    as Arctic ozone hole, 162–162
    chemical loss in Arctic
      vortex, 155–160
    and denitrification, 129
    impact on the troposphere, 191,
      207–208
    interannual variability in ozone
      loss, 160–164
    and natural variability in
      stratospheric ozone, 154–155
    and PSC formation, 112
    in springtime, 271
  basic mechanism leading to, 146

Polar ozone loss (*continued*)
  chemical mechanisms of, 18–21
  and increase in water
    vapor, 264–265
  in springtime, 269–271
Polar regions
  climate evolution in, 191
  EESC recovery in, 260
  heterogeneous chlorine activation
    in, 289
  mean stratospheric age in, 56
  mixing ratios in, 8
  springtime global ozone depletion
    in, 172–173
  sulfate aerosol loading and ozone
    destruction in, 284
  surface cooling in, 286
  uncertainty about ozone recovery
    over, 216
  in "world avoided" simulation,
    24
Polar stratosphere
  denitrification of, 18
  expected temperature changes
    in, 240–241, 262
  important heterogeneous reactions
    in, 133
  ozone depletion in (*See* polar
    ozone loss)
  reactions on cloud particles in, 6
Polar stratospheric cloud formation
  potential (PFP), 284, 285, 290
Polar stratospheric clouds (PSCs), 6
  Antarctic diversity in, 115–117
  Antarctic formation of, 122–123
  in Arctic, 154, 207
  Arctic STS and NAT
    measurements, 117–119
  and chemistry of lower
    stratosphere, 256
  and chlorine activation, 135–138,
    164
  as crystalline nitric acid
    trihydrate, 18
  distribution and composition
    of, 112–119

  and enhanced ClO
    concentrations, 91
  expected increase in, 262
  formation potential, 284, 285, 290
  growth stages of, 111
  heterogeneous reaction rates
    on, 131–135
  historical overview, 108–112
  LiDAR-based classification
    of, 113–115
  liquid, 111–114
  mixed-phase, 113–115, 118, 121,
    125, 126
  NAT nucleation, 119–128
    ice-assisted, 121–123, 130–131
    without ice, 124–125
  and ozone loss, 146
  and ozone recovery, 216
  polar vortices containing, 146, 147
  potential formation
    pathways, 119–120, 122
  reaction of NCl and ClONO$_2$
    on, 18
  and seasonal polar
    temperatures, 19
  simulation of NAT-rock formation
    and denitrification, 128–131
  Type-I, 111, 113, 114
  Type Ia, 113, 114, 118
  Type-Ia, 121, 124
  Type Ia-enh, 113, 114, 118, 122
  Type Ib, 113, 114, 118
  Type-Ib, 121
  Type-II, 111, 113, 114
  volumes of, 161, 163–164
Polar vortices
  Antarctic, 15, 16, 91, 145–146,
    150–154
  Arctic, 16, 146, 155–160
  and enhanced aerosol loading, 287
  and extreme weather events, 235
  and mid-latitude ozone
    depletion, 178–180
  and rate of ozone destruction, 21
Polvani, L., 237
Pope, F. D., 94, 95

Porter, George, 80
Potential for chlorine activation (PACl), 285, 290, 292–294
Potential vorticity gradient, 203, 205, 236
Poulet, G., 83
Ppb (parts per billion), 2
Ppbm (mass mixing ratios), 2
Ppbv (parts per billion by volume), 2
Ppm (parts per million), 2
Ppmm (mass mixing ratios), 2
Ppmv (parts per million by volume), 2
Ppt (parts per trillion), 2
Pptm (mass mixing ratios), 2
Pptv (parts per trillion by volume), 2
Precipitation
    impact of geo-engineering on, 284–286
    and poleward shift of westerly jet, 199
    and volcanic eruptions, 282
Pressures, 2
Process-based model evaluation, 184
Product gas injection (PGI), 84, 97, 98
PSCs. *See* polar stratospheric clouds

Quasi-biennial oscillation (QBO), 170, 171, 218, 224
    and eleven-year solar cycle, 220
    and tropical upwelling, 233–234
Quasi-isentropic motion, 234
Quasi-liquid layer, 112

Radiation
    impact of enhanced greenhouse gas concentrations on, 216–222
    infrared, 216, 282
    long-wave
        and concentrations of radiatively active gases, 216
        downwelling, 190, 202
        forcing of, 281
        upwelling, 192

short-wave
    absorption by ozone, 3
    and concentrations of radiatively active gases, 216
    downwelling, 190, 192, 202
    scattering and absorption of, 282
    and solar radiation management, 280–281
solar
    and concentrations of radiatively active gases, 216
    and ozone production, 169
terrestrial, 216
ultraviolet (*See* ultraviolet (UV) radiation)
Radiative forcing, 68–69, 216–217
    seasonal, 224
    and solar radiation management, 281
Ramanathan, V., 202
Ramaswamy, V., 220, 235
Randel, W. J., 181, 289
Rasch, P. J., 286
Ravishankara, A. R., 134–137
RDF (reverse domain filling), 178
Reaction rates (halogens), 86, 87
Reactive chlorine (ClO_x), 174, 176
Reactive intermediates, 80
Reactive nitrogen (NO_y), 122
    conversion to HNO_3, 177
    expected increases in, 264
    NO_x/NO_y equilibrium ratio, 282, 284, 289, 290
    removal of (*See* denitrification)
    source of, 183
    vertical redistribution of, 128, 130, 131
    vertical transport of, 129
Reactive uptake coefficient (γ), 133–135
Recovery. *See* ozone recovery
Reid, G., 228
Reid, S. J., 181

Remote sensing (RS)
measurement, 81–83
NAT nucleation without ice,
124
NAT-PSC nucleation, 121
of PSC occurrence and types,
113
Renitrification, 129
REPROBUS CTM, 156, 157
Reservoir species, 4, 11
Reverse domain filling (RDF),
178
Rex, M., 112
RICH radiosonde data, 217
Richter, A., 97
Rind, D., 232, 237, 240
Robock, A., 286
Rosenlof, K., 228
Rossby waves. *See* planetary waves
Rowland, F. S., 10, 13, 34,
80, 176
RS measurement. *See* remote sensing
measurement
Rusin, N., 279

SAD. *See* surface area densities
SAGE. *See* Stratospheric Aerosol and
Gas Experiment
Salawitch, R. J., 98
Salby, M., 228
Salby, M. L., 181
Salts, naturally emitted, 38
SAM. *See* Southern Annular Mode
SAM II (Stratospheric Aerosol
Measurement II), 110, 129
Santacesaria, V., 124
SAOZ instruments, 156, 157
Scaife, A., 240
Scaife, A. A., 237
Scheele, C. W., 79
Schofield, R., 98
Schumacher, H. J., 80
Schwarzkopf, M., 235
Sea ice concentrations (SICs), 255,
286
Sea level pressure, 198–199, 205

Seasons
and Antarctic ozone hole
variation, 15–16, 149, 152–153
and Arctic ozone depletion, 16–18
chemical loss, 157–159
natural variability, 160–164
and Brewer-Dobson
circulation, 225
and cooling in Antarctic, 192–193,
198
and extra-tropical tropopause layer
ozone, 239
and heterogeneous chlorine
activation, 132, 135
and mid-latitude ozone
depletion, 174, 180–181
and mid-latitude ozone
variation, 174, 175
and polar ozone loss cycles, 20, 21
and polar stratospheric clouds, 19
and radiative forcing, 224
and southern hemisphere mean
ozone, 197
and total column ozone, 9
and vortex stability, 146
Sea surface temperatures (SSTs), 215,
230–233
changes in, 32
and chemistry-climate
interactions, 255
and increased upwelling, 263
simulations for 21st century, 257
Seidel, D. J., 238
Sensitivity simulations, 257
SGI (source gas injection), 97
SGs. *See* source gases
SH. *See* southern hemisphere
Shanklin, J. D., 15
Shi, Q., 134
Shindell, D. T., 237, 271
Short-lived chemicals, 36. *See also*
very short-lived substances (VSLS)
Short-wave (SW) radiation
absorption by ozone, 3
and concentrations of radiatively
active gases, 216

downwelling, 190, 192, 202
scattering and absorption of, 282
and solar radiation management, 280–281
SICs (sea ice concentrations), 255, 286
Sigmond, M. C., 199
Simpson, I. R., 203
Simulations
past reference and future reference, 257
sensitivity, 257
of stratospheric cooling, 192
of stratospheric ozone in the 21st century, 256–258
Sinks
balance of sources and, 42
for long-lived halogenated trace gases, 38
methane as, 63
for methyl bromide, 51
Southern Ocean CO_2 sink, 200
Slab ocean model (SOM), 285
SLIMCAT model, 130–132, 135, 154, 156
Slusser, J. R., 89
Solar cycle
11-year, 169–171, 183, 218–220, 235
and mid-latitude ozone depletion, 183
Solar radiation
and concentrations of radiatively active gases, 216
and ozone production, 169
Solar radiation management (SRM), 280, 284–286
Solid PSCs, 112–114
Solomon, S., 18, 111, 192, 194, 197, 228, 236, 237
SOLVE/THESEO campaigns, 121, 124, 157–159
SOM (slab ocean model), 285
Son, S. W., 202, 203, 206
Soukharev, B. E., 181

Source gases (SGs), 33–69
brominated, 56
and climate change, 67–69
emissions histories for, 41
halogen
"banks" of, 10
increase in, 10–13
legally binding controls on, 13
as ozone-depleting substances, 10
past and projected abundance of, 12–13
photochemical breakdown of, 10, 11
recovery from effect of, 21–24
longer-lived halogenated, 33–34, 38–54
human *vs.* natural sources of, 40–46
measuring/interpreting changes in abundance of, 46–50
from natural processes, 50–52
systematic changes in total tropospheric chlorine and bromine from, 52–54
timescales and processes that remove, 38–40
mid-latitude ozone depletion and changes in, 183
non-halogenated, 62–67
methane, 63–65
nitrous oxide, 65–66
sulfur compounds, 66–67
in stratospheric halogen chemistry, 84–85
and stronger BD circulation, 227
total atmospheric halogen loading changes, 59–62
very short-lived substances, 54–59
Source gas injection (SGI), 97
Southern Annular Mode (SAM), 191, 197–200, 234–236
Southern hemisphere (SH). *See also* Antarctic
air mix between northern hemisphere and, 50–52

Southern hemisphere (SH)
(*continued*)
   Brewer-Dobson circulation
      in, 225, 226
   climate change models, 194–195
   cold pole problem in, 259
   planetary waves in, 8, 9
   sulfate aerosol loading and ozone
      destruction in, 284
   tropospheric circumpolar
      circulation in, 236
   tropospheric geopotential height
      variability in, 191–194
   tropospheric jet shift in, 16
   winter stratosphere in, 146
Southern high latitudes, chlorine
   inventory for, 87
Southern mid-latitudes
   chlorine inventory for, 87
   climate in, 199
   column ozone in, 269
   ozone depletion in, 173–174
   total ozone and ESC for, 22–24
Southern Ocean, 191, 199, 200
Sowers, T., 45
SPARC. *See* Stratospheric Processes
   and their Role in Climate
Spectrograph, 1
Spectroscopy, 80
Spence, P., 199, 200
Spinks, J. W. T., 80
SRM (solar radiation
   management), 280, 284–286
SSTs. *See* sea surface temperatures
SSU (Stratospheric Sounding
   Unit), 218, 219
Steady-state condition, in determining
   lifetimes, 43
Steinbrecht, W., 181, 239
Stenchikov, G., 237
Stolarski, R. S., 4, 15, 34, 176
Stratosphere
   aerosol layer in, 109–110
   CFC transport through, 39
   cooling in, 192, 193
   frozen particles in, 112–113

greenhouse gas effect in, 216
halogens and ozone chemistry
   of, 78
liquid particles in, 113
water vapor in, 109
Stratosphere-troposphere
   coupling, 203, 233–241. *See also*
upper troposphere/lowermost
   stratosphere (UTLS)
expected tropopause
   changes, 240–241
modelling, 225
tropical and extra-tropical
   tropopause layer, 238–240
Stratospheric Aerosol and Gas
   Experiment (SAGE), 172, 185, 229
Stratospheric Aerosol Measurement
   II (SAM II), 110, 129
Stratospheric dynamics
   and aerosol injection, 287–288
   estimates of future changes of, 215
   impact of enhanced greenhouse gas
      concentrations on, 222–233
      Brewer-Dobson circulation and
         mean age of air, 225–231
      importance of atmospheric
         waves, 223–225
      role of sea surface
         temperatures, 231–233
   impact of volcanic eruptions
      on, 282, 283
   and polar ozone loss, 146, 160–161
Stratospheric halogen
   chemistry, 78–99
   abundances, 83–85
   chlorine *vs.* bromine
      chemistry, 87–89
   halogen catalyzed ozone loss
      cycles, 89–93
      ClO/BrO cycle, 93
      ClO dimer cycle, 92
      ClOOCl photolysis, 93–95
   history of halogen
      chemistry, 79–80
   measurement of stratospheric
      halogen species, 81–83

partitioning, 85–87
in UTLS, 95–99
   chlorine chemistry in
      tropopause region, 98–99
   inorganic bromine
      budget, 96–98
Stratospheric halogens, 21st century
   changes in, 260–261
Stratospheric ozone
   altitude profile of, 1, 3
   chemistry of, 3–7
   climate change influence on, 21
   distribution of, 7–9, 169, 226, 227
   early observations of, 1, 3
   and geo-engineering
      impact of geo-
         engineering, 287–295
      motivation for proposed
         approaches, 279–281
   global abundance of, 170, 171
   impact of volcanic eruptions
      on, 282, 284
   natural variability in, 154–155
   observations of, 172
   polar, 290–295
   transport changes, 262–264
   in the 21st century, 253–273
      changes in stratospheric
         halogens, 260–261
      chemistry-climate
         models, 254–256
      evaluation, 259–260
      factors affecting, 260–265
      mid-latitude ozone, 267–269
      simulations, 256–258
      springtime polar
         ozone, 269–271
      temperature changes, 261–262
      tropical ozone, 265–268
Stratospheric ozone depletion
   in the Antarctic, 192–207
      future changes in, 206–207
      impact on the
         troposphere, 192–200
      mechanisms of tropospheric
         response to, 201–206

   and ocean, 199–200
   and surface climate, 197–199
   and tropospheric
      circulation, 192–197
   in the Arctic, 207–208
Stratospheric ozone layer, 1–9
   anthropogenic influence on, 9–15
      increase in halogen source
         gases, 10–13
      upper stratospheric ozone
         depletion, 13–15
   early observations, 1, 3
   future of, 21–26
      and Montreal Protocol, 24–26
      recovery projections, 21–24
   impact of climate change
      on, 214–243
      coupling of stratosphere and
         troposphere, 203, 233–241
      radiation and
         chemistry, 216–222
      stratospheric
         dynamics, 222–233
   polar stratospheric ozone
      depletion, 15–21
      Antarctic ozone hole, 15–16
      Arctic ozone depletion, 16–18
      chemical mechanisms of, 18–21
   units of atmospheric
      measurement, 2–3
Stratospheric Processes and their
   Role in Climate (SPARC), 183,
   184, 227, 231, 254, 259. *See also*
   Chemistry-Climate Model
   Validation (CCMVal) Activity
Stratospheric Sounding Unit
   (SSU), 218, 219
Stratospheric variability, 224
STS. *See* supercooled ternary
   solutions
Subtropics
   expected ozone decrease in, 267,
      269
   polar expansion of Hadley cell
      in, 199
   summer moistening trend in, 209

Subtropics (*continued*)
  transport of chemical species and
    particles, 234
  upper troposphere temperatures
    in, 232
  zonal wind in, 203
Sulfate aerosols
  chlorine activation on, 19–20
  $ClONO_2$ heterogeneous reaction
    on, 88
  cold
    and chlorine activation, 177
    heterogeneous reactions
      on, 131–135
  heterogeneous reactions
    on, 19–20, 88, 131–135, 177
  liquid, 281–282
  northern hemisphere loading,
    284
  particles
    chlorine transformation from
      PSCs, 108
    heterogeneous reaction rates
      on, 131–135
    lower stratospheric
      heterogeneous reactions
        on, 177
    and NO_x/NO_y equilibrium
      ratio, 282, 284, 289, 290
    solar radiation management
      using, 280
Sulfur compounds, 66–67
Sulfur dioxide (SO_2), 67
  injection into
    stratosphere, 286–289
  from volcanic eruptions, 281
Sulfur gases
  human-influences emissions of, 37
  from volcanic eruptions, 281
Sulfuric acid (H_2SO_4)
  hydrates of, 111
  in PSC particles, 112, 113
  volcanic and non-volcanic, 67
  from volcanic eruptions, 281
Sulphur hexafluoride (SF_6), 68, 69,
  230

Sunlight
  and Arctic photolysis rates, 164
  in Arctic vortex, 154
  artificial reflection of, 280
  and polar ozone depletion, 21
  polar vortices exposed to,
    146, 147
Supercooled ternary solutions
  (STS), 18–19
  Arctic measurements of, 117–119
  in chlorine activation, 135–138
  coexistence of NAT and, 113, 115
  formation of, 113
  liquid STS clouds, 115
  in NAT nucleation, 121
"super-recovery," 216
Supersonic transport, impact of, 9–10
Surface area densities (SAD)
  and chemical ozone
    depletion, 284, 285, 290, 292
  of liquid sulfate aerosols, 281–282
  and sulfuric aerosol
    injection, 287–289
Surface climate
  Antarctic, 197–199
  Arctic, 207
Surface temperature
  Antarctic, 191
  associated with stratospheric ozone
    depletion, 190–191
  impact of geo-engineering
    on, 284–286
Surf zone, 223–224, 256
SW radiation. *See* short-wave
  radiation
Synoptic waves, 225

Tabazadeh, A., 128, 131
Temperature, 181, 182. *See also*
  climate change; cooling; warming
  and aerosol injection, 286, 287
  and chemistry of upper
    stratosphere, 176–177
  expected changes in, 220–221,
    261–262
  heating in stratosphere, 3

and heterogeneous reaction rates, 135–137

impact of geo-engineering on atmospheric temperature, 286–287

surface temperature, 284–286

influence on ozone content, 221–222

NAT temperature, 112, 115, 135, 136

ozone-destroying reactions dependent on, 215

past changes in, 217–220

polar-cap changes in, 204

and polar heterogeneous reactions, 19–20

and polar ozone holes, 162–163

and polar stratospheric clouds, 19

and rate of ozone destruction, 21

for solid water ice particles, 112

stratospheric, 192, 193

tropospheric, 202, 215

and upper stratospheric ozone, 14

in upper troposphere/lowermost stratosphere, 191

and UV absorption by ozone, 201, 202

and volcanic eruptions, 282, 283

in "world avoided" simulation, 24–26

Terrestrial ecosystems

nitrous oxide in, 65–66

as sinks for long-lived halogenated trace gases, 38

Terrestrial radiation, 216

Thompson, D., 228, 234, 236, 237

Thompson, D. W. J., 192, 194, 197, 203, 234, 235

3D Chemical Transport Models, 131, 183

of Arctic ozone loss, 156, 159–160

connection between polar and mid-latitude ozone loss, 179, 180

dynamical changes and mid-latitude ozone, 181

3D chemistry-climate models, 183–185

3D models, 183–184

3-stage model (PSC development), 120

Tilmes, S., 286, 287, 289, 290

Timescales

and chlorine activation in the Arctic, 163–164

for decomposition of halogenated source gases, 38–40

in lower stratosphere, 256

for sulfur dioxide removal, 67

for transport/removal of inorganic halogen, 56

TOMS (Total Ozone Mapping Spectrometer), 172

Toon, O. B., 18, 128

Total available stratospheric bromine, 85

Total available stratospheric chlorine, 84, 85

Total ozone, 163

changes in, 1960 to 2100, 22–24

measurement of, 1

in "world avoided" simulation, 25

Total Ozone Mapping Spectrometer (TOMS), 172

Total tropospheric bromine, 52–54

Total tropospheric chlorine, 52–54

Transport. *See also* Brewer-Dobson circulation

in advection of air into mid-latitudes, 178

and climate change, 215

and distribution of stratospheric ozone, 7–9

expected changes in, 262–264

halogen, 56

and mid-latitude ozone changes, 170

mixing, 7, 256

and mid-latitude ozone changes, 170

for polar and tropical latitudes, 8

Transport (*continued*)
  nitric acid, 122
  reactive nitrogen, 129
  and residence times of molecules in
    stratosphere, 56
  and seasonal/interannual ozone
    variability, 180–182
  in stratosphere, 256
  and stronger BD circulation, 228
  in subtropics and extra-tropics,
    234
  in tropical-tropopause layer, 239
Trichloroethylene (C_2HCl_3), 55–56
Tropical ozone, in the 21st
  century, 265–268
Tropical stratosphere
  drop in water vapor concentrations
    after 2001, 229
  quasi-biennial oscillation in, 224
Tropical tropopause layer (TTL)
  defined, 238
  effect of geo-engineering on, 289
  expected warming of, 265
  and sea surface
    temperatures, 232–233
  and stratosphere-troposphere
    coupling, 238–240
  vertical mixing in, 234
  VSLS in, 97
Tropical upper troposphere, 56, 58
Tropical upwelling, 232–233,
  263–264
  expected changes in, 241, 266
  and geo-engineering, 289
Tropics
  expected ozone decrease in, 267, 269
  methane emissions in, 65
  mixing ratios in, 8
  ozone mixing ratios in, 289
  sea surface temperature anomalies
    in, 232
  simulated lower stratospheric total
    ozone in, 25
  and stronger BD circulation, 227
  total ozone and ESC, 1960 to
    2100, 22, 24

Tropopause
  change in downwelling SW
    radiation across, 190
  chlorine chemistry in, 98–99
  defined, 238
  expected changes in, 240–241
  height of, and mid-latitude ozone
    depletion, 181, 182
  ozone trends in, 95, 96
  sensitivity to ozone changes
    in, 217
Troposphere. *See also* upper
  troposphere/lowermost
  stratosphere (UTLS)
  halogens and ozone chemistry
    of, 78
  impact of polar ozone loss
    on, 190–209
    Antarctic, 191–207
    Arctic, 207–208
  methane and ozone formation
    in, 63
  mixing times and ODS loss rates
    in, 59
  reaction to sea surface temperature
    anomalies, 232
  stratosphere-troposphere
    coupling, 203, 233–241
  temperatures in, 202
    and aerosol injection, 286,
      287
    and mid-latitude ozone
      depletion, 181, 182
  warming effect of greenhouse
    gases, 216
Tropospheric climate, 16
Tropospheric source gases
  chlorine and bromine from
    long-lived ODSs, 52–54
  concentrations and emissions, 37
Tsias, A., 113
TTL. *See* tropical tropopause layer
Tung, K., 228
21st century
  Antarctic erythemal UV in, 202
  stratospheric ozone in, 253–273